I0888378

Advanced Geospatial Practices in Natural Environment Resource Management

Rubeena Vohra
Bharati Vidyapeeth's College of Engineering, India

Ashish Kumar
Bennett University, India

A volume in the Advances in Geospatial Technologies (AGT) Book Series

IGI Global
PUBLISHER of TIMELY KNOWLEDGE

Published in the United States of America by
IGI Global
Engineering Science Reference (an imprint of IGI Global)
701 E. Chocolate Avenue
Hershey PA, USA 17033
Tel: 717-533-8845
Fax: 717-533-8661
E-mail: cust@igi-global.com
Web site: http://www.igi-global.com

Library of Congress Cataloging-in-Publication Data

Names: Vohra, Rubeena, 1981- editor. | Kumar, Ashish, 1986- editor.
Title: Advanced geospatial practices in natural environment resource management / edited by Rubeena Vohra, Ashish Kumar.
Description: Hershey PA : Engineering Science Reference, [2024] | Includes bibliographical references. | Summary: "The contents in the book emphasize not only on the issues pertaining to the environment but also presents the advanced management practices for the protection of varied fields related to environment"-- Provided by publisher.
Identifiers: LCCN 2023043759 (print) | LCCN 2023043760 (ebook) | ISBN 9798369313961 (hardcover) | ISBN 9798369313978 (ebook)
Subjects: LCSH: Natural resources--Management--Geographic information systems--Case studies. | Natural resources--Management--Data processing--Case studies. | Geospatial data--Computer processing--Case studies
Classification: LCC HC85 .A363 2024 (print) | LCC HC85 (ebook) | DDC 333.70285--dc23/eng/20231128
LC record available at https://lccn.loc.gov/2023043759
LC ebook record available at https://lccn.loc.gov/2023043760

This book is published in the IGI Global book series Advances in Geospatial Technologies (AGT) (ISSN: 2327-5715; eISSN: 2327-5723)

British Cataloguing in Publication Data
A Cataloguing in Publication record for this book is available from the British Library.
All work contributed to this book is new, previously-unpublished material.
The views expressed in this book are those of the authors, but not necessarily of the publisher.
For electronic access to this publication, please contact: eresources@igi-global.com.

Advances in Geospatial Technologies (AGT) Book Series

Nilanjan Dey
Techno India College of Technology, India

ISSN:2327-5715
EISSN:2327-5723

MISSION

The geospatial technology field is a fast-paced, high growth industry that is involved in a variety of fields, including military planning, public health, land use, environmental protection, and Google Earth mapping. With such a diverse body of applications, the research in geospatial technologies is always evolving and new theories, methodologies, tools, and applications are being developed. **Advances in Geospatial Technologies (AGT) Book Series** is a reference source and outlet for research that discusses all aspects of geographic information, including areas such as geomatics, geodesy, GIS, cartography, remote sensing, and other areas. Because geospatial technologies are so pervasive in such a variety of areas, AGT also includes books that address interdisciplinary applications of the technologies.

COVERAGE

- Disaster Informatics
- Public Health Applications of Geospatial Technologies
- Cartography
- Geographical information systems
- Digital Terrain Modeling
- Environmental Monitoring
- Military Security and Geospatial Technologies
- Public Sector Use of Geospatial Technologies
- Surveying Technologies
- Spatial Reference Systems

IGI Global is currently accepting manuscripts for publication within this series. To submit a proposal for a volume in this series, please contact our Acquisition Editors at Acquisitions@igi-global.com or visit: http://www.igi-global.com/publish/.

The Advances in Geospatial Technologies (AGT) Book Series (ISSN 2327-5715) is published by IGI Global, 701 E. Chocolate Avenue, Hershey, PA 17033-1240, USA, www.igi-global.com. This series is composed of titles available for purchase individually; each title is edited to be contextually exclusive from any other title within the series. For pricing and ordering information please visit http://www.igi-global.com/book-series/advances-geospatial-technologies/73686. Postmaster: Send all address changes to above address.

Titles in this Series

For a list of additional titles in this series, please visit:
http://www.igi-global.com/book-series/advances-geospatial-technologies/73686

Emerging Trends, Techniques, and Applications in Geospatial Data Science
Loveleen Gaur (Amity University, India & Taylor's University, Malaysia & University of the South Pacific, Fiji) and P.K. Garg (Indian Institute of Technology, Roorkee, ndia)
Engineering Science Reference • © 2023 • 309pp • H/C (ISBN: 9781668473191) • US $240.00

Addressing Environmental Challenges Through Spatial Planning
Athar Hussain (Netaji Subhas University of Technology, India) Kailash Chandra Tiwari (Delhi Technological University, India) and Alpana Gupta (Indian Institute of Technology, Delhi, ndia)
Engineering Science Reference • © 2022 • 329pp • H/C (ISBN: 9781799883319) • US $215.00

Methods and Applications of Geospatial Technology in Sustainable Urbanism
José António Tenedório (Universidade NOVA de Lisboa, Portugal) Rossana Estanqueiro (Universidade NOVA de Lisboa, Portugal) and Cristina Delgado Henriques (Universidade de Lisboa, Portugal)
Engineering Science Reference • © 2021 • 684pp • H/C (ISBN: 9781799822493) • US $215.00

Interdisciplinary Approaches to Spatial Optimization Issues
Sami Faiz (University of Tunis El Manar, Tunis, Tunisia) and Soumaya Elhosni (University of Tunis El Manar, Tunis, Tunisia)
Engineering Science Reference • © 2021 • 271pp • H/C (ISBN: 9781799819547) • US $215.00

Geospatial Technologies for Effective Land Governance
Moha El-Ayachi (Institut Agronomique et Vétérinaire Hassan II, Morocco) and Loubna El Mansouri (Institut Agronomique et Vétérinaire Hassan II, Morocco)
Engineering Science Reference • © 2019 • 302pp • H/C (ISBN: 9781522559399) • US $205.00

IGI Global
PUBLISHER of TIMELY KNOWLEDGE

701 East Chocolate Avenue, Hershey, PA 17033, USA
Tel: 717-533-8845 x100 • Fax: 717-533-8661
E-Mail: cust@igi-global.com • www.igi-global.com

Table of Contents

Detailed Table of Contents

The atmosphere has a significant impact on the physical environment and the people. The atmospheric processes go hand in hand with surface processes and affect the climate and contribute to the changing climate. The adversely altering environment is not the only issue that comes with climate change, but the impact it has on human and other biological populations is also worrisome and needs to be addressed. Geospatial technologies, which combine remote sensing and geographic information systems, provide flexible cross-scale tools for examining climate change, the long-term evolution of the climate system, and its effects on social and ecological systems. In order to better understand how the Earth, ocean, biosphere, cryosphere, and atmosphere have changed over time, scientists have used long-term satellite data sets. The chapter discusses the various atmospheric processes, their implications, their inter-relationship with land processes, and the application of geospatial methods, in particular remote sensing, for the observation and monitoring of the atmosphere.

Water and wastewater engineering are essential for public health and environmental protection. It involves the design, construction, and operation of systems to treat water and wastewater. The basic principles of water and wastewater treatment include physical, chemical, and biological processes. Engineers consider the nature and concentration of contaminants, the desired level of treatment, the cost of treatment, the availability of land, the climate, and environmental regulations when designing water and wastewater treatment systems. Common types of water treatment systems include conventional treatment, reverse osmosis, electrodialysis, and ion exchange. Common types of wastewater treatment systems include primary treatment, secondary treatment, and tertiary treatment. Critical tasks involved in operating and maintaining water and wastewater treatment systems include monitoring and adjusting the operation of treatment processes, conducting maintenance and repairs, and sampling and testing water or wastewater.

Water demand and quality management are vital components of ensuring universal access to clean water and safeguarding water resources for future generations. It involves strategies to reduce water consumption, including initiatives like water pricing, public education, and water-efficient technologies, safeguarding water quality through pollution prevention, wastewater treatment, and watershed management. Decreasing water demand can enhance water quality by reducing pollution entering water bodies. The goal of management is to cut back on water use without compromising social or economic advancement. Municipalities, industry, and agriculture are just a few of the industries that can use water quality management techniques. Integrated water demand and quality management is an approach to water resource management that considers the interconnectedness of water quantity and quality. This approach recognizes that water demand management and water quality management are complementary and must be considered together to achieve sustainable water resource management.

With the fast growth of aquatic data, machine learning is essential for data analysis, categorization, and prediction. Data-driven models using machine learning may effectively handle complicated nonlinear problems in water research, unlike conventional approaches. Machine learning models and findings have been used to build, monitor, simulate, evaluate, and optimize water treatment and management systems in water environment research. Machine learning may also enhance water quality, pollution control, and watershed ecosystem security. This chapter discusses how ML approaches were used to assess water quality in surface, ground, drinking, sewage, and ocean. The authors also suggest potential machine learning applications in aquatic situations.

Weather forecasting is the utilization of science and technology to foresee the conditions of the atmosphere for a given location and time. Weather forecasting is high priority since it helps to settle future climate changes and provides information on critical weather conditions. As the weather has a great impact on various aspects of human life, aquatic life, aviation industry, and others, efforts have been made for decades to improve the efficiency of weather forecasting to ensure a better life and to reduce economic loss, but the result is not more precise than expected. The present research focuses on improving the efficiency of weather forecasting, focusing on various forms of precipitation such as rain, snow, hailstorms, and snowflakes by making use of historical numerical weather datasets across the globe. The efficiency in terms of performance measures has been compared with existing models.

Glacial lake outburst floods (GLOFs) are increasingly threatening the sustainability and infrastructure of high-mountain regions globally. This study specifically addresses the GLOFs risk in the Kinnaur district of Himachal Pradesh. Using a comprehensive analytical framework, the authors explore factors affecting GLOF initiation, map potential downstream impact zones, and assess societal vulnerability to these climate-induced calamities. For precise analysis, they utilize Landsat images (1989-2018) and Cartosat DEM (30 m) data to map glacial lakes. ArcGIS 10.8 generates geospatial characteristics, while the maximum likelihood classification (MLC) algorithm is employed for land use and land cover (LULC) mapping. Indices like NDSI, NDWI, and NDPI contribute additional insights. Identifying 39 glacial lakes, 10 deemed hazardous due to size (>0.1 km2), the study underscores the specific threat of cross-border GLOFs from China's Upper Satluj River Basin to Eastern Himachal Pradesh. In a prospective deglaciated scenario, it is envisaged that a substantial escalation in glacial lake outburst flood (GLOF) hazard levels will prevail across the majority of administrative units. Given the increasing encroachment of glacial lakes towards precipitous headwalls, the potential for glacial lake outburst flood initiation is significantly amplified. This elevates the risk to surrounding areas, as the displacement of ice and rock can initiate substantial GLOFs. In light of this prognosis, an essential initial action for adaptation in these circumstances involves strengthening resilience and implementing capacity-building measures designed to reduce the risk of GLOFs.

Gopinadh Rongali, National Centre for Medium Range Weather Forecasting, MoES, Noida, India
Ashok K. Keshari, Indian Institute of Technology Delhi, New Delhi, India
Ashwani K. Gosain, Indian Institute of Technology Delhi, New Delhi, India
R. Khosa, Indian Institute of Technology Delhi, New Delhi, India
Ashish Kumar, School of Computer Science Engineering and Technology, Bennett University, Greater Noida, India

In the study of snow melt runoff, the temperature lapse rate (TLR) is an essential variable. For the Beas River Basin in the Western Himalayas, it has been approximated in the current study. In this work, the split-window (SW) technique for TLR estimate has been used to recover the land surface temperature (LST). LST in the study area has a negative correlation with elevation values, and the trend shows that LST and elevation have an inverse relationship, according to data from the United States Geological Survey's (USGS) advanced spaceborne thermal emission and reflection radiometer (ASTER) and global digital elevation model (GDEM). The TLRs for the Beas River Basin region vary from 0.71°C/100 m to 0.87°C/100 m during the time period of 18 April 2013 to 27 June 2015. The findings were calculated using lapse rates calculated from maps produced by the moderate resolution image spectroradiometer (MODIS). There is excellent agreement between the MODIS-Terra data and the air temperature and LST from Landsat-8. The modelling of snow and glacier melt flow in the Himalayan region will benefit from the current work.

The availability of reliable power without interruption is critical for a fast and sustained economic growth. No country can develop and sustain itself beyond subsistence levels unless it has access to modern energy (electrical) services at a minimum. According to studies, despite the country's vast renewable energy potential, access to alternate unconventional types of energy is extremely limited. More than half of the world lack access to electricity and continue to rely primarily on traditional energy sources. This may be significantly advanced by incorporating prosumers in present smart-grid infrastructure to generate power with renewable distributed generation technologies to meet expanding demand while minimising environmental risk. The authors assessed the current state of awareness of the potential positive implications of the energy transition for various communities and socioeconomic groups on the frontlines of the change using a fast growing corpus of academic and policy research.

Hyperspectral imagery holds the essence of spectral and spatial attributes, making it a rich source of information for applications like target detection in the field of agriculture, urban areas, etc. The chapter provides an insight of extraction of shape-based features fused with the spectral characteristics for detection of urban targets particularly roads and roofs. Roads are considered as linear structures whereas roofs occur in clusters. Therefore, to detect them, spectral information is not sufficient. The work proposes integration of both spectral and shape aspects of a target, which has given better results using SVM. The experimentation involves results in the form of spectral analysis, shape analysis, fusion of spectral and shape-based features along with the detection rate and supporting results. Delineating the roads and roofs of urban topography is an essential fragment for city planning, urban sprawl estimation, population studies, sustainable development, etc. The work done here may help numerous governmental and non-governmental organizations for conducting related studies.

This chapter used a systematic review to organize the data by a systematic review of existing literature. The chapter documented climate smart agriculture (CSA) practices that exist in the country to cope with various production risks. These included the adaption of better seed varieties, application of weather information, efficient fertilizer use, intercropping, and crop residue and canal irrigation. Food system issues, affecting food production, processing, distribution, and consumption, are among the main causes of food insecurity and hunger in Tanzania. Tanzanian rural households are particularly susceptible to environmental pressures.

By integrating automated driving systems (ADS) and AI-driven advanced driver assistance systems (ADAS) like the traffic sign detection (TSD) technology, the automotive sector can develop smart and self-driving cars. Traffic signs (TS) play a crucial role in avoiding accidents and traffic congestion. Motorists need to understand the visual representations of various data elements incorporated in traffic symbols. There are often instances where drivers neglect TS located ahead of their vehicles, resulting in severe outcomes. This research offers an automatic TSD forecast utilising the hybrid deep belief network (HDBN) model for classification to address this issue. When it comes to forecasting the future world of smart urban cities, the given HDBN model primarily focuses on high-precision traffic prediction. The rider sunflower optimization (RSFO) technique is utilised to improve the hyper parameter tuning, which improves the overall effectiveness of the traffic flow prediction process. Overall, the suggested TSD system is found to be a highly efficient method of detecting TS, performing exceptionally well in relation to precision, recall, accuracy, and F1. The suggested solution under evaluation appears to perform better in terms of accuracy than other current methods stated in this chapter.

Preface

As editors of *Advanced Geospatial Practices in Natural Environment Resource Management*, we are pleased to present this comprehensive reference book that delves into the intricate realms of environmental science, engineering, and technology. Our goal is to conscientiously enlighten students, scholars, and researchers about the critical areas demanding protection within our existing environment. This volume not only addresses environmental issues but also emphasizes advanced management practices essential for safeguarding diverse fields related to the environment.

The focus on managing environmental issues through advanced scientific solutions is a profound area of research, contributing to the ongoing development of solutions and measurements on the ground. The deeper understanding of environmental processes and the evaluation of management strategies through advanced geospatial and AI/ML technology aim to enhance the environment, ultimately aligning with established sustainable goals.

Specifically, the engineering studies in environmental science provide insights into problems related to water, soil, and air. Environmental engineering and technology underscore the development of technical solutions needed to address these problems in a manner compatible with legislative, economic, social, and political concerns. The holistic technical solutions presented in this book will significantly aid governments, businesses, and customers in the broader societal community.

Improvements in natural environmental systems contribute to the general well-being of all humans. The application and development of advanced engineering techniques, coupled with geospatial and AI/ML knowledge, present a promising route map for dealing with future hazardous risks in public health sectors. Environmental engineering studies in this book investigate, analyze, and predict how the exposure of hazardous materials in the environment leads to the destruction and contamination of vital planetary resources. The advanced practices outlined here can contribute to solving challenges posed to public health by providing valuable analytical insights and workable engineering solutions, ultimately reducing disease and preventing hazard exposure for human well-being.

The book is structured into several key sections, covering fundamental aspects, historical perspectives, reasons, and challenges for the depletion of natural resources. It delves into the concepts and analysis of advanced solutions versus conventional approaches for sustainable development, exploring the impact of scientific solutions on the environment and people. Additionally, it addresses strategies and planning for the retention of natural resources, considering economical and political concerns for sustainable development. Real-world case studies and success stories provide practical insights, and the book concludes with a focus on research and future prospects of advanced technology in geospatial practices.

The chapters in *Advanced Geospatial Practices in Natural Environment Resource Management* collectively offer a comprehensive exploration of diverse facets within the realms of environmental science, engineering, and technology. Each chapter provides a focused lens on critical topics, ranging from atmospheric processes and climate change to water and wastewater engineering to temperature lapse rates for snowmelt runoff. This anthology serves as a curated guide for educators, researchers, and students, offering insights into cutting-edge geospatial practices essential for the sustainable management of natural resources and the preservation of our environment.

Chapter 1 provides a comprehensive exploration of the intricate interplay between atmospheric processes, surface processes, and climate, elucidating their impact on the physical environment and human populations. Focusing on the changing climate, the chapter delves into the adverse effects on both the environment and biological populations. Geospatial technologies, particularly the integration of remote sensing and geographic information systems, offer versatile tools for examining climate change, the long-term evolution of the climate system, and their effects on social and ecological systems. The discussion extends to the application of geospatial methods, specifically remote sensing, for observing and monitoring atmospheric processes, emphasizing the importance of long-term satellite data sets in understanding the evolution of Earth, oceans, biosphere, cryosphere, and atmosphere.

Essential for public health and environmental protection, Chapter 2 explores the core principles underlying water and wastewater engineering. Encompassing the design, construction, and operation of systems for treating water and wastewater, the chapter covers physical, chemical, and biological processes crucial to these treatment systems. Considerations such as the nature and concentration of contaminants, treatment cost, land availability, climate, and environmental regulations are highlighted in the design process. Various types of water treatment systems, including conventional treatment, reverse osmosis, electrodialysis, and ion exchange, are discussed, along with wastewater treatment systems. The chapter also outlines critical tasks involved in operating and maintaining these systems, emphasizing monitoring, adjustment of treatment processes, maintenance, repairs, and water or wastewater sampling and testing.

Chapter 3 underscores the importance of water demand and quality management in ensuring universal access to clean water and safeguarding water resources for future generations. Strategies to reduce water consumption, including water pricing, public education, and water-efficient technologies, are explored alongside initiatives for safeguarding water quality through pollution prevention, wastewater treatment, and watershed management. The chapter emphasizes the interconnectedness of water quantity and quality in integrated water demand and quality management, offering an approach that considers both aspects together for sustainable water resource management. The goal is to reduce water use without compromising social or economic advancement, benefiting municipalities, industries, and agriculture.

Addressing the fast growth of aquatic data, Chapter 4 delves into the indispensable role of machine learning in data analysis, categorization, and prediction in water research. Machine learning models are presented as effective tools for handling complex nonlinear problems in water treatment and management systems, surpassing conventional approaches. The chapter highlights the application of machine learning models and findings in building, monitoring, simulating, evaluating, and optimizing water treatment and management systems, contributing to water quality improvement, pollution control, and watershed ecosystem security. Practical applications of machine learning in assessing water quality in various environments, including surface, ground, drinking, sewage, and ocean, are discussed, offering potential avenues for further research in aquatic situations.

Focusing on the critical role of weather forecasting in predicting atmospheric conditions, Chapter 5 delves into efforts to improve the efficiency of weather forecasting. Recognizing the significant impact of weather on various aspects of human life, aquatic life, and industries such as aviation, the chapter explores the use of historical numerical weather datasets to enhance the precision of forecasting. Emphasis is placed on improving the efficiency of weather forecasting, with a particular focus on various forms of precipitation, including rain, snow, hail storms, and snowflakes. The research evaluates the performance of existing models and proposes advancements for more accurate and reliable weather forecasting.

Chapter 6 addresses the escalating threat of Glacial Lake Outburst Floods (GLOFs) in high-mountain regions globally, with a specific focus on the Kinnaur district of Himachal Pradesh. Employing a comprehensive analytical framework, the chapter assesses factors contributing to GLOF initiation, delineates potentially impacted areas downstream, and evaluates societal vulnerability to these climatically-induced calamities. The utilization of Landsat images and Cartosat DEM data, combined with ArcGIS, facilitates glacial lake mapping and provides topographic information for the assessment. The findings contribute to the understanding of GLOF risks, providing valuable insights for risk mitigation and management in high-mountain regions.

Chapter 7 focuses on the essential variable of temperature lapse rate (TLR) in the study of snowmelt runoff, specifically for the Beas river basin in the Western Himalayas. The split-window (SW) technique is employed to estimate TLR by recovering land surface temperature (LST). The study establishes a negative correlation between LST and elevation values, revealing an inverse relationship between LST and elevation. Using data from the United States Geological Survey's (USGS) advanced spaceborne thermal emission and reflection radiometer (ASTER) and global digital elevation model (GDEM), the chapter provides TLR values for the Beas River Basin region during a specific time period. The findings, supported by lapse rates calculated from maps produced by the moderate resolution image spectroradiometer (MODIS), demonstrate excellent agreement between MODIS-Terra data, air temperature, and LST from Landsat-8.

Highlighting the critical importance of reliable power for sustained economic growth, Chapter 8 addresses the challenge of limited access to alternative energy sources despite vast renewable energy potential. The integration of prosumers in smart-grid infrastructure is proposed as a solution to generate power with renewable distributed generation technologies. The chapter evaluates the current state of awareness regarding the potential positive implications of the energy transition, emphasizing the need to incorporate renewable energy sources into existing infrastructure. Drawing on academic and policy research, the chapter explores the role of energy transition in achieving economic growth while minimizing environmental risk.

Focusing on hyperspectral imagery, Chapter 9 explores its significance in providing rich information for applications such as target detection in urban areas. The extraction of shape-based features fused with spectral characteristics is presented as a valuable method for detecting urban targets, specifically roads and roofs. The integration of both spectral and shape aspects using support vector machines (SVM) yields improved results. The chapter discusses the experimentation process, including spectral analysis, shape analysis, fusion of spectral and shape-based features, and detection rate. The findings hold significance for city planning, urban sprawl estimation, population studies, and sustainable development, offering valuable insights for governmental and non-governmental organizations conducting related studies.

Utilizing a systematic review of existing literature, Chapter 10 documents Climate Smart Agriculture (CSA) practices in Tanzania aimed at coping with various production risks. The identified practices include the adoption of better seed varieties, application of weather information, efficient fertilizer use, intercropping, and crop residue and canal irrigation. The chapter emphasizes the role of food system issues in contributing to food insecurity and hunger in Tanzania, particularly affecting rural households. The comprehensive review provides insights into the current landscape of CSA practices, contributing to the understanding of sustainable agricultural approaches in Tanzania.

Chapter 11 explores the integration of Automated Driving Systems (ADS) and AI-driven Advanced Driver Assistance Systems (ADAS) to develop smart and self-driving cars, with a specific focus on Traffic Sign Detection (TSD) technology. Recognizing the crucial role of traffic signs in avoiding accidents and traffic congestion, the chapter presents an automatic TSD forecast utilizing the hybrid deep belief network (HDBN) model for classification. The research emphasizes the importance of understanding visual representations in traffic symbols and proposes the use of the rider sunflower optimization (RSFO) technique for hyperparameter tuning. The findings contribute to the advancement of smart urban cities by enhancing the precision of traffic flow prediction.

The target audience for this book includes educators, mentors, training professionals, researchers, and students. The content is tailored to cater to a diverse readership, spanning across educators, school students, teaching assistants, researchers, Ph.D. students, faculties of computer science, school educators, management, and information technology professionals, as well as students pursuing MBA, B.Tech, B.Ed, M.Tech, MCA, and BCA.

We believe that the topics covered, ranging from the introduction to environmental science and engineering to GIS-aided environmental planning and management, will serve as a valuable resource for individuals across various disciplines. The hands-on chapters contributed by academicians and professionals from reputed organizations further enrich the content, making this book a comprehensive guide for anyone interested in advancing geospatial practices in natural environment resource management.

Rubeena Vohra
Bharati Vidyapeeth's College of Engineering, India

Ashish Kumar
Bennett University, India

Chapter 1
Atmospheric Processes and Climate Change

Jaya Yadav
Delhi Technological University, India

Dyvavani Krishna Kapuganti
Delhi Technological University, India

ABSTRACT

The atmosphere has a significant impact on the physical environment and the people. The atmospheric processes go hand in hand with surface processes and affect the climate and contribute to the changing climate. The adversely altering environment is not the only issue that comes with climate change, but the impact it has on human and other biological populations is also worrisome and needs to be addressed. Geospatial technologies, which combine remote sensing and geographic information systems, provide flexible cross-scale tools for examining climate change, the long-term evolution of the climate system, and its effects on social and ecological systems. In order to better understand how the Earth, ocean, biosphere, cryosphere, and atmosphere have changed over time, scientists have used long-term satellite data sets. The chapter discusses the various atmospheric processes, their implications, their inter-relationship with land processes, and the application of geospatial methods, in particular remote sensing, for the observation and monitoring of the atmosphere.

1. INTRODUCTION

The atmosphere's state at any given location and time is known as weather. The long-term average range of weather, or the accumulation of daily and seasonal weather events, is what is meant by climate. However, the idea of climate encompasses much more than this, since it also refers to the weather extremes that are specific to a region,

DOI: 10.4018/979-8-3693-1396-1.ch001

such as summer heat waves and winter cold spells. The physical environment and how it functions are greatly impacted by the weather. Atmospheric processes, which include the solar influx, scattering, cloud formation, condensation, heat retention, and wind patterns, individually and jointly affect the climate and contribute to the changing climate (Ahrens & Henson, 2018).

The structure of atmosphere changes with elevation. With increasing height from Earth surface, the layers of atmosphere are troposphere (~upto 10 km), stratosphere (from 10 km to 52 km), mesosphere (from 52 km to 85 km), thermosphere (from 85 km to 692 km), and exosphere (from 692 km to 9,978 km) (Ahrens & Henson, 2018). Each of these layers have different composition and concentration of gaseous elements, and with height and concentration, their properties and functioning also changes (Khvostikov, 1965). There are different gas molecules present in all the strata of the atmosphere and based on the gravitational energy, these gases move about and interact with each other. The gravitational fractionation is however, obliterated by turbulence and convective winds (Fleagle & Businger, 1981). The major components of atmosphere are nitrogen and its di-oxides (78.08%), oxygen (20.95%), and argon (0.93%) and the minor components are water (0-4%), carbon-di-oxide, neon, helium, methane, hydrogen, carbon monoxide, ozone and halogen derivatives of organochlorine compounds, with trace components such as ammonia and nitrous oxide (Thompson R., 1998). The concentration of these gases varies with the increasing strata of atmosphere.

Troposphere, the lowest layer of the atmosphere, is primarily made up of N_2, O_2, H_2O and Ar. Majority of the atmospheric processes leading to the formation of weather and climate occur in this layer. One of the main factors affecting global mean climate of the Earth is its energy budget, the equilibrium between the energy the Earth is receiving from the Sun and the energy it is radiating back into space (Krishnan, et al., 2020). The composition of the atmosphere varies the incoming and outgoing radiative flow at the surface thereby affecting the climate. It is clear that variable trace gases contribute significantly to both short-term climate change and how weather systems work. Recently, it is found that role of carbon dioxide in the greenhouse effect and global warming has increased significantly, but other gases like methane, nitrous oxide, and chlorofluoro carbons (CFCs) also play a significant role (Malhi, Kaur, & Kaushik, 2021).

The main causes of the current climate change are human-caused (anthropogenic) emissions of greenhouse gases (GHGs), aerosols, and changes in land use and land cover (LULC) (Krishnan, et al., 2020). GHGs warm the surface by limiting much of the terrestrial radiation of Earth that is lost to space. According to ice core records, atmospheric concentrations of the main GHGs—carbon dioxide, methane, and nitrous oxide—are now higher than they have ever been in the last 800,000 years and there is a high degree of certainty that their mean growth rates over the past century are unheard of in the previous 22,000 years (Krishnan, et al., 2020).

Aerosols are microscopic particles or droplets that come from both natural and man-made sources and are suspended in the atmosphere (Li, et al., 2022). Mineral dust from soil erosion, sea salt, and volcanic eruptions are examples of natural sources while transport, industrial air pollution, biomass burning, which releases airborne sulphates, nitrates, ammonium, and black carbon, and dust from land degradation processes like desertification are the main anthropogenic sources (Mushtaq, Sharma, Bangotra, Gautam, & Gautam, 2022). Aerosols tend to cool the surface either directly (by scattering or absorbing solar radiation) or indirectly (by promoting cloud formation) and thus, human-caused aerosol pollution has partially offset the warming brought on by anthropogenic GHG emissions (Krishnan, et al., 2020).

Over the past few centuries (and even earlier), significant changes in land use and land cover (LULC) have affected a large portion of the Earth's land surface, primarily due to deforestation and the expansion of agriculture (Singh, 2014). Deforested areas are less effective as carbon dioxide sinks and, if biomass burning occurs alongside them, they are a direct source of GHGs. The terrestrial albedo is also changed when land is turned from natural vegetation to agriculture or pasturage, which affects the surface radiative balance. Such drastic change of surface composition alters the energy and particle exchange between lithosphere/hydrosphere and atmosphere, ultimately influencing the atmospheric processes. The net effect of human-induced climate forcing has been an increase in the global average near-surface air temperature by approximately 1°C since pre-industrial times (Krishnan, et al., 2020). Each of the last three decades has been successively warmer at the Earth's surface than any preceding decade since 1850 and 2001–2018 have been 18 of the 19 warmest years in the observational record (Krishnan, et al., 2020). The response anticipated from a warming planet is consistent with trends in other significant global climate indicators, such as the rate and patterns of precipitation, temperature and precipitation extremes, atmospheric water vapor concentration, melting of continental ice, sea level rise, ocean heat content, ocean acidification, and the frequency of strong cyclones.

Besides, the modern climate models predict that anthropogenic global warming and related climate change will persist throughout the twenty-first century, with significant ramifications (Malhi, Kaur, & Kaushik, 2021). However, there are still a lot of areas where research is needed, especially for climate projections at smaller spatial and temporal scales for instance, CMIP5 simulations of past and projected changes in monsoon rainfall show significant regional variations. The range of predictions made by different climate models regarding future changes in the Indian Ocean warming, regional sea-level rise, tropical cyclone activity, extreme weather and climate, changes in the Himalayan snow cover, etc. must also be narrowed (Krishnan, et al., 2020). Emphasis is to be made on key processes in climate models, increase the understanding of the science behind climate change (e.g. clouds, aerosol–cloud interactions, vegetation–atmosphere feedbacks, etc.), further, increase human capacity to deal with these issues.

There is need for continued monitoring of the meteorological process to entirely understand the atmospheric processes. Geospatial Earth Observation application proved to be very valuable in studying climate change; changes in land use and land cover, snow and glaciers, ocean dynamics, and dynamics of atmosphere. It is even easy to assess the damages caused by some natural hazards, and efforts are being made to use satellite data for monitoring or early warning of the hazards. In order to better understand Earth's vulnerability and risk, scientists are working to combine ground and satellite observations. In order to better understand how the Earth, ocean, biosphere, cryosphere, and atmosphere have changed over time, scientists have used long-term satellite data sets and improved their understanding of global climate change.

The further sections of this chapter will discuss the physics of atmosphere, atmospheric processes, land processes and geospatial advances in climate change studies regarding a few atmospheric processes.

2. PHYSICS OF THE ATMOSPHERE

Radiation from the Sun is the only source of energy that enters the atmosphere. This energy initiates and maintains numerous atmospheric processes. A large flux and short wavelength are used to represent the solar radiation that the Earth's surface receives from the sun (known as insolation). The radiation is also called electromagnetic radiation with a wavelength range of 10^{-12} to 10^{3} m. Comprised of particles called photons that carry electromagnetic energy, the EMR has a magnetic field and an electric field, which propagate perpendicular to each other in a wave-nature and reach the earth surface (Andrews, 2010).

As radiation travels through the atmosphere, it is diminished by absorption, reflection, and scattering. Absorption is the phenomenon where matter takes in the energy of the photon that comes in contact with it. In this case, the energy and momentum of photons is transferred to matter that it has come in contact with. This type of radiation may emit out of the particle later on, in the form of dissipating heat or as thermal energy. The second common phenomenon is reflection. Reflection is a non-selective process that affects the entire spectrum's range of wavelengths. It is the bouncing back of the incident photons when they come in contact with surface of a matter. In the atmosphere, it primarily originates from larger aerosols and cloud tops (and is influenced by cloud thickness). Upon reaching Earth surface, it occurs from all earth surface features. The third common phenomenon, scattering, is the spreading of light in all directions when it bounces back after contact with the surface of matter. Scattering mostly occurs in the atmospheric layers when light comes in contact with clouds, aerosols and dust particles. Based on the wavelength of the incident radiation and the size of the particle that causes scattering, it is of three distinct types. Rayleigh scattering occurs when the particles that scatter light are smaller than the wavelengths. The shorter blue wavelengths are scattered more readily than the longer red wavelengths in, which is generally referred to as Rayleigh scattering. Shorter wavelengths are primarily absorbed by clouds and atmospheric gases. The majority of gamma rays, X-rays, and harmful UV radiation are absorbed by oxygen and ozone. Mie scattering occurs when the particles that scatter light are about the same size as the wavelength of the radiation. Finally, the non-selective scattering occurs when particles are larger than the wavelength of the radiation. Here, all wavelengths are scattered about equally, a reason why fog and clouds appear white. The remaining of the radiation that reaches the earth's surface and absorbed by it is converted to thermal energy, which heats the surface and causes it to emit long-wave radiation (Caballero, 2014).

Atmospheric physics, the radiation theories and the different phenomena occurring, are vital for understanding the atmospheric processes. They are a fundamental part of weather forecast models and are integrated into modelling as physical parameterization.

3. ATMOSPHERIC PROCESSES

The entirety of Earth and the space around it is in constant exchange of energy between various entities that form the atmosphere and land. The exchange of energy and particles leads to occurrence of several physical processes. Atmosphere is an envelope where many such processes take place. This is aided by both, the solar radiation and Earth's rotation. The radiant fluxes, air pressure, wind patterns and regulation of chemicals between the biosphere and geosphere for the baseline for all the other physical processes. The major atmospheric processes that have garnered much of research focus are wind patterns, temperature formation and precipitation (Thompson R. D., 1998).

Air pressure pays an important role in the formation of wind patterns. Wind patterns, in turn, are associated with condensation and precipitation. Surpluses and deficits in solar radiation cause pressure differences in the atmosphere, such as polar anticyclones and equatorial low pressure, which cause the winds to blow and encourage the formation of precipitation and aridity (Huntley, 2023). Formation of precipitation is possible from the regulation of water through the soil, plants and air, in the form of evaporation and evapotranspiration. These processes dissipate heat of their own, which contributes to the larger source of temperature formation. Temperature formation is aided by conduction and convection processes.

Without an efficient atmospheric transfer by non-radiative energy fluxes, the Earth's annual radiant energy surplus, or available net radiation, would raise surface temperatures. Conduction and convection processes help move the energy away from the Earth's surface as needed. Heat is transferred within a substance through conduction, which involves the collision of quickly moving molecules. It is understood that this transfer works best in solids and least well in gases. The layer of air closest to the Earth's surface, or the atmospheric boundary layer, is therefore the only one where molecular conduction is significant. Convection, on the other hand, is the turbulent vertical movement of energy that can only happen in liquids and gases. Local air movements, or eddies, in the atmosphere causes turbulent motions brought on by free or forced convection to transport energy and water vapor upward. Free and forced convective systems, in which air parcels rise through the atmosphere with distinct temperature changes, result in energy transfers within the troposphere. Since it results in stable or unstable conditions, this relationship regulates the amount of vertical movement of air parcels within the troposphere (Caballero, 2014; Andrews, 2010).

The operation of atmospheric processes and systems at all scales depends critically on atmospheric water vapor. Water constantly circulates throughout the atmosphere. Here, evaporation occurs whereby the sun's energy turns vast amounts of liquid water into water vapor. The moist air is then carried by winds to other locations, where condensation causes the water vapor to condense back into liquid (or ice) and form clouds. The solid ice crystals or liquid cloud particles may enlarge and descend to the surface as rain, snow, or hail under specific circumstances. When precipitation enters an ocean, the water cycle restarts. In contrast, a significant portion of the water only makes its way back to the ocean after a difficult journey if the precipitation falls on a continent. The hydrologic (or water) cycle is this movement and transformation of water molecules from a liquid state to a vapor state and back again.

4. ADVANCES IN GEOSPATIAL APPLICATION TO STUDY ATMOSPHERIC PROCESSES

The previous couple of decades have seen a great deal of application of the geospatial sector, as a whole, for research in atmospheric and Earth observations. This was possible because of the scientific advancement of the technologies employed in satellite studies and other remote sensing fields such as airborne and field-based cameras. With the growing importance of climate studies and emphasis on global physical processes, this section will discuss a detailed information on geospatial advances in three of the major processes that influence the global climate change, viz., greenhouse gases, aerosols and the thermal environment due to the changing LULC.

4.1. Greenhouse Gases

Credible information on GHGs, including source of the emissions and their volume, is crucial for combating climate change and this data can be used to assess the usefulness of GHG mitigation strategies, policies, and initiatives. The functioning of GHG trading schemes is made possible by the potential to evaluate GHG reductions at various source levels such as the global, national, sectoral, or individual point. Without trustworthy trading units, these schemes would be incoherent (Sun, et al., 2022). The confidence of buyers and sellers in such trading schemes is based on this credibility. Accurate GHG data is also necessary for studies on the interactions between GHGs and global warming. It is possible to speculate the sources and sinks of key GHGs, identify the physical and chemical processes that cause their variations, and forecast their future trends. Accurately capture the diurnal, monthly, seasonal, and inter-annual variability of important GHGs to further understand how GHG emissions interact with the atmosphere and how the climate responds to both natural and anthropogenic GHG emissions. In addition to promoting research into the carbon cycle, a trustworthy GHG monitoring system could aid in the worldwide development of chemical transport models (CTMs) and emission inventory compilation technologies (Sun, et al., 2022).

According to estimates from the United Nations Intergovernmental Panel on Climate Change (IPCC), net-zero carbon dioxide (CO2) emissions, or "neutrality," must be achieved by the integrated Earth system by 2050. Carbon neutrality is a crucial strategy in the fight against climate change. It is the global goal where annual CO2 emissions are equal to the amount of CO2 reduction through strategies that can either increase carbon sinks or reduce carbon sources, such as decarbonization, carbon offset, and energy conservation (Sun, et al., 2022).

In-situ networks provide a wealth of accurate surface CO2 concentration measurements, but one major limitation of these networks is their limited coverage. Additionally, there remain substantial uncertainties surrounding the magnitude, distribution, and duration of both natural and man-made emissions (Boesch, et al., 2021). In particular, measurements of atmospheric CO2 from space can offer the necessary global coverage with the temporal and spatial resolution required to enhance our comprehension of regional carbon fluxes, especially in regions like the Tropics or high latitudes that are not well sampled by surface networks.

The development of optical remote sensing methods sampling the entire atmospheric column over the past two decades has proven to be very beneficial for tracking GHGs in the atmosphere. Using satellite measurements of vertical columns or column averaged CO_2 and CH_4 mixing ratios may ameliorate this situation; satellite will cover the entire globe more efficiently, in cost effective manner, while providing large sets of data. To determine the global distributions of GHGs, a number of cutting-edge satellites with various spatiotemporal resolutions are launched, including SCIAMACHY from the European Space Agency (ESA), GOSAT and GOSAT-2 from Japan, OCO-2 and OCO-3 from the United States, TanSat and Gaofen-5 (GF-5) series satellites from China, GHGSat from Canada, etc. These satellites primarily use infrared grating or Fourier transform infrared (FTIR) spectrometers to measure total columns of GHGs through atmospheric limb or nadir observations. The lowest layer of the atmosphere, where the sourced sink signals are strongest, is also sensitive to changes in greenhouse gas concentrations in these measurements.

High-resolution FTIR spectrometers that are mounted on the ground are effective tools for determining GHG total columns and profiles. The NDACC-IRWG (Network for Detection of Atmospheric Composition Change-Infrared Working Group) and the TCCON (Total Carbon Column Observing Network) both use high-resolution FTIR spectrometers to monitor total columns and profiles of GHGs and atmospheric pollutants. The high-resolution IFS120/125HR FTIR spectrometers have some exceptional qualities, such as high precision and stability, but they also have some drawbacks. They are bulky, expensive spectrometers that require a lot of infrastructure to operate, and maintaining their optical alignments is challenging and time-consuming. The COCCON (COllaborative Carbon Column Observing Network) are made using EM27/SUN spectrometers. Figure 1 shows network of TCCON, NDACC and COCCON.

Examples of monitoring platforms include tower-based, balloon, tethered balloon, ship-borne, vehicle-borne, airborne, space-based, and ground-based remote sensing platforms like ground-based FTIR. Manual sampling analysis and surface *in-situ* measurement sites (e.g., laboratory measurement or surface monitoring network) are also examples of monitoring platforms (Figure 2). Because no single monitoring platform can fully satisfy the requirements of stereoscopic monitoring of GHG emissions due to its limited coverage or spatial resolution, scientists usually combine a number of observation platforms to create a stereoscopic monitoring system.

Specific *in-situ* monitoring and remote sensing technologies have to be compared, verified, and optimized in depth in order to learn from and complement one another in the development of a reliable GHG stereoscopic monitoring network. In order to

Figure 1. Global FTIR observation networks
Sun et al. (2022)

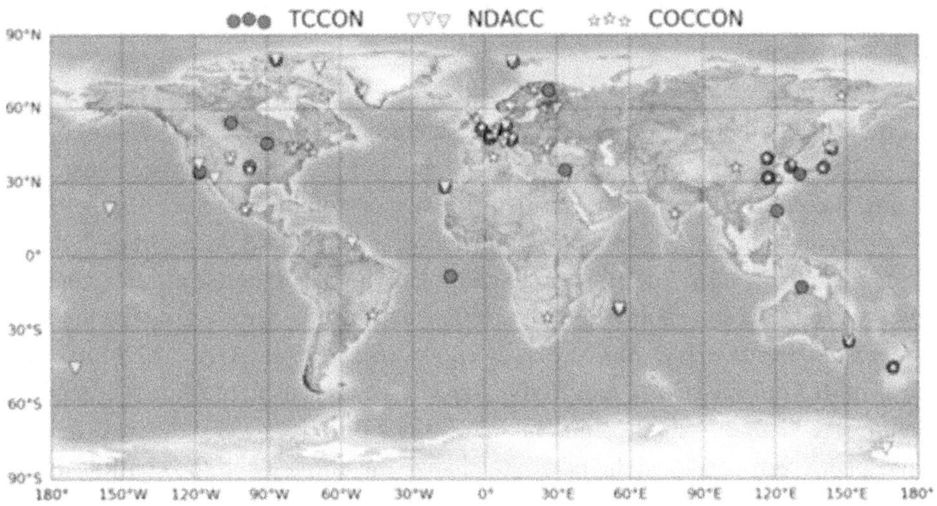

Figure 2. Global GHG monitoring network coordinated by NOAA and WMO
Sun et al. (2022)

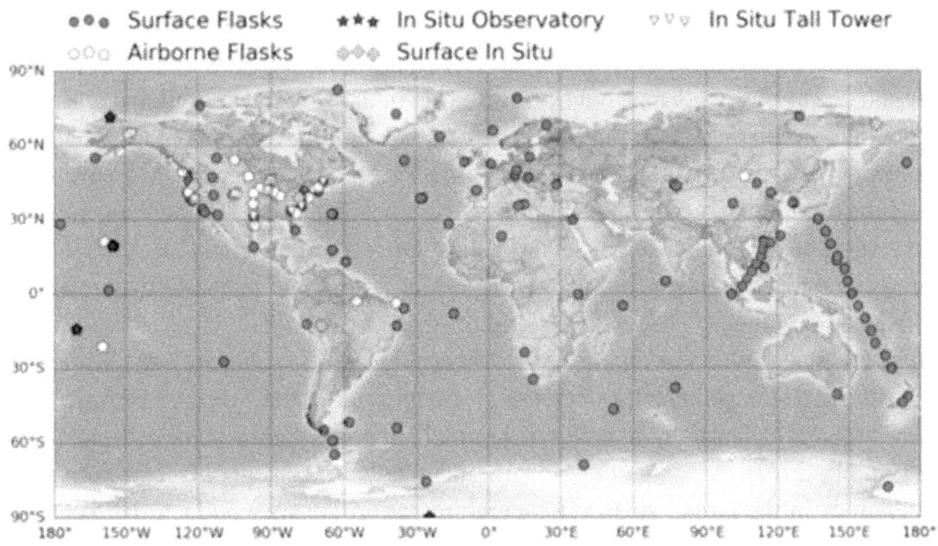

achieve stereoscopic GHG monitoring, full use of the various monitoring platforms and technologies should be made, and play to their strengths while avoiding their flaws.

4.2. Aerosols

An integral component of Earth's atmosphere are atmospheric aerosols. Aerosols are made up of solid, liquid, and gaseous materials that differ in size, phase, and chemical complexity and are described as a suspension of liquid or solid particles dispersed throughout the atmosphere that are present in a gaseous medium (Lee, Li, Kim, & Kokhanovsky, 2009). Particulate Matter (PM), sulphates, nitrates, mineral dust, sea salt, black carbon, organic carbon, etc. are all generally grouped together as aerosols. Further, PM can be classified as coarse (10 μm), fine (2.5 μm), and ultrafine (0.1 μm) based on their aerodynamic diameter (Mushtaq, Sharma, Bangotra, Gautam, & Gautam, 2022; Zhang, et al., 2021). Both anthropogenic and natural atmospheric aerosols have the power to change the climate and the equilibrium of the earth's atmosphere.

Globally, aerosols have the greatest concentrations and instantaneous forcing in proximity to their sources; however, their effects on the energy balance and climate can be more profound and widespread geographically. They cause perturbations in the spatially inhomogeneous radiative energy balance at the surface and at the top of the atmosphere. Aerosols cause a heterogeneous energy perturbation that includes a strong interhemispheric asymmetry with specific implications for both hemispheres' climates. On hemispheric scales, they modify the vertical and horizontal thermal structure, which has an immediate impact on the circulation of air and ocean. Aerosols may have played a role in the slowdown of global warming from 2000 to 2015 by causing a sea surface temperature (SST) response that was opposite in sign but similar in pattern to that of greenhouse gases. Moreover, the Atlantic meridional overturning circulation may be impacted by the aerosol-induced SST variations, which also cause anomalies in ocean circulation (Li, et al., 2022). Aerosols tend to weaken the hydrologic cycle, in contrast to the climate-altering effects of greenhouse gases, which tend to strengthen it.

Large-scale climate anomalies like the warming of the Arctic, the harsh winters in the Northern Hemisphere, the decrease in precipitation in the Northern Hemisphere, and the hemi-spherically asymmetric rainfall trends are also caused by aerosols (Li, et al., 2022). Aerosols that absorb and scatter light generally cause discrete alterations in the global surface temperature pattern, energy budget, hydrological cycle, and climate responses (Krishnan, et al., 2020; Ranjan, Patra, & Gorai, 2021).

At a regional scale, the local warming or cooling brought on by aerosols can create a temperature gradient between contaminated areas and their surroundings, and this can disrupt or change the regional circulation (Li, et al., 2022). The monsoon scenario, particularly in South and East Asia, is the most representative case. The South and East Asian monsoons are weakened when aerosol cooling at the surface lessens the temperature difference between the land and the ocean. One important factor influencing the monsoon shifts is the fraction of absorbing aerosols. The so-called atmospheric brown clouds, which were first identified over the Indian Ocean in aircraft carried from the Indian subcontinent, are formed when black carbon (BC) combines with organic carbon (Li, et al., 2022). By raising the temperature of the atmosphere above the ocean, these aerosols lessen the strength of the South Asian monsoon and the temperature differential between the ocean and land. Additionally, BC has the potential to significantly weaken the monsoons in South America and Africa. Aerosols, both nearby and far away, can have an impact on regional variations in the hydrologic cycle in Asia (Fadnavis, et al., 2020).

Aerosols lower atmospheric stability and cool the surface on even smaller scales. Aerosols that absorb are more effective in this process; however, because of their higher loading, aerosols that scatter typically have a greater overall effect. The planetary boundary layer height may be lowered by this process, leading to the so-called aerosol–boundary–layer feedback, which is believed to play a major role in the severe pollution events that occur in China.

For the past twenty years, satellite remote sensing has been a huge help with a variety of studies, including aerosol modelling, weather forecasting, climate change, and the investigation of global pollution issues (Mushtaq, Sharma, Bangotra, Gautam, & Gautam, 2022). The products for satellite remote sensing aerosol monitoring are sufficiently accessible despite their low spatial resolution. Limiting them to regional and global scopes, though, makes them unsuitable for monitoring at the city level. Pollutant concentration has frequently been measured by sparsely distributed regulatory control networks. Installing inexpensive sensors was done in addition to routine monitoring, which offers concentration areas for air pollutants, as they became more widely available. Figure 3 & 4 give some of the operational satellite sensors that provide aerosol information.

A worldwide network of sun photometers at ground level called AERONET stations makes more precise calculations (Verma, Prakash, Soni, & Ram, 2019). Aerosol optical depth (AOD) and aerosol properties are utilized for the validation and calibration of satellite AOD retrieval. However, because these satellites are not widely dispersed throughout the world, there is a need for improved coverage of AOD retrieved

from satellites. In comparison to ground-based aerosol observing sensors, satellite remote sensing has inherent demand because it provides high spatial resolution data and nearly continuous spatial coverage over large areas (Verma, Prakash, Soni, & Ram, 2019). But the most precise AOD concentrations for estimating aerosol effects on a regional and global scale come from sparsely distributed ground-based remote

Figure 3. Few operational satellite sensors products providing aerosol information
Ranjan, Patra, and Gorai (2021)

AOD product	Satellite	Orbit	Wavelength	Data available since
MODIS	Terra	Sun-synchronous	VIS, IR	1999
MISR			VIS	1999
MODIS	Aqua		VIS, IR	2002
MAIAC	Terra & Aqua (MODIS)		VIS	2000
OMI	Aura		VIS, UV	2003
CALIOP	CALIPSO		LiDAR (532 nm, 1064 nm)	2006
GOCI	COMS	Geostationary		2010
VIIRS	Soumi NPP	Sun-synchronous	VIS, IR	2011
AHI	Himawari-8	Geostationary		2014
ABI	GOES-16			2016

Note: *ABI* Advanced Baseline Imager, *AHI* Advanced Himawari Imager, *GOCI* Geostationary Ocean Colour Imager, *IR* Infrared, *MAIAC* Multi-Angle Implementation of Atmospheric Correction, *UV* Ultraviolet, *VIS* Visible light

Figure 4. The main satellite platforms and sensors for global aerosol observations
Zhang et al. (2021)

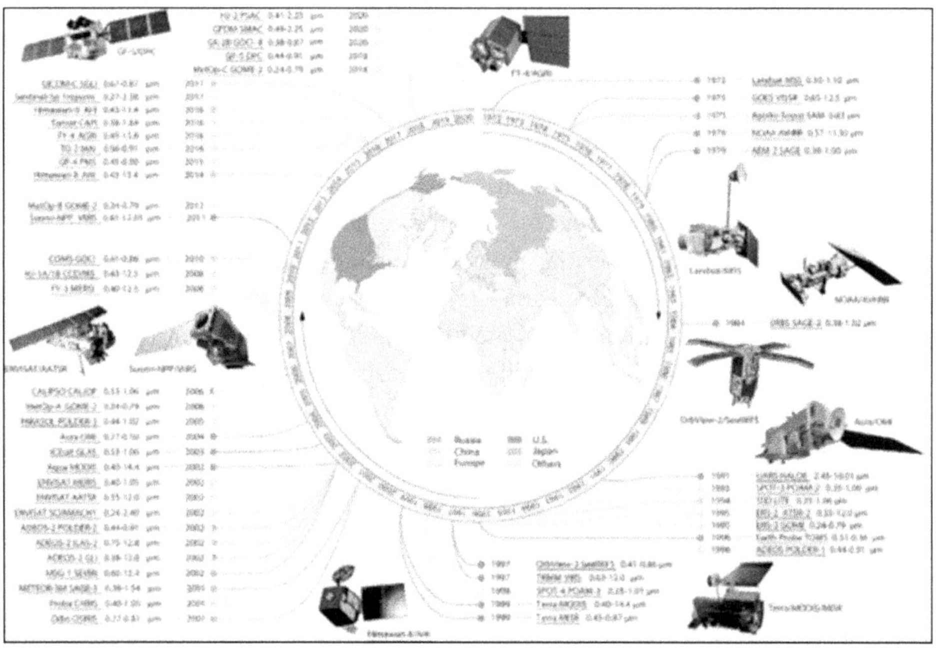

sensing, like aerosol data from the Moderate Resolution Imaging Spectroradiometer (MODIS) and the Aerosol Robotic network (AERONET) (Mushtaq, Sharma, Bangotra, Gautam, & Gautam, 2022).

The most widely used datasets for estimating ground level PM concentration are those based on MODIS-based AOD data. However, compared to MODIS data in previous studies, MISR and VIIRS based AOD datasets have shown better consistency in PM estimation (Garrigues, et al., 2022). Several satellite sensors, such as Terra, MISR, and SeaWiFS, when used in combination can improve the spectral and temporal capabilities for AOD retrieval and more precise PM predictions and also satellite data with medium to high resolution (Landsat, Sentinel, among others) can improve the precision of PM estimation (Ranjan, Patra, & Gorai, 2021). These days, a more accurate estimation of PM concentrations is being considered using the vertical profile of aerosol based on LIDAR data. Unfortunately, due to higher costs and other complications, these data are only available for a limited number of spatial regions.

For PM estimation, statistical regression-based models have been used extensively worldwide over the past 20 years. The MEM model is the most widely used statistical regression-based model and has been found to perform optimally; on the other hand, the GWR model has demonstrated better consistency for PM prediction at the local or regional scale. Compared to linear regression models, the multiple regression (MR) models that were created by adding different auxiliary variables, such as land-use, PBLH, and meteorological factors, have demonstrated a somewhat higher level of consistency. However, MR models are found to perform relatively poorly when compared to models such as MEM, GWR, LUR, etc.

These days, PM concentration estimation is also done using a variety of machine learning and deep learning techniques, such as artificial neural networks (ANN), two-stage models (TSM), random forests (RF), support vector regression (SVR), decision trees (DT), and gauss neural networks (GNN) (Huntingford, et al., 2019; Ranjan, Patra, & Gorai, 2021). However, further validation is required for these techniques on a national and international level.

Figure 5 shows the approaches for the utilisation of ground and satellite data to study aerosols. In order to predict PM concentration more accurately and reliably, it is important to investigate the high-resolution (temporal and spectral) satellite data for AOD retrieval (Verma, Prakash, Soni, & Ram, 2019). To further increase the accuracy level of PM concentration prediction, more auxiliary factors must be identified and taken into account in the model and the main goals of upcoming aerosol observation techniques are to increase measurement accuracy and broaden the range of parameters that satellite sensors can retrieve (Li, et al., 2022). To obtain crucial parameters that cannot be obtained from remote sensing alone, and to achieve the necessary accuracy of various parameters for constraining aerosol forcing, satellite remote sensing must be combined with in situ observations (Ranjan, Patra, & Gorai, 2021).

4.3. Thermal and Vegetation Processes From LULC Change

All of the features covering the surface of Earth are classified into any one of the land cover or land use (LULC) category. LULC entails the information about an

Figure 5. AOD measurement through ground and satellite
Mushtaq, Sharma, Bangotra, Gautam, and Gautam (2022)

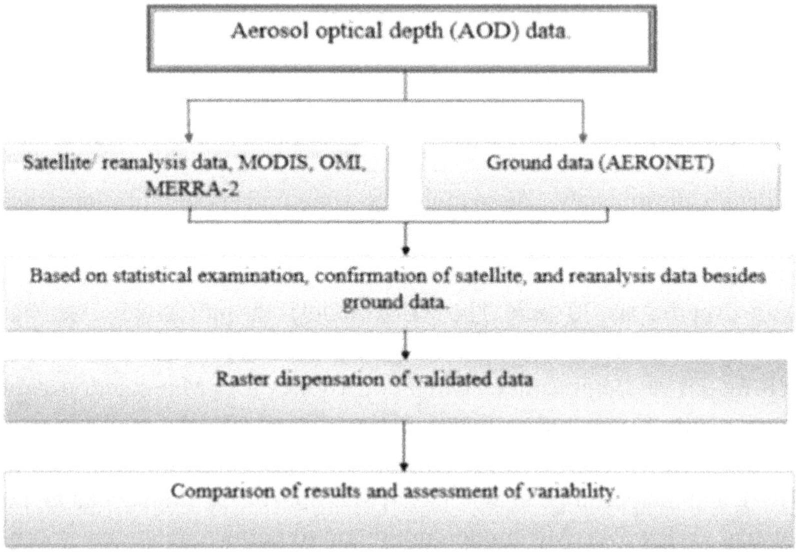

Earth feature, in time, occurring as either a cover or as a practice (Yirsaw, Wu, Shi, Temesgen, & Bekele, 2017). For this purpose, a forest class is land cover where as an agricultural class is a land use. Based on the features and physiological functioning of the features, each category has its own microclimate and meteorological processes (Gallo, Easterling, & Peterson, 1996). The changes in LULC exert a profound impact on the overlaying atmosphere. The regional meteorological conditions in the atmosphere affect the physiology of land underneath and, as a feedback process, the land changes impart some altering pressure on the atmospheric processes above. Majority of the atmospheric processes occurring in the atmosphere are somehow inter-related to the alterations of the land underneath, be it emission of GHGs, loss of carbon to the atmosphere, increasing concentration of aerosols, or the regulation of water linking soil moisture, evapotranspiration and planetary boundary layer to form clouds or precipitation (Ito & Hajima, 2020). One of the most problematic outcomes of this interaction is the heat exchange and emissivity. The incident radiation that the earth absorbs at any point in time is emitted at a later time, especially night time, in the form of thermal energy. The long wave radiation that comes from the land surface is calculated from land surface emissivity, which is the ratio of the energy emitted from the land to the energy emitted from an ideal black body at the same temperature, using the Stefan-Boltzmann law. Emissivity is of importance because the closer the value of emissivity is to one, the better that surface is as an emitter. The ability of land surface to reflect the shortwave incoming radiation into the atmosphere depends upon emissivity. This is denoted by surface albedo, which is the fraction of sunlight reflected by Earth, and it is also important to model the land surface energy balance (Dickinson, 1983). The other important phenomenon is heat formation, denoted by temperature measures. Temperature can be measured at the level of surface as Land Surface Temperature (LST) or 2m above the surface and upwards as air temperature (Ta).

Several variations of land surface play a critical role in maintaining the atmosphere as it is. To start with, vegetation plays a major role in denoting how gases, especially carbon, is sequestered in terrestrial zones, and only little amounts of it is escaping into the space. Besides, vegetation has lower albedo and reduces the reflecting thermal radiation, thus, reducing the temperature. Additionally, vegetation has foliage and an active crown. Based on the Leaf Area Index (LAI), and the sky view factor, part of the incoming radiation passes through canopy and reaches ground. The remaining radiation is either absorbed by the foliage, scattered off the foliage or backscattered off the foliage (Hardwick, et al., 2015). Now since thermal energy is the radiation that is emitted from the Earth surface, this leaves little scope for thermal reflectance from the vegetated surfaces. LAI, which is the area of leaf per unit ground area, partitions the surface heat fluxes, regulates light intake by the leaf and directly affects the stomatal conductance, which in turn impacts the flow of carbon and water through the plant body (Tao, et al., 2013). Hence, finally, vegetation suffices regulation of water in the form of evapotranspiration. Evapotranspiration (ET) is a phenomenon where the transpiration of water from soil to atmosphere occurs through the plant body and evaporation of water occurs from the surface of the plant body in a combined manner. ET functions under the conductance of stomata but is strongly influenced by availability of moisture in soil and the vapor pressure deficit outside the leaf. However, ET has been termed as the necessary evil of the plant physics since it drives flux of carbon alongside (Krishna, Watham, Padalia, Srinet, & Nandy, 2023). This extensive process of carbon and ET regulation releases the chemical products into the atmosphere, where they contribute to GHGs and precipitation. In the process of ET, water moves from higher moisture surroundings to dryer air than saturated air. Hence, upon release, it imparts a cooling effect and reduces temperature (Tan, Wong, Tan, Jusuf, & Chiam, 2015).

There are numerous methods and techniques to study the thermal behaviour of land and the heat flux across terrestrial land and atmosphere. Starting from in-situ field data collection, the wide array ranges to remote sensing studies. Amongst the remote sensing techniques, the geospatial sector has seen great advancements in these applications. Using remote sensing data and geospatial technologies it is possible to map and analyse the changes in land cover, visualize a time series change in LULC, and compute LST, which underlines the radiative load of the Earth's surface. Additionally, it is also possible to model the physiology of vegetation processes in view of the changing meteorology. Spatial analysis allows differentiating between the temperature from leaves of the canopy in dense vegetation and temperature from the whole canopy, various parts of the trees and ground surface in sparse vegetation. These changes may not be wholly captured by meteorological stations. The study of LULC and relating LST is mostly done using multispectral (MS) and thermal satellite imagery, respectively. Thermal sensors, the main source for thermal reflectance data of Earth surface, detect emitted/reflected terrestrial radiation. Microwave sensors are also known to capture the terrestrial radiation. While thermal datasets have challenges of clouds present in the atmosphere at the time of satellite overpass, microwave overcomes this challenge through its cloud penetrative ability. However, microwave does not give as high spatial resolution as thermal datasets (Almeida, Teodoro, & Goncalves, 2021). The details sensors and their specifications given in the later portion of this section are taken from their respective websites.

Thermal Sensors

Their operation range is 8-14 µm. Radiation to and from Earth's surface can enter and exit within this atmospheric window. Mostly Landsat, Advanced Spaceborne Thermal Emission and Reflection radiometer (ASTER) and MODerate resolution Imaging and Spectroradiometer (MODIS) are used for thermal studies.

Landsat Mission

Although Landsat mission started in 1972, thermal band was included only in the Landsat 4 satellite. Currently, Thematic Mapper (TM), Enhanced Thematic Mapper (ETM+, Thermal InfraRed Sensor (TIRS) (series 4-9) offer thermal imagery. However, Landsat 4 and 5 decommissioned in 2011. At a height of 705 km, Landsat operates in near-polar orbit and has a revisit time between 8 days and 16 days. TM on Landsat 4 started its operation in July 1982 and decommissioned it in June 2001. TM on Landsat 5 started its operation in March 1984 to January 2013. It is a multispectral sensor with seven bands, and band 6 collects thermal data. The MX bands have a spatial resolution of 30 m and the thermal band have a spatial resolution of 120 m. Often for thermal studies, the thermal band is resampled to match the 30 m MS resolution. ETM+ on Landsat 7 started in April 1999 and the mission is still ongoing. It has eight spectral bands, which include panchromatic (PAN) and thermal bands as well. The spatial resolution of MX bands is 30 m, thermal band is 60 m and PAN is 15 m. The Landsat 8 mission, launched in 2013, has two sensors, OLI – Operational Land Imager and TIRS- Thermal InfraRed Sensor. The OLI has nine spectral bands which includes 1 PAN at 15 m resolution and 8 MX bands with 30 m resolution. The TIRS has two thermal bands of different wavelengths, with

100 m resolution. The Landsat 8 mission was launched in 2021 with similar features to Landsat 8 and both sensors renamed to OLI2 and TIRS2. The heliosynchronus orbital path of Landsat series allows it to collect temporal data of the same area. The United States Geological Survey (USGS) disseminates free Landsat data in level 1(radiometric corrected) and level 2 (atmospheric corrected). Landsat has its predetermined algorithm for calculation of LST using the emissivity and vegetation spectral indices.

ASTER Mission

ASTER on-board NASA's (National Aeronautics Space Administration) Terra satellite was launched in December 1999. It has three subsets of different wavelength regions. The visible and near-infrared (VNIR) are the three MS bands 15 m spatial resolution, six shortwave infrared (SWIR) bands with 30 m spatial resolution and five bands with 90 m resolution in the thermal infrared region. The VNIR bands also form stereoscopic imagery with the rare-view and nadir-view telescopes. It has a revisit time of 16 days. ASTER provides data in Level 1A that is the raw data without any pre-processing and calibration, Level 1B with geometric and radiometric correction, Level 2, which has various products such as surface temperature, surface emissivity, surface reflectance etc., and Level 3A which has digital elevation models.

MODIS Mission

NASA launched the MODIS sensor on-board the Terra satellite in 1999 and on-board the Aqua satellite in 2003. The sensor is made with 36 bands in the VNIR and infrared wavelengths. It is one of the low spatial resolution sensors, with the resolution ranging between 250 m to 1 km in different bands. Alongside physical atmospheric and land features, MODIS can capture biological properties of ocean and vegetation. The surface reflectance dataset from MODIS is available at 250 m and the thermal dataset is available at 1 km. the other bands are pre-processed products of surface temperature, gross and net primary productivity, vegetation indices, evaporation precipitation, LST etc. MODIS calculates the LST using Temperature Emissivity Separation algorithm. Studies have shown that integration of MODIS data from both the satellites provides valuable results for daytime and night-time LST.

Other Polar and Geostationary Satellites

Other than the above-mentioned satellites, which are mostly used for LST studies, there are other satellites that are stationed in polar orbits or are geostationary and provide temporal data. These satellites include:

Geostationary Operational Environmental System (GOES) is a NASA launched satellite controlled by National Ocean and Atmospheric Administration (NOAA), which is a geosynchronous equatorial orbiter. It was purposed to cover southern Canada, Central and South America, continental United States, and the Pacific and Atlantic oceans. Spinning Enhanced Visible InfraRed Imager (SEVIRI), on-board Meteosat Second Generation (MGS), operates in a geostationary orbit with twelve bands and is used for predictive weather models. Out of the twelve bands, eight bands are in the thermal infrared channel. National Polar-orbiting Partnership (NPP), working as a collaboration of NASA and Joint Polar Satellite System (JPSS) is used to monitor climate trend and biological productivity.

Additionally, passive microwave radiometers are also used to compute LST. They emit the natural emissions that come from the Earth's atmosphere and land. These datasets have their limitations regarding spatial resolution and are known to be unreliable for temperature measurements on frozen surfaces. Some of the microwave radiometers are Advanced Microwave Scanning Radiometer-Earth Observation System (AMSR-E) on-board NASA's Aqua, ASMR2 on-board Japan Aerospace eXploration Agency's Global Change Observation Mission-Water1 (JAXA's GCOM-W1), and Microwave Radiation Imager (MRI) on-board the FengYun-3B.

LULC classification is mostly done using multispectral or hyperspectral imagery. The mapping is done on different scales based on the purpose. The annual maps are made at a scale of 1:250000 and for five years, it is made on a scale of 1:50000. Maps of the scale of 1:10000-1:25000 are made for urban applications, whereas, 1:5000-1:10000 is used for vegetation mapping. For comprehensive understanding of the land surface and associated land processes, various satellite images are used. One particular review found that approximately 77.03% of the LULC data in their reviewed literature came from single source, 19.69% from two sources, and others had more than two sources (Almeida, Teodoro, & Goncalves, 2021). Much of the LULC analysis is done using Landsat exclusively or in integration with other data sources. Following this, the low-resolution MODIS and medium resolution ASTER datasets are used. High resolution and very-high resolution satellites such as SPOT, IKONOS, Worldview, and Orbview are also used for LULC studies. Other than satellite data, airborne sensors such as Unmanned Arial Vehicles (UAV), drones, and cameras aboard airplanes are also used (Izuka, et al., 2018). Usage of various geospatial datasets gives improvements in spatial and temporal resolution. However, data from airborne sensors, high resolution, and very-high resolution are commercial products.

The other parameters crucial for studying the interrelationship between land processes and atmospheric processes are meteorological variables (Lauwaet, et al., 2016). Weather and climate are geographical phenomenon. It is important to keep in mind that meteorology or climatic factors vary location to location. Much of the meteorological parameters are retrieved from weather station situated at ground. These stations are monitored either by the national government, the state government or city municipality. These provide close-to-ground values but lack in spatial coverage as they have minimal coverage diameter. Meteorological data can also be collected using satellite sensors. However, it is not easy to capture local variations in meteorology from the elevation of hundreds of kilometres. The most commonly used satellite datasets are the 2.5 degrees spatial resolution NCEP Reanalysis produced by National Centre for Environmental Prediction (NCEP) and National Centre for Atmospheric Research (NCAR) for every 6-hours, and the 9 km spatial resolution ECMWF Atmospheric Reanalysis (ERA5) by European Centre for Medium-range Weather Forecast (ECMWF) for every hour.

There a number of softwares in the geospatial market for analysing the land processes. Some of these are commercial and some are open source softwares. These include ArcGIS and ENVI as most widely used softwares, followed by ERDAS. The supplementary parameters are studied using FRAGSTATS for landscape analysis and SPSS for statistical analysis. The major statistics applied here are correlation and regression, to understand the relationship between temperature and various land process parameters. Machine learning techniques to study the correlation are also well documented (Mohammad, Goswami, Chauhan, & Nayak, 2022) and with geospatial advancement in big data, ML has a greater scope to answer questions in the field of atmospheric-land processes and their implications on climate change. The methodological approach for processing these datasets is also widely varying. Much of the literature that studies land processes and their interrelationship with surface and atmospheric temperature focuses on 3D representation of the Earth (Derdouri, Wang, Mutayama, & Osaragi, 2021), impacts of landscape metrics on formation of LST (Guo, Wu, Cao, Chen, & Yang, 2020), spatial autocorrelation of temperature (Kowe, et al., 2022), and digital elevation models (Khandelwal, Goyal, Kaul, & Mathew, 2018). For physiological processing focus is made on aerosol concentration because of land use features, photosynthesis, fractional vegetation cover and spectral indices (Zhang, Odeh, & Ramadan, 2013), carbon and other GHG emission from vegetation and urban spaces (Velasco, Roth, Norford, & Molina, 2016), land barrenness, albedo, and spatial extent of water bodies. Table 1 gives a list of the various satellite datasets, parameters and statistics used for the applications mentioned above.

To analyse the historical land processes and predict the future trends, it is important to take a modelling approach. The empirical modelling, such as correlation and regression between variables, will allow for simple statistical understanding about the relationship between different variables of the study (Lu, Lu, Horton, & Ren, 2014).

Meanwhile, a process-based modelling approach, such as Biome BGC (White, Thornton, Running, & Nemani, 2000), Urbclim (Lauwaet, et al., 2016), terrestrial biogeochemical models etc., analyses the entire physiological processing of a system,

Table 1. Satellites used to study LULC and LST, most used indices and parameters, and most applied statistical techniques

<table>
<tr><th colspan="2">Satellites</th><th rowspan="2">Parameters</th><th rowspan="2">Statistical Techniques</th></tr>
<tr><th>Variable</th><th>Satellite/Sensor</th></tr>
<tr><td>Land Surface Temperature</td><td>Landsat TM/ ETM+/TIRS
MODIS
ASTER
SEVIRI
Meteosat
GOES
NPP
AVHRR
IRMSS</td><td rowspan="3">LULC
Vegetation Spectral Indices
Vegetation Fractional Cover
Green Volume
Photosynthesis
Impervious Surfaces
Green house gases
Normalised Difference Building Index
Water bodies
Built-cover
Digital Elevation models
Barrenness
Vegetation spatial configuration
Vegetation composition
Geographical climate classification
Humidity
Evapotranspiration
Albedo
Demographic data
Rainfall/clouds
Air pressure</td><td rowspan="3">Correlation
Regression
Descriptive Statistics
Spatial Statistics
Spatial Autocorrelation
Variance
Standard Error/Deviation
Numerical modelling
Predictive modelling
Machine learning
Geographically weighted regression</td></tr>
<tr><td>Land Use/Land Cover</td><td>Landsat TM/ ETM+/OLI
MODIS
ASTER
LISS
IKONOS
SPOT
Worldview
Orbview
UAV</td></tr>
<tr><td>Climatic Factors</td><td>NCEP/NCAR Reanalysis
ERA5 Reanalysis
In-situ Weather Station</td></tr>
</table>

which includes a large horde of parameters. Novel approaches are required, which will function on the knowledge of the physiology of a system and at the same time make clear-cut relationships between the parameters being assessed. Such semi-empirical models are possible with the use of relevant data, integrated ground and spatial parameters, and advanced approaches of ML.

5. CONCLUSION

With the global increase in population and its impact on land change, the land processes are also dynamically changing, which in turn influence the atmospheric processes. Advancements in modelling these changes, which will include all the driving forces, is essential to understand long-term climate change. Geospatial technologies have found their way in mapping and analysing these processes with efficiency. However, to handle the large number of datasets and parameters of climatic and biophysical nature there is a need for advanced techniques such as machine learning. For improved monitoring of atmosphere and climatic events, it is crucial to integrate the high accuracy of ground data and large spatial coverage of satellite data.

REFERENCES

Ahrens, C. D., & Henson, R. (2018). *Essentials of Meteorology An Invitation to the Atmosphere* (8th ed.). Cengage Learning.

Almeida, C. R., Teodoro, A. C., & Goncalves, A. (2021). Study of the urban heat island (UHI) using remote sensing data/techniques: A systematic review. *Environments (Basel, Switzerland)*, *8*(10), 105. doi:10.3390/environments8100105

Andrews, D. G. (2010). *An introduction to atmospheric physics*. Cambridge University Press. doi:10.1017/CBO9780511800788

Boesch, H., Liu, Y., Tamminen, J., Yang, D., Palmer, P. I., Lindqvist, H., Cai, Z., Che, K., Di Noia, A., Feng, L., Hakkarainen, J., Ialongo, I., Kalaitzi, N., Karppinen, T., Kivi, R., Kivimäki, E., Parker, R. J., Preval, S., Wang, J., ... Chen, H. (2021). Monitoring Greenhouse Gases from Space. *Remote Sensing (Basel)*, *13*(14), 2700. Advance online publication. doi:10.3390/rs13142700

Caballero, R. (2014). *Physics of the Atmosphere*. IOP Publishing. doi:10.1088/978-0-7503-1052-9

Derdouri, A., Wang, R., Mutayama, Y., & Osaragi, T. (2021). Understanding the links between LULC changes and SUHI in cities: Insights from two-decadal studies (2001–2020). *Remote Sensing (Basel)*, *13*(18), 3654. doi:10.3390/rs13183654

Dickinson, R. E. (1983). Land surface processes and climate—Surface albedos and energy balance. *Advances in Geophysics*, *25*, 305–353. doi:10.1016/S0065-2687(08)60176-4

Fadnavis, S., Mahajan, A. S., Choudhury, A. D., Roy, C., Singh, M., Biswas, M. S., & Sarin, M. M. (2020). Atmospheric Aerosols and Trace Gases. In R. Krishnan, J. Sanjay, C. Gnanaseelan, M. Mujumdar, A. Kulkarni, & S. Chakraborty (Eds.), *Assessment of Climate Change over the Indian Region: A Report of the Ministry of Earth Sciences (MoES), Government of India* (pp. 93–116). Springer Singapore. doi:10.1007/978-981-15-4327-2_5

Fleagle, R. G., & Businger, J. A. (1981). *An introduction to atmospheric physics*. Academic Press Inc.

Gallo, K. P., Easterling, D. R., & Peterson, T. (1996). The influence of land use/land cover on climatological values of the diurnal temperature range. *Journal of Climate*, *9*(11), 2941–2944. doi:10.1175/1520-0442(1996)009<2941:TIOLUC>2.0.CO;2

Garrigues, S., Remy, S., Chimot, J., Ades, M., Inness, A., Flemming, J., Kipling, Z., Laszlo, I., Benedetti, A., Ribas, R., Jafariserajehlou, S., Fougnie, B., Kondragunta, S., Engelen, R., Peuch, V.-H., Parrington, M., Bousserez, N., Vazquez Navarro, M., & Agusti-Panareda, A. (2022). Monitoring multiple satellite aerosol optical depth (AOD) products within the Copernicus Atmosphere Monitoring Service (CAMS) data assimilation system. *Atmospheric Chemistry and Physics*, *22*(22), 14657–14692. doi:10.5194/acp-22-14657-2022

Guo, G., Wu, Z., Cao, Z., Chen, Y., & Yang, Z. (2020). A multilevel statistical technique to identify the dominant landscape metrics of greenspace for determining land surface temperature. *Sustainable Cities and Society*, *61*, 102263. doi:10.1016/j.scs.2020.102263

Hardwick, S. R., Toumi, R., Pfeifer, M., Turner, E. C., Nilus, R., & Ewers, R. M. (2015). The relationship between leaf area index and microclimate in tropical forest and oil palm plantation: Forest disturbance drives changes in microclimate. *Agricultural and Forest Meteorology*, *201*, 187–195. doi:10.1016/j.agrformet.2014.11.010 PMID:28148995

Huntingford, C., Jeffers, E. S., Bonsall, M. B., Christensen, H. M., Lees, T., & Yang, H. (2019). Machine learning and artificial intelligence to aid climate change research and preparedness. *Environmental Research Letters*, *14*(12), 124007. Advance online publication. doi:10.1088/1748-9326/ab4e55

Huntley, B. J. (2023). Solar Energy, Temperature and Rainfall. In B. J. Huntley (Ed.), *Ecology of Angola: Terrestrial Biomes and Ecoregions* (pp. 95–125). Springer. doi:10.1007/978-3-031-18923-4_5

Ito, A., & Hajima, T. (2020). Biogeophysical and biogeochemical impacts of land-use change simulated by MIROC-ES2L. *Progress in Earth and Planetary Science*, *7*(45), 1–15. doi:10.1186/s40645-020-00372-w

Izuka, K., Itoh, M., Shiodera, S., Matsubara, T., Dohar, M., & Watanabe, K. (2018). Advantages of unmanned aerial vehicle (UAV) photogrammetry for landscape analysis compared with satellite data: A case study of postmining sites in Indonesia. *Cogent Geoscience*, *4*(1), 1498180. doi:10.1080/23312041.2018.1498180

Khandelwal, S., Goyal, R., Kaul, N., & Mathew, A. (2018). Assessment of land surface temperature variation due to change in elevation of area surrounding Jaipur, India. *The Egyptian Journal of Remote Sensing and Space Sciences*, *21*(1), 87–94. doi:10.1016/j.ejrs.2017.01.005

Khvostikov, I. A. (1965). *The upper Layers of the Atmosphere*. National Aeronautics and Space Administration.

Kowe, P., Dube, T., Mushore, T. D., Ncube, A., Nyenda, T., Mutowo, G., & Kizilirmak, G. (2022). Impacts of the spatial configuration of built-up areas and urban vegetation on land surface temperature using spectral and local spatial autocorrelation indices. *Remote Sensing Letters*, *13*(12), 1222–1235. doi:10.1080/2150704X.2022.2142073

Krishna, D. K., Watham, T., Padalia, H., Srinet, R., & Nandy, S. (2023). Improved gross primary productivity estimation using semi empirical (PRELES) model for moist Indian sal forest. *Ecological Modelling*, *475*, 110175. doi:10.1016/j.ecolmodel.2022.110175

Krishnan, R., Gnanaseelan, C., Sanjay, J., Swapna, P., Dhara, C., Sabin, T. P., & Niyogi, D. (2020). Introduction to Climate Change Over the Indian Region. In R. Krishnan, J. Sanjay, C. Gnanaseelan, M. Mujumdar, A. Kulkarni, & S. Chakraborty (Eds.), *Assessment of Climate Change over the Indian Region: A Report of the Ministry of Earth Sciences (MoES), Government of India* (pp. 1–20). Springer Singapore. doi:10.1007/978-981-15-4327-2_1

Lauwaet, D., De Ridder, K., Saeed, S., Brisson, E., Chatterjee, F., Van, L. N., & Hooyberghs, H. (2016). Assessing the current and future urban heat island of Brussels. *Urban Climate*, *15*, 1–15. doi:10.1016/j.uclim.2015.11.008

Lee, K. H., Li, Z., Kim, Y. J., & Kokhanovsky, A. (2009). Atmospheric Aerosol Monitoring from Satellite Observations: A History of Three Decades. In Y. J. Kim, U. Platt, U. Platt, M. B. Gu, & H. Iwahashi (Eds.), *Atmospheric and Biological Environmental Monitoring* (pp. 13–38). Springer Netherlands. doi:10.1007/978-1-4020-9674-7_2

Li, J., Carlson, B. E., Yung, Y. L., Lv, D., Hansen, J., Penner, J. E., Liao, H., Ramaswamy, V., Kahn, R. A., Zhang, P., Dubovik, O., Ding, A., Lacis, A. A., Zhang, L., & Dong, Y. (2022). Scattering and absorbing aerosols in the climate system. *Nature Reviews. Earth & Environment*, *3*(6), 363–379. doi:10.1038/s43017-022-00296-7

Lu, Y., Lu, S., Horton, R., & Ren, T. (2014). An empirical model for estimating soil thermal conductivity from texture, water content, and bulk density. *Soil Science Society of America Journal*, *78*(6), 1859–1868. doi:10.2136/sssaj2014.05.0218

Malhi, G. S., Kaur, M., & Kaushik, P. (2021). Impact of Climate Change on Agriculture and Its Mitigation Strategies: A Review. *Sustainability (Basel), 13*(3), 1318. Advance online publication. doi:10.3390/su13031318

Mohammad, P., Goswami, A., Chauhan, S., & Nayak, S. (2022). Machine learning algorithm based prediction of land use land cover and land surface temperature changes to characterize the surface urban heat island phenomena over Ahmedabad city, India. *Urban Climate*, *42*, 101–116. doi:10.1016/j.uclim.2022.101116

Mushtaq, Z., Sharma, M., Bangotra, P., Gautam, A. S., & Gautam, S. (2022). Atmospheric Aerosols: Some Highlights and Highlighters, Past to Recent Years. *Aerosol Science and Engineering*, *6*(2), 135–145. doi:10.1007/s41810-022-00133-w

Ranjan, A. K., Patra, A. K., & Gorai, A. K. (2021). A Review on Estimation of Particulate Matter from Satellite-Based Aerosol Optical Depth: Data, Methods, and Challenges. *Asia-Pacific Journal of Atmospheric Sciences*, *57*(3), 679–699. doi:10.1007/s13143-020-00215-0

Singh, R. P. (2014). Dust Storms and Their Influence on Atmospheric Parameters over the Indo-Gangetic Plains. In J. Sundaresan, K. M. Santosh, A. Déri, R. Roggema, & R. Singh (Eds.), *Geospatial Technologies and Climate Change* (pp. 21–35). Springer International Publishing. doi:10.1007/978-3-319-01689-4_2

Sun, Y., Yin, H., Wang, W., Shan, C., Notholt, J., Palm, M., Liu, K., Chen, Z., & Liu, C. (2022). Monitoring greenhouse gases (GHGs) in China: Status and perspective. *Atmospheric Measurement Techniques*, *15*(16), 4819–4834. doi:10.5194/amt-15-4819-2022

Tan, C. L., Wong, N. H., Tan, P. Y., Jusuf, S. K., & Chiam, Z. Q. (2015). Impact of plant evapotranspiration rate and shrub albedo on temperature reduction in the tropical outdoor environment. *Building and Environment*, *94*, 206–217. doi:10.1016/j.buildenv.2015.08.001

Tao, Z., Santanello, J. A., Chin, M., Zhou, S., Tan, Q., Kemp, E. M., & Peters-Lidard, C. D. (2013). Effect of land cover on atmospheric processes and air quality over the continental United States–a NASA Unified WRF (NU-WRF) model study. *Atmospheric Chemistry and Physics*, *13*(13), 6207–6226. doi:10.5194/acp-13-6207-2013

Thompson, R. (1998). *Atmospheric Processes and Systems* (1st ed.). Routledge. doi:10.4324/9780203268643

Thompson, R. D. (1998). *Atmospheric processes and systems*. Psychology Press. doi:10.4324/9780203268643

Velasco, E., Roth, M., Norford, L., & Molina, L. T. (2016). Does urban vegetation enhance carbon sequestration? *Landscape and Urban Planning, 148*, 99–107. doi:10.1016/j.landurbplan.2015.12.003

Verma, S., Prakash, D., Soni, M., & Ram, K. (2019). Atmospheric Aerosols Monitoring: Ground and Satellite-Based Instruments. In S. Sarvajayakesavalu (Ed.), *Advances in Environmental Monitoring and Assessment*. IntechOpen. doi:10.5772/intechopen.80489

White, M. A., Thornton, P. E., Running, S. W., & Nemani, R. R. (2000). Parameterization and sensitivity analysis of the BIOME–BGC terrestrial ecosystem model: Net primary production controls. *Earth Interactions, 4*(3), 1–85. doi:10.1175/1087-3562(2000)004<0003:PASAOT>2.0.CO;2

Yirsaw, E., Wu, W., Shi, X., Temesgen, H., & Bekele, B. (2017). Land Use/Land Cover Change Modelling and the Prediction of Subsequent Changes in Ecosystem Service Values in a Coastal Area of China, the Su-Xi-Chang Region. *Sustainability (Basel), 9*(7), 1204. doi:10.3390/su9071204

Zhang, Y., Li, Z., Bai, K., Wei, Y., Xie, Y., Zhang, Y., Ou, Y., Cohen, J., Zhang, Y., Peng, Z., Zhang, X., Chen, C., Hong, J., Xu, H., Guang, J., Lv, Y., Li, K., & Li, D. (2021). Satellite remote sensing of atmospheric particulate matter mass concentration: Advances, challenges, and perspectives. *Fundamental Research, 1*(3), 240–258. doi:10.1016/j.fmre.2021.04.007

Zhang, Y., Odeh, I. O., & Ramadan, E. (2013). Assessment of land surface temperature in relation to landscape metrics and fractional vegetation cover in an urban/peri-urban region using Landsat data. *International Journal of Remote Sensing, 34*(1), 168-189. doi:10.1080/01431161.2012.712227

Chapter 2
Water and Wastewater Engineering

Avinash Kumar
Delhi Technological University, India

Jaya Yadav
Delhi Technological University, India

Rubeena Vohra
Bharati Vidyapeeth's College of Engineering, India

Anand Sebastian
Centre for Geoinformatics Integrated Rural Technology Centre, Palakkad, India

ABSTRACT

Water and wastewater engineering are essential for public health and environmental protection. It involves the design, construction, and operation of systems to treat water and wastewater. The basic principles of water and wastewater treatment include physical, chemical, and biological processes. Engineers consider the nature and concentration of contaminants, the desired level of treatment, the cost of treatment, the availability of land, the climate, and environmental regulations when designing water and wastewater treatment systems. Common types of water treatment systems include conventional treatment, reverse osmosis, electrodialysis, and ion exchange. Common types of wastewater treatment systems include primary treatment, secondary treatment, and tertiary treatment. Critical tasks involved in operating and maintaining water and wastewater treatment systems include monitoring and adjusting the operation of treatment processes, conducting maintenance and repairs, and sampling and testing water or wastewater.

DOI: 10.4018/979-8-3693-1396-1.ch002

INTRODUCTION

Water and wastewater engineering is the discipline that deals with the provision, treatment, and disposal of Water and wastewater. Water supply and distribution, wastewater collection and treatment, water quality management, environmental impact assessment, and sustainable water resources management are all part of a broad field. Various organizations, including water utilities, environmental consulting companies, and government agencies, can employ water and wastewater engineers. Their responsibilities include designing, constructing, and operating water and wastewater treatment systems, monitoring water quality, and developing strategies to protect water resources. According to J. Paul Guyer (2005), they play a crucial role in ensuring the availability of clean Water for communities and protecting our environment.

HISTORY OF WATER AND WASTEWATER ENGINEERING

Throughout history, Water and wastewater engineering has been crucial in the progress of civilizations. Since ancient times, early societies have devised basic systems for the collection and storage of Water, as well as the disposal of wastewater. For example, the ancient Egyptians built canals and dams to irrigate their crops. They also developed latrines and sewers to gather and dispose of human waste. Water and wastewater engineering development accelerated in the 19th century due to the Industrial Revolution. The growth of cities and the increasing use of Water and wastewater led to several problems, including waterborne diseases and pollution. Engineers responded by developing new technologies for water treatment and wastewater disposal. In the 20th century, water and wastewater engineering continued to evolve J. Paul Guyer (2005). New technologies were developed, such as the activated sludge process for wastewater treatment and the reverse osmosis process for desalination. These technologies helped to improve the quality of water and wastewater treatment, and they also helped to protect the environment.

IMPORTANCE OF WATER AND WASTEWATER ENGINEERING

Environmental and public health protection depends on Water and wastewater engineering. Drinking, cooking, and taking a bath all require clean Water. It is also essential for industrial processes and agricultural irrigation. If not adequately treated, wastewater can pollute rivers, lakes, and oceans. It can also spread disease. Water and wastewater engineers play a vital role in ensuring that we have access to clean Water and that wastewater is treated correctly. They work to design, construct, and operate water and wastewater treatment systems that meet the needs of our communities. They also work to develop new technologies that can improve the efficiency and effectiveness of Water and wastewater treatment.

Water and wastewater engineering is becoming increasingly important in today's world, according to The World Bank (2020). With the growing population and demand for Water, new solutions are needed to meet these challenges. Apart from public health and environmental protection, Water and wastewater engineering also plays a crucial role in economic development. Clean Water is essential for businesses and industries, and proper wastewater treatment can help prevent pollution and protect natural resources. Water and wastewater engineering is a challenging yet rewarding field where engineers work to solve some of the most pressing problems of our time. Their work is making a significant difference in people's lives worldwide.

Water and wastewater treatment fundamentals include physical, chemical, and biological treatment methods (Stensel, 2009).

PHYSICAL TREATMENT PROCESSES

Physical treatment processes are used to remove solids, suspended particles, and other large contaminants from Water and wastewater. These processes include:

Screening: This procedure eliminates big objects from wastewater, including rags, sticks, and toys.
Grit removal: Removes sand, gravel, and other heavy solids from wastewater.
Sedimentation: This process allows suspended solids to settle out of wastewater.
Flocculation: This process helps clump together small particles so that sedimentation can more easily remove them.
Filtration: This process removes suspended particles from Water by passing them through a filter. Chemical treatment processes

Chemical treatment processes are used to remove dissolved contaminants from Water and wastewater. These processes include:

Coagulation: This process uses chemicals to clump together small particles so that they can be more easily removed by sedimentation or filtration.
Neutralization: This process uses chemicals to adjust the pH of Water or wastewater.
Oxidation: This process uses chemicals to oxidize (remove electrons from) contaminants.

Chlorination: This process uses chlorine to kill bacteria and other microorganisms in Water or wastewater.

Ozonation: This process uses ozone to kill bacteria and other microorganisms in Water or wastewater.

BIOLOGICAL TREATMENT PROCESSES

Biological treatment processes use microorganisms to remove organic matter from Water and wastewater. These processes include, according to Burton, F. L. (2014):

Activated sludge: Activated sludge is a method for decomposing organic materials in wastewater that combines bacteria and other microorganisms.

Trickling filters: In this method, microorganisms that break down organic debris in wastewater are grown on a bed of pebbles or other medium.

Aerobic digestion: In the presence of oxygen, bacteria are used in the aerobic digestion process to break down organic matter in wastewater.

Anaerobic digestion: In the absence of oxygen, microorganisms are used in this procedure to break down organic matter in wastewater.

The specific treatment processes used for a particular water or wastewater application will depend on the nature and concentration of the contaminants, the desired level of treatment, and the cost of the treatment. In addition to the physical, chemical, and biological treatment processes described above, many other treatment processes are used for specific applications Burton, F. L., (2014). For example, membrane filtration is used to remove dissolved solids from Water, and ion exchange is used to remove dissolved ions from Water.

DESIGN OF WATER AND WASTEWATER TREATMENT SYSTEMS

Factors to Consider in Designing Water and Wastewater Treatment Systems

The nature and concentration of the contaminants: The type and level of treatment required will depend on the nature and concentration of the contaminants in the Water or wastewater WHO (2005). For example, if the Water is contaminated with heavy metals, then a different treatment process will be required than if the Water is contaminated with organic matter.

The desired level of treatment: The desired level of treatment will also affect the design of the treatment system. For example, if the Water is only required to be treated to a level that is safe for drinking, then a less complex treatment system can be used than if the Water is required to be treated to a level that is safe for industrial use.

The cost of treatment: According to the WHO Report, the cost of treatment is an important factor to consider, especially in the case of large water or wastewater treatment systems. The engineer will need to balance the cost of treatment with the desired level of treatment.

The availability of land: The availability of land is also an important factor to consider, especially for large water or wastewater treatment systems. The engineer will need to make sure that there is enough land available to accommodate the treatment system.

The climate: The climate can also affect the design of the treatment system. For example, if the climate is cold, then the treatment system will need to be designed to prevent the Water or wastewater from freezing.

The environmental regulations: The environmental regulations in the area where the treatment system will be located will also need to be considered. David M. Reckhow (2012). The engineer will need to make sure that the treatment system meets all of the applicable environmental regulations.

TYPES OF WATER AND WASTEWATER TREATMENT SYSTEMS

There are many different types of water and wastewater treatment systems. Some of the most common types of water treatment systems include:

Conventional treatment: The most popular kind of water treatment technology is conventional treatment. Removing impurities from Water combines chemical, biological, and physical techniques.

Reverse osmosis: One kind of membrane filtration that is used to remove dissolved solids from Water is reverse osmosis.

Electrodialysis: Water can have its ions removed using a membrane technique called electrodialysis.

Ion exchange: This is a type of process that uses resins to remove ions from Water.

Some of the most common types of wastewater treatment systems include:

Primary treatment: The initial step in the treatment of wastewater is called primary treatment. Wastewater undergoes the elimination of solids, suspended particles, and other major impurities.

Secondary treatment: The second step in the treatment of wastewater is called secondary treatment. It eliminates organic materials from wastewater via biological means.

Tertiary treatment: The third stage of treating wastewater is called tertiary treatment. It cleans wastewater of fertilizers and other impurities Metcalf & Eddy, Inc. (2019). The specific type of water or wastewater treatment system that is used will depend on the nature and concentration of the contaminants, the desired level of treatment, and the cost of treatment.

OPERATION AND MAINTENANCE OF WATER AND WASTEWATER TREATMENT SYSTEMS

Key tasks involved in operating and maintaining water and wastewater treatment systems:

Monitoring and adjusting the operation of treatment processes: This includes checking the levels of chemicals, the flow rates of Water or wastewater, and the temperatures of the Water or wastewater.

Conducting maintenance and repairs: This includes repairing or replacing worn or damaged equipment and performing preventive maintenance tasks such as lubrication and cleaning, Metcalf & Eddy, Inc. (2019).

Sampling and testing water or wastewater: This is done to ensure that the Water or wastewater meets the required standards for quality and safety.

Recording and reporting data: This includes recording the results of monitoring, testing, and maintenance activities.

Training and supervising staff: Ensuring that all personnel have the necessary training to operate and maintain the treatment system is a crucial aspect of staff supervision and training Metcalf & Eddy, Inc. (2019).

CHALLENGES AND TROUBLESHOOTING

Some of the challenges that may arise in operating and maintaining water and wastewater treatment systems include Stensel, H. D. (2014):

Equipment failures: Breakdowns of equipment: Several things, including operator error, overloading, and wear and tear, can cause equipment to malfunction.

Changes in Water or wastewater quality: Changes in Water or wastewater quality can make it difficult to maintain the treatment system at its designed level of performance.

Environmental conditions: Environmental conditions such as temperature, rainfall, and humidity can also affect the treatment system's performance.

Regulatory compliance: Water and wastewater treatment systems must comply with various regulations, which can change over time.

Troubleshooting problems in water and wastewater treatment systems can be complex and time-consuming. The first step is to identify the cause of the problem. This may involve collecting data, conducting tests, and inspecting the equipment. Once the cause of the problem has been identified, the appropriate corrective action can be taken. The operation and maintenance of water and wastewater treatment systems is a critical task that requires a high level of skill and knowledge Stensel, H. D. (2014). By following the key tasks outlined above and being aware of the challenges and troubleshooting techniques, operators can help ensure that these systems operate safely and efficiently.

ECONOMICS OF WATER AND WASTEWATER TREATMENT

Costs of water and wastewater treatment financial management of water and wastewater treatment systems.

The following are some aspects of the economics of treating Water and wastewater, including treatment costs and system financial management:

COSTS OF WATER AND WASTEWATER TREATMENT

The size and complexity of the treatment system, the kind of treatment procedures utilized, the cost of labor and materials, and the environmental rules that must be followed are just a few of the variables that can cause the costs of treating Water and wastewater to vary greatly. In general, the costs of wastewater treatment are higher than the costs of water treatment. This is because wastewater treatment typically requires more complex treatment processes to remove a wider range of contaminants.

The following are some of the major cost components of Water and wastewater treatment:

1. Capital costs: These are the costs of constructing the treatment system, including land, equipment, and buildings.
2. Operating costs: These are the costs of operating the treatment system, including labor, chemicals, and energy.
3. Maintenance costs are the costs of repairing and replacing worn or damaged equipment.
4. Regulatory compliance costs: These are the costs of complying with environmental regulations, such as the cost of monitoring and testing Water and wastewater quality.

FINANCIAL MANAGEMENT OF WATER AND WASTEWATER TREATMENT SYSTEMS

The financial management of water and wastewater treatment systems is important to ensure that these systems are operated and maintained cost-effectively. The financial management process typically includes the following steps: Metcalf & Eddy. (2013):

1. Establishing financial goals: The first step is to establish financial goals for the water and wastewater treatment system. These goals may include minimizing costs, maximizing efficiency, or improving environmental performance.
2. Developing a financial plan: The next stage is to create a financial plan that details the strategy for achieving the financial objectives. This strategy should include a budget, a cash flow projection, and a debt payback schedule.
3. Managing financial resources: The third step is to manage the financial resources of the water and wastewater treatment system. This includes collecting revenue, paying bills, and investing funds.
4. Monitoring financial performance: The fourth step is to monitor the financial performance of the water and wastewater treatment system. This involves comparing actual results to the budget and making adjustments as needed.

The financial management of water and wastewater treatment systems is a complex and challenging task. However, by following the steps outlined above, water and wastewater utilities can help to ensure that these systems are operated and maintained cost-effectively and sustainably.

ENVIRONMENTAL IMPACTS OF WATER AND WASTEWATER TREATMENT

The impacts of Water and wastewater treatment on the environment are:

According to Metcalf & Eddy. (2013), Water and wastewater treatment can have several environmental impacts, both positive and negative. Some of the positive impacts include:

1. Reduction of pollution: Water and wastewater treatment can help to reduce pollution by removing contaminants from Water and wastewater. This can help to protect water bodies, such as rivers, lakes, and oceans, and it can also help to protect human health.
2. Improved water quality: Water and wastewater treatment can help improve water quality by removing contaminants. This can make Water safe for drinking, bathing, and other uses.
3. Reclaimed Water: Water and wastewater treatment can be used to produce reclaimed Water, which can be used for irrigation, industrial processes, and other purposes. This can help to conserve freshwater resources.

Some of the negative impacts of Water and wastewater treatment include:

1. Discharge of treated wastewater: Treated wastewater is often discharged into rivers, lakes, or oceans. This can contain residual contaminants, which can harm the environment.
2. Use of chemicals: Water and wastewater treatment often use chemicals, such as chlorine and coagulants. These chemicals can hurt the environment, particularly if they are released into the environment.
3. Energy consumption: Water and wastewater treatment requires energy, which can contribute to greenhouse gas emissions.
4. Land use: Water and wastewater treatment plants require land, which can take away from other uses, such as agriculture or conservation.

MITIGATION MEASURES

There are several strategies that can be implemented to reduce the negative effects of Water and wastewater treatment on the environment (P.A., 2009). These actions include, among others:

1. Using more efficient treatment processes: Several more efficient treatment processes can be used to reduce the amount of energy and chemicals used in Water and wastewater treatment.
2. Reusing and recycling water: Reusing and recycling water can help to reduce the amount of wastewater that needs to be treated.

3. Treating wastewater to a higher standard: Treating wastewater to a higher standard can remove more contaminants, which can reduce the environmental impact of the discharge.
4. Using less land: Water and wastewater treatment plants can be designed to use less land. This can be done by using taller treatment tanks or by locating the plant in a less sensitive area, the Water Environment Federation (WEF,2013). The environmental impacts of Water and wastewater treatment are a complex issue. By taking the appropriate mitigation measures, it is possible to reduce these impacts and protect the environment.

EMERGING TECHNOLOGIES IN WATER AND WASTEWATER TREATMENT

New Technologies for Water and Wastewater Treatment

Several new technologies are being developed for Water and wastewater treatment. Some of these technologies include the Environmental Protection Agency (EPA,2016):

1. Membrane filtration: Membrane filtration is a method that uses membranes to remove contaminants from Water and wastewater. Membranes are thin, semi-permeable barriers that allow water molecules but not contaminants to pass through.
2. Advanced oxidation processes (AOPs): Advanced oxidation processes (AOP) are a group of chemical procedures that remove pollutants from Water and wastewater by using strong oxidizing agents. Organic debris, heavy metals, and viruses are just a few of the impurities that AOPs may remove.
3. Microbial fuel cells (MFCs): MFCs Microbial fuel cells (MFCs) are devices that generate electricity from wastewater using bacteria. MFCs have the potential to clean wastewater as well as generate renewable energy.
4. Solar photocatalysis: Solar photocatalysis is a method that employs sunlight to degrade pollutants in Water and wastewater, Showkatfar, M. (2018). Solar photocatalysis can remove a variety of pollutants, including organic debris, heavy metals, and pathogens.
5. Nanotechnology: Nanotechnology is the application of materials at the nanoscale (1-100 nano-meters). Nanotechnology, such as nanofiltration membranes and nano-sorbents, can be used to develop innovative water and wastewater treatment methods.

POTENTIAL BENEFITS OF EMERGING TECHNOLOGIES

The potential benefits of emerging technologies in Water and wastewater treatment include Stensel, H. D. (2014):

1. Improved water quality: Emerging technologies can be used to remove a wider range of contaminants from Water and wastewater, resulting in improved water quality.
2. Reduced energy consumption: Emerging technologies can be more energy-efficient than traditional treatment processes, which can help to reduce greenhouse gas emissions.
3. Reduced use of chemicals: Emerging technologies can use fewer chemicals than traditional treatment processes, which can reduce the environmental impact of Water and wastewater treatment.
4. Reduced cost: Emerging technologies can be more cost-effective than traditional treatment processes, which can make water and wastewater treatment more affordable.
5. Increased sustainability: Emerging technologies can be more sustainable than traditional treatment processes, which can help to protect the environment. The development and adoption of emerging technologies in Water and wastewater treatment is a promising area of research. These technologies have the potential to improve water quality, reduce energy consumption, and protect the environment.

REFERENCES

American Water Works Association. (2016). *Water treatment plant design* (5th ed.). McGraw-Hill Education.

Chobanoglous, G., Burton, F. L., & Stensel, H. D. (2014). *Wastewater Treatment* (5th ed.). McGraw-Hill Education.

Emerging Technologies for Wastewater Treatment and In-Plant Wet Weather Management Fact Sheet. (n.d.). Environmental Protection Agency (EPA). https://www.epa.gov/sustainable-water-infrastructure/emerging-technologies-wastewater-treatment-and-plant-wet-weather

Eswaran, V., & Mohan, M. (2018). Emerging technologies for wastewater treatment: A review. *Journal of Chemical Technology and Biotechnology*, *93*(10), 2747–2767.

Friedler, E., & Wilderer, P. A. (2009). *Introduction to wastewater treatment.* John Wiley & Sons.

Judd, S. J. (2017). The future of desalination: Technology, economics, and the environment. *Science*, *357*(6347), aad3527. https://www.science.org/doi/10.1126/science.1200488

Lee, J., Esmaeili, A., & Showkatfar, M. (2018). Emerging technologies in wastewater treatment: A review. *Chemosphere*, *199*, 2740. https://www.sciencedirect.com/topics/earth-and-planetary-sciences/wastewater-treatment

Metcalf & Eddy. (2013). *Wastewater engineering: Treatment and reuse* (5th ed.). McGraw-Hill Education.

Metcalf & Eddy, Inc. (2019). *Wastewater Engineering: Treatment and Resource Recovery* (6th ed.). McGraw-Hill Education.

New Technology Trends in The Wastewater Management Industry. (2022). Forbes. https://www.forbes.com/sites/theyec/2022/10/25/new-technology-trends-in-the-wastewater-management-industry/

Processes. (n.d.). *Special Issue: Emerging Technologies for Water and Wastewater Treatment*. https://www.mdpi.com/journal/water/special_issues/Wastewater_Techniques

Shannon, M. A., Bohn, P. W., Elimelech, M., Georgiou, J. G., Gitis, V., Jones, R. R., & Xu, Y. (2008). Science and technology for water purification in the coming decades. *Nature*, *452*(7185), 301–310.

Tchobanoglous, G., Burton, F. L., & Stensel, H. D. (2014). *Wastewater treatment* (2nd ed.). McGraw-Hill Education.

The United States Environmental Protection Agency (EPA). (n.d.). https://www.epa.gov/ground-water-and-drinking-water/basic-information-about-your-drinking-water

The Water Environment Federation (WEF). (n.d.). https://www.wef.org/

The World Bank. (2020, July 15). *Water and Sanitation*. https://www.worldbank.org/en/topic/water

The World Health Organization (WHO). (n.d.). https://www.who.int/teams/environment-climate-change-and-health/water-sanitation-and-health

United States Environmental Protection Agency (EPA). (2010). https://www.epa.gov/

Water Environment Federation. (2020). *Operation of municipal wastewater treatment plants (Manual of Practice No. 11)*. Water Environment Federation.

Chapter 3
Water Demand and Quality Management

Avinash Kumar
Delhi Technological University, India

Dyvavani Krishna Kapuganti
Delhi Technological University, India

Rubeena Vohra
Bharati Vidyapeeth's College of Engineering, India

ABSTRACT

Water demand and quality management are vital components of ensuring universal access to clean water and safeguarding water resources for future generations. It involves strategies to reduce water consumption, including initiatives like water pricing, public education, and water-efficient technologies, safeguarding water quality through pollution prevention, wastewater treatment, and watershed management. Decreasing water demand can enhance water quality by reducing pollution entering water bodies. The goal of management is to cut back on water use without compromising social or economic advancement. Municipalities, industry, and agriculture are just a few of the industries that can use water quality management techniques. Integrated water demand and quality management is an approach to water resource management that considers the interconnectedness of water quantity and quality. This approach recognizes that water demand management and water quality management are complementary and must be considered together to achieve sustainable water resource management.

DOI: 10.4018/979-8-3693-1396-1.ch003

1. INTRODUCTION

Water demand and quality control are crucial to ensure everyone can access clean and safe drinking water. Water demand management is the process of reducing water usage to decrease the demand for water resources. The act of conserving and improving the quality of water resources is known as water quality management. Water demand and quality control are closely related, as reducing water demand can help enhance water quality by reducing the pollution that enters water bodies (Gleick, 2003). Improving water quality can make it more appealing to customers and encourage them to conserve water. Managing water demand and quality is crucial for adapting to climate change. Sea level rise and changes in precipitation patterns are two significant effects of climate change that are expected to impact water resources significantly. Effective management of water demand and quality can assist communities in adapting to these changes and safeguarding their water resources (P. H., 2003). Here are some of the critical challenges to water demand and quality management:

1. Population expansion and urbanization are putting further strain on water resources.
2. Climate change is expected to impact water supplies due to changes in precipitation patterns and rising sea levels.
3. Domestic, industrial, and agricultural water pollution is a severe problem in many parts of the world.
4. Water is wasted due to inefficient water use and infrastructural leaks.

Water demand and quality management can be improved by taking various actions to address these issues such as investing in water infrastructure to improve efficiency and reduce leaks (P. H., 2003).

1. Implement water-saving measures such as water-efficient appliances and fixtures, as well as public education initiatives.
2. Reducing agricultural, industrial, and residential water pollution through legislation, incentives, and best practices.
3. Developing integrated water management strategies that consider the environmental, social, and economic elements of water consumption.
4. Water demand and quality management are critical to ensure everyone can access clean, safe water and preserve our water resources for future generations.

Background of Global Water Challenges: Water is not only necessary for life, but it is also an important resource for economic development and social well-being.

Water supplies, on the other hand, are under rising strain due to a variety of issues, including WHO 2017:

1. Population growth: The world's population is expected to exceed 9.7 billion by 2050, putting further demand on water resources.
2. Urbanisation: As more people move to cities, there is an increase in water demand and wastewater generation WHO (2018).
3. Climate change: The increased frequency of extreme weather events, such as floods and droughts, can impact water availability and quality.
4. Water pollution: Water pollution is a big issue in many regions of the world, and it can make drinking water and other uses unhealthy.

Water Demand Management

Water demand management is lowering water usage without jeopardizing economic or social development. There are numerous water demand management measures that can be adopted, including (WHO 2017):

1. Water pricing: Water pricing can deter water waste. Increased water prices, for example, can encourage high-volume customers to lower their consumption.
2. Public education and awareness: These activities can help educate the public daily about the need for water conservation and practical techniques to save water.
3. Water-saving fixtures and appliances: Low-flow shower heads and toilets are examples of fixtures and appliances that can help reduce water use in homes and offices.
4. Rainwater harvesting is collecting and storing precipitation for future use. This can be an important supply of water for non-potable uses such as gardening and irrigation.
5. Greywater is wastewater from baths, showers, and sinks that don't contain any focal matter. Greywater can be reused for gardening, irrigation, and other non-potable applications.

Water Quality Management

Protecting and improving water quality is known as water quality management. Numerous water quality management measures can be adopted, including:

1. Point source pollution is pollution that originates in a particular area, such as a factory or sewage treatment plant. Regulations and permits are two methods for managing pollution at its source.

2. Non-point source pollution can come from various sources, such as agricultural and stormwater runoff. It can be reduced by implementing best management practices (BMPs) (Gleick, 2003).
3. Water treatment is the process of purifying water so that humans can consume it by removing impurities. Water treatment facilities employ a variety of treatment procedures, including coagulation, filtration, and disinfection.
4. Watershed management: The management of water resources in a specific area, known as watershed management, can help to protect water quality by reducing pollutants and improving the health of the ecosystem (P. H., 2003).

Integrated Water Demand and Quality Management

Integrated water demand and quality management is a water resource management technique that takes into account the interdependence of water quantity and quality. Recognizing the interdependence of water demand and quality is crucial for sustainable water resource management. By evaluating these two factors together, we can unlock many benefits, as expertly pointed out by Gleick, P. H. in 2003. This approach acknowledges that managing water demand and quality go hand-in-hand and should be evaluated together for sustainable water resource management. Integrating water demand and quality management offers numerous benefits, including those highlighted by Gleick, P. H. (2003).

1. Increased water efficiency: By reducing water use and the requirement for water treatment, integrated water demand and quality management can help to increase water efficiency.
2. Pollution reduction: Integrated water demand and quality management can help to reduce pollution by regulating both point source and non-point source pollution.
3. Improved water quality: By lowering the levels of contaminants in water, integrated water demand and quality management can help to improve water quality.
4. Improved water security: Integrated water demand and quality management can help improve water security by assuring a consistent supply of clean water to meet society's and the environment's needs (WHO, 2018).

Importance of Water Demand and Quality Management: Water demand and quality management are essential for a number of reasons, including:

1. To maintain a consistent supply of safe drinking water and other uses. Water is necessary for life, but it is also used for various other reasons, including agriculture, industry, and enjoyment. Water consumption is increasing as

the world's population expands and urbanizes. Water demand management solutions can help minimize water use while ensuring a consistent supply of clean water for everyone.

2. To safeguard public health. Water contamination can lead to a number of diseases, including cholera, typhoid, and diarrhea. Water quality management strategies can aid in the reduction of pollution and the protection of public health.
3. To safeguard ecosystems. Water is necessary for all life. Therefore, water pollution can harm ecosystems and diminish biodiversity. Strategies for managing water quality can help protect ecosystems and their services.

To aid in economic progress. Many economic activities, including as agriculture, industry, and tourism, rely on water. Water demand and quality management solutions can aid economic development by assuring a consistent supply of clean water and conserving ecosystems. In addition to these broad benefits, water demand and quality management can aid in the resolution of specific water issues, such as:

1. Climate change: The increased frequency of extreme weather events like floods and droughts can have an impact on water supply and quality. Water quality and demand management strategies can help water systems become more resilient to climate change.
2. Water scarcity: Water scarcity is becoming increasingly problematic in many places of the world. Water demand management solutions can aid in the reduction of water usage and the conservation of water resources.
3. Water contamination: Water pollution is a major issue in many regions of the world. Water quality management strategies can aid in the reduction of pollution and the improvement of water quality. Overall water demand and quality management, we can say, are critical for guaranteeing a reliable supply of clean water, preserving public health and ecosystems, promoting economic development, and addressing specific water concerns WHO (2021).

Chapter Objectives and Scope: The goals of this chapter on water demand and quality management are as follows:

1. Provide an overview of the key concepts of water demand and quality management, including the interdependence of water quantity and quality.
2. Discuss the major worldwide water concerns driving the need for water demand and quality management.
3. Emphasise the significance of water demand and quality management in terms of human health, the environment, and the economy.

4. Examine various water demand and quality management methods and procedures that can be put in place to address global water concerns.
5. Explain the advantages of integrated water demand and quality management.

Water demand and quality management in its broadest sense

1. Strategies for managing water demand include price, public education and awareness, water-saving appliances and fixtures, rainwater harvesting, and greywater reuse (Gleick, 2003).
2. Strategies for water quality management, such as point source pollution control, non-point source pollution control, water treatment, and watershed management.
3. Integrated water demand and quality management approaches like water reuse and nature-based solutions.
4. The use of water demand and quality management techniques and measures to address unique global water concerns such as climate change, scarcity, and pollution WHO (2018).

2. UNDERSTANDING WATER DEMAND

Water demand is the amount of water that customers want to meet their needs. It is divided into three major sections:

1. Domestic water demand: This includes drinking, cooking, bathing, and other home activities.
2. Agricultural water demand, which includes irrigation and livestock production.
3. The industrial water demand category includes Water required for production, manufacturing, and other industrial operations.

A variety of factors influence water demand, including:

1. Increased population: As the population expands, so does the water demand.
2. Economic development: Economic development increases water demand for agriculture, industry, and household uses.
3. Climate change: As a result of climate change, more extreme weather events, such as droughts and floods, are occurring, impacting water availability and demand.
4. Water pricing: Pricing for water can be used to influence demand. Higher water pricing, for example, can discourage wasteful water consumption.

According to the Ministry of Jal Shakti, Government of India. (2021). Water demand management is the process of lowering water usage without jeopardizing economic or social development. There are numerous water demand management measures that can be adopted, including:

1. Water prices can be used to discourage excessive water use. Increased water prices, for example, can encourage high-volume customers to lower their consumption.
2. Public education and awareness: These initiatives can assist individuals in realizing the value of water conservation and provide them with practical advice on how to do so.
3. Water-saving fixtures and appliances: Low-flow showerheads and toilets are two examples of fixtures and appliances that can help reduce water use in homes and workplaces.
4. Rainwater harvesting is collecting and storing precipitation for future use. This can be an important supply of water for non-potable uses such as gardening and irrigation.
5. Greywater is wastewater that does not include feces and is collected from bathtubs, sinks, and showers. Greywater can be recycled for gardening, irrigation, and other non-potable uses.

Water Demand Influencing Factors: Water demand factors can be broadly categorized into two categories:

- Natural factors: These factors include climate, population growth, and water availability.
- Human factors: These factors include economic development, urbanization, water pricing, and public education and awareness

Natural factors:

1. Climate: Temperature, precipitation, and evaporation are all elements that influence water demand. Hotter regions, for example, have higher water demand due to increased evaporation and irrigation needs.
2. Population growth: As the population expands, so does the water demand.
3. Water availability: Water availability in a location can also influence water demand. Water demand is often lower in locations with limited water resources due to the need to save water. National Water Policy (2012).

Human factors: Economic development: Economic development leads to increased water consumption for industrial, agricultural, and commercial purposes.

1. Urbanization: More people are moving to cities, leading to increased water demand and wastewater generation in urban areas.
2. Water pricing: The price of water can influence water consumption. For example, higher water prices can discourage wasteful water use.
3. Public education and awareness: These initiatives can assist individuals in realizing the value of water conservation and provide them with practical advice on how to do so. These factors must be interconnected. For example, economic development can lead to urbanization, which can, in turn, lead to increased water demand. Climate change can also impact all of the factors listed above.

Population Growth and Urbanization: Urbanisation and population growth are two of the most important demographic trends of the twenty-first century. According to UN projections, the world population will reach 9.7 billion by 2050, with metropolitan regions housing more than half of the population. The consequences of urbanization and population growth on water quality and demand management are numerous. Water consumption will rise as the population expands and more people move to cities. This will place a strain on water resources and result in water scarcity in many parts of the world. Urbanization also has a significant impact on water quality. Cities generate large amounts of wastewater, which can pollute water resources if not properly treated. Additionally, urbanization can lead to increased stormwater runoff, which can carry pollutants into water bodies. To effectively manage water demand and quality, it's crucial to address challenges posed by population growth and urbanization.These strategies should focus on reducing water consumption, improving water quality, and ensuring a reliable supply of clean water for all. According to the Ministry of Urban Development, Government of India. (2015).

Here are some specific examples of water demand and quality management measures that can be adapted to meet population growth and urbanization challenges:

1. Water pricing: Water pricing can penalize wasteful water consumption while encouraging conservation.
2. Public education and awareness initiatives can assist people in understanding the need for water conservation and how to conserve water in their daily lives.
3. Water-efficient appliances and fixtures: Low-flow toilets and showerheads, for example, can help reduce water consumption in homes and workplaces.

4. Rainwater harvesting: Rainwater harvesting is the technique of collecting and storing rainwater for future use. This can be a useful source of water for landscaping, irrigation, and other non-potable needs.
5. Greywater reuse: If you have wastewater from your bathtub, shower, or washbasin that does not include any feces', reusing greywater is a smart option. Greywater can be reused for non-potable activities such as gardening and irrigation.
6. Wastewater treatment: Wastewater treatment plants can aid in the removal of pollutants from wastewater and the preservation of water quality.
7. Stormwater management measures can help to limit the quantity of stormwater runoff and contaminants entering aquatic bodies. We can assist in securing a reliable supply of clean water for all by implementing these and other water demand and quality management techniques, even in the face of population expansion and urbanization.

Industrial and Agricultural Water Needs: Industrial and agricultural water requirements differ based on the industry or type of agriculture. However, some frequent water uses are as follows:

1. Industry: Water is employed in many industrial operations, including manufacturing, mining, and energy generation. For example, water is used to cool machinery, create power, and convey raw materials and finished products.
2. Water is required for irrigation, which is the process of watering crops. Irrigation is required in many regions of the world to produce crops since there is insufficient rainfall to meet plant needs. Water is also used in the production of cattle, food processing, and other agricultural activities.

Industrial and agricultural water requirements are substantial. Agriculture accounts for over 70% of worldwide water withdrawals, according to the United Nations Food and Agriculture Organisation. The industry contributes to around 20% of global water withdrawals. Water scarcity is becoming more common in many parts of the world due to increased demand for water from industry and agriculture. Water demand management measures must be implemented to reduce water consumption in these industries. Here are some ways for managing water demand that can be used in industry and agriculture Ministry of Urban Development, Government of India. (2015):

1. Industry: Water recycling and reuse, water-efficient technologies, and process optimization.
2. Agriculture: Irrigation efficiency improvements, drought-tolerant crops, and precision agriculture.

Climate Change Impacts on Water Demand: Climate change has a range of effects on water demand, including:

1. Increased evaporation: Warmer temperatures lead to increased evaporation, which reduces water availability.
2. Changes in precipitation patterns: Climate change is causing more frequent droughts and floods in some locations. This could lead to a water deficit in certain regions and an increase in water demand in others.
3. Increased need for irrigation: As farmers try to adjust to shifting weather patterns, there is an increase in demand for irrigation. These effects impose a strain on water supplies and make meeting water demands more challenging. To overcome these difficulties, it is critical to employ water demand management and climate adaptation techniques.

Water Demand Estimation Ways: According to Sharma et al. (2009). There are several ways to assess water demand, including:

1. Trend analysis: This method entails extrapolating previous trends in water consumption into the future. It is a straightforward strategy, but it can be wrong if the factors that drive water demand, such as population increase, economic development, and climatic change, alter.
2. End-use analysis: This method estimates water demand for various end applications such as home, agricultural, and industrial water consumption. It's a more complicated method than trend analysis, but it's more accurate if you know what factors drive water demand for different end uses.
3. Water balance modeling: In this method, a water balance model is created to simulate the movement of water through a system. Water balance models can be used to forecast water demand under many scenarios, including population increase, economic development, and climate change. Ministry of Urban Development, Government of India. (2015).

Per Capita Water Consumption: Per capita water consumption is the average amount of water utilized by a person over a given time period, usually per day or year. It is calculated by dividing the total amount of water utilized by all people by the total number of people. There are substantial differences in per capita water use around the world due to a range of factors such as cultural customs, economic development, and climate. In general, wealthy countries consume more water per capita than poor countries. According to the United Nations Food and Agriculture Organisation (FAO), the global average per capita water usage is 3,800 liters per day. However, water consumption varies significantly across countries. For example, in the United States, the average per capita water use is 1,100 liters per day, whereas in Ethiopia, the average per capita water usage is approximately 60 liters per day. Water consumption per capita is an important indication of water use efficiency and water security. Water managers use it to plan and implement water-saving strategies. Here are some ideas for lowering per capita water consumption: Smart Cities Mission (2015)

1. Take shorter showers.
2. Fix any leaks in your plumbing.
3. Water your lawn less often and more deeply.
4. Collect rainwater to use for watering plants or washing your car.
5. Use water-efficient appliances and fixtures.

Water Demand Forecasting Models: Water demand forecasting models employ historical data and other criteria to anticipate future water demand. Water utilities, water resource managers, and other stakeholders utilize these models to make informed decisions about water planning and management. The following are some of the most frequent types of water demand forecasting models:

1. Time series models: These models forecast future water demand using historical water demand data. Depending on the parameters included in the model, time series models can be simple or complex.
2. End-use models: These models break down water demand into several sectors, such as domestic, agricultural, and industrial. End-use models can be used to forecast future water demand for each industry based on population increase, economic development, and other variables.
3. Water balance models: These models replicate the flow of water through a system while accounting for elements such as precipitation, evaporation, runoff, and groundwater recharge. Water balance models can be used to forecast future water demand under various scenarios of climate change According to World Resources Institute. (2021). The optimal water demand forecasting model to adopt will be determined by the specific situation and the available data.

Case Studies on Water Demand Management: Here are some case studies on water demand management:

Case Study 1: Singapore

Singapore, with a population of nearly 6 million people, is a water-scarce country. The country has put in place a variety of water demand control methods, including:

1. Singapore has a tiered water pricing scheme, with higher prices for higher water consumption. This motivates individuals to utilize water more wisely.
2. Public education and awareness: To promote water conservation, Singapore runs a number of public education and awareness campaigns. For example, the country's "Public Utilities Board" runs a program called "Every Drop Counts" that educates people on the significance of water conservation and offers advice on how to save water.
3. Water-efficient appliances and fixtures: Singapore mandates that all new buildings include water-efficient appliances and fixtures. People can also get government help to upgrade their homes with water-saving equipment and fixtures. Singapore has been able to lower its per capita water use by more than 30% as a result of these water demand management methods during the 1990s.

Case Study 2: Los Angeles, California

Los Angeles is a large city in a semi-arid region. The city has implemented a number of water demand management strategies, including:

1. Water pricing: Los Angeles has a tiered water pricing system, with higher prices for higher water consumption. This encourages people to use water more efficiently.
2. Public education and awareness: To promote water conservation, Los Angeles conducts a number of public education and awareness campaigns. For example, the city's "Water and Power Department" runs a "WaterWise" campaign that educates people on the significance of water conservation and offers tips on how to save water.
3. Rebate programs: Los Angeles offers a number of rebate programs for water-efficient appliances and fixtures. For example, the city offers a rebate for people who install a low-flow toilet.
4. Water recycling and reuse: The city of Los Angeles recycles and repurposes wastewater for irrigation and other non-potable uses. Since the 1990s, Los Angeles has been able to lower its per capita water use by more than 20% as a result of these water demand management methods Falkenmark, M., & Rockström, J. (2006).

Water Demand Management in India's Water-Scarce Regions: Water demand management (WDM) is the practice of lowering water consumption while maintaining vital human requirements and economic development. It is an essential component of long-term water resource management in water-stressed areas. Over 40% of India's population lives in water-stressed areas, making it a water-scarce country National Water Policy (2012). Due to population expansion and economic development, the country's water demand is predicted to increase by 50% by 2050. There are several WDM solutions that can be adopted in water-stressed areas of India. These are some examples:

Water prices can be used to penalize inefficient water consumption and encourage conservation. Higher water tariffs, for example, can be levied for higher water consumption. Water-efficient fixtures and appliances: Low-flow showerheads and toilets, among other fixtures and appliances, can help reduce water consumption in homes and businesses. Rainwater harvesting is the process of collecting and storing rainwater for later use. This can be an excellent source of water for irrigation, landscaping, and other non-potable applications. Greywater reuse: Greywater is wastewater that does not contain feces from sinks, showers, and baths. Greywater can be recycled for irrigation, landscaping, and other non-potable applications.

Agricultural water conservation practices, such as drip irrigation and precision agriculture, can help to minimize agricultural water usage. In addition to these fundamental WDM techniques, a number of specialized WDM efforts have been launched in India's water-scarce regions. Here are a couple of such examples: Jal Shakti Abhiyan. The Jal Shakti Abhiyan is a countrywide water conservation and management effort initiated by the Government of India in 2019. Rainwater harvesting, groundwater recharge, and watershed development are the primary goals of the initiative. The Atal Mission for Rejuvenation and Urban Transformation (AMRUT): The Atal Mission for Rejuvenation and Urban Transformation (AMRUT) is a national mission launched by the Government of India in 2015 to improve urban infrastructure and services. The mission emphasizes water conservation and management. The Smart Cities Mission is a nationwide initiative established by the Government of India in 2015 to develop 100 smart cities across the country. The mission emphasizes water conservation and management.

3. ENSURING WATER QUALITY

Important Water Quality Parameters: The following are important parameters for determining water quality:

1. Temperature, turbidity, color, odor, taste, and electrical conductivity are physical properties.
2. pH, acidity, alkalinity, hardness, salinity, dissolved oxygen, nutrients, and contaminants are examples of chemical parameters.
3. Biological parameters, which include bacteria, viruses, and parasites.

Physical Parameters:

Three physical characteristics often used to measure water quality are turbidity, color, and odor. The cloudiness or haziness of water is referred to as turbidity. Suspended particles like clay, silt, and algae cause it. Water with high turbidity can be ugly and can also interfere with water treatment operations. Nephelometric turbidity units (NTU) are used to measure turbidity. The hue or color of the water is referred to as color. Dissolved substances such as organic debris, iron, and manganese produce it. The colors of natural waters can range from clear to blue, green, brown, and yellow. Color does not always indicate poor water quality; however, it might be an indicator of organic waste or metal pollution. Platinum-cobalt units (PCU) are used to measure color.

Odor is the smell of water. It is caused by volatile substances, such as hydrogen sulfide and gasoline. Natural waters can have a variety of odors, including earthy, fishy, and musty. Odors can be a sign of pollution from organic matter or chemicals. Odors are described using qualitative terms, such as "earthy," "fishy," and "chemical." Turbidity, color, and odor can be measured using a variety of methods, including field kits and laboratory analysis. The specific techniques used will depend on the accuracy and sensitivity required. These three physical parameters are important indicators of water quality. High turbidity, color, or odor can indicate pollution or other problems. By monitoring these parameters, we can better understand the water quality and identify any potential issues. Here are some specific examples of how turbidity, color, and odor can be used to assess water quality:

1. Turbidity: High turbidity can make water unsafe to drink, as it can interfere with the disinfection process. Turbidity can also cause water to be unsuitable for irrigation, as it can clog irrigation systems.
2. Colour: Water with a high color can be unpleasant and interfere with water treatment operations. Color can also be an indicator of organic matter or metal pollution.
3. Odour: Odours might be a symptom of organic materials or chemical pollution. Some odors, like hydrogen sulfide, are toxic.

Chemical Parameters (pH, Dissolved Oxygen, Contaminants): pH, dissolved oxygen (DO), and contaminants are three chemical parameters commonly used to assess water quality. Water's acidity or alkalinity is determined by its pH. A neutral score of seven is assigned to it in a range of 0 to 14. Water is considered acidic if its pH is less than seven and alkaline if it is greater than 7. The majority of aquatic life favors pH levels between 6.5 and 8.0. The amount of oxygen dissolved in water is measured as dissolved oxygen, or DO. In marine life, DO is important. Low DO levels can cause fish and other marine organisms to suffocate. Contaminants are substances that are present in water that can harm human health or the environment. Common contaminants include bacteria, viruses, parasites, metals, pesticides, and herbicides. These three chemical parameters are important indicators of water quality. High pH, low DO, and high levels of contaminants can make water unsafe to drink, swim in, or use for irrigation. Here are some specific examples of how pH, DO, and contaminants can be used to assess water quality:

1. pH: Water below 6.5 or above 8.5 can be corrosive to pipes and plumbing fixtures. Acidic water can also leach metals out of pipes and into the water.
2. DO: Low DO levels can indicate pollution from organic matter. Organic matter consumes oxygen as it decomposes. Low DO levels can also be a sign of stagnant water.
3. Contaminants: Bacteria, viruses, and parasites can all cause waterborne ailments such as diarrhea, vomiting, and cramping. Metals and pesticides can also cause a wide range of health issues, including cancer, neurological issues, and reproductive issues. We can better understand water quality and identify potential concerns by monitoring pH, DO, and pollutants. This data can help people make informed decisions about using and managing water resources. It is crucial to remember that the particular pH, DO, and pollutant values regarded as acceptable will vary depending on the intended use of the water. For example, the pH limits for drinking water are stricter than the pH standards for irrigation water.

Biological Parameters (Bacteria, Viruses): Bacteria and viruses are two biological indicators widely used to measure water quality. Bacteria are single-celled microorganisms found in water, soil, and the atmosphere. Some bacteria are useful, while others are harmful. Viruses are even smaller than bacteria and can replicate exclusively within live cells. Viruses are responsible for a wide range of disorders, including diarrhea, vomiting, and respiratory infections. Because bacteria and viruses can cause a range of waterborne ailments, biological factors are essential indicators of water quality. Waterborne infections are especially dangerous for children, the elderly, and those with compromised immune systems. Here are a few examples of how bacteria and viruses can be used to evaluate water quality:

1. Bacteria: The presence of certain types of bacteria, such as fecal coliforms, can be a sign of pollution from sewage. Fecal coliforms are bacteria found in the feces of humans and other animals.
2. Viruses: The presence of viruses in water can be a sign of pollution from sewage or other sources. Viruses can also be spread through contact with infected people or animals. A qualified professional should perform water quality testing for

bacteria and viruses. Water quality testing kits are also available for purchase but may not be as accurate as laboratory tests.

3. It is critical to understand that the presence of bacteria and viruses in water does not always imply that it is hazardous to drink. However, it is critical to disinfect water before ingesting it, particularly if it comes from a private well or another untreated source. Water can be disinfected using several processes, such as boiling, chlorination, and ultraviolet (UV) light. The optimum strategy will be determined by the sort of bacteria and viruses found in the water.

Water Quality Regulations and Standards: Water quality standards and regulations are critical tools for ensuring public health and environmental protection. They establish limitations on the amount of pollutants and other contaminants that can be present in water. Government agencies such as the Environmental Protection Agency (EPA) in the United States typically develop water quality criteria. These standards are based on scientific facts on the effects of pollutants on the environment and human health. Water quality regulations are policies put in place to ensure that water quality criteria are met. Businesses and industries may be required by these regulations to clean their effluent before dumping it into public waterways. They may also order communities to test their drinking water and, if necessary, treat it. Water quality standards and regulations are important for a number of reasons. They help to:

1. Protect public health from waterborne illnesses, such as cholera, typhoid, and dysentery.
2. Protect aquatic life from the harmful effects of pollutants.
3. Preserve water quality for recreation, such as swimming, fishing, and boating.
4. Support economic development by ensuring that businesses have access to clean water.

International and National Standards for Drinking Water: To assure the safety and purity of drinking water for human use, international and national drinking water standards are produced. These guidelines often specify limitations on the levels of various contaminants allowed in drinking water, such as bacteria, chemicals, and physical factors. The World Health Organisation (WHO) Guidelines for Drinking Water Quality are among the most extensively utilized international drinking water standards. These guidelines are based on the best current scientific data and are revised regularly to reflect new information. Over 100 pollutants are limited under the WHO Guidelines for Drinking Water Quality, including:

1. Microorganisms: bacteria, viruses, and parasites

2. Chemicals: heavy metals, pesticides, and herbicides
3. Physical parameters: turbidity, color, and odor

The WHO Guidelines also include advice on the design and operation of water treatment and distribution systems. National drinking water standards are frequently based on WHO Guidelines, although they may also be adjusted to each country's individual environment and needs. In India, for example, the Bureau of Indian Standards (BIS) produced the IS 10500:2012 drinking water standard. This standard bans the presence of over 100 pollutants, including microorganisms such as bacteria, viruses, and parasites. Heavy metals, insecticides, and herbicides are examples of chemicals. Turbidity, color, and odor are physical qualities.

The IS 10500:2012 standard also includes suggestions for water treatment and distribution system design and operation. It is crucial to emphasize that international and national drinking water standards are only one component of a holistic approach to assuring drinking water safety and quality. Other critical measures include:

1. Source water protection: protecting the quality of water at its source, such as rivers and lakes
2. Water treatment: removing contaminants from water using a variety of methods, such as coagulation, filtration, and disinfection
3. Water distribution: delivering water to consumers in a safe and sanitary manner
4. Water quality monitoring: regularly testing water to ensure that it meets all applicable standards

Monitoring and Compliance Procedures: Monitoring and compliance procedures are essential for ensuring that water quality standards and regulations are met. According to WRM (2019) Reports, These procedures typically involve collecting samples of water and testing them for contaminants. Monitoring procedures may vary depending on the type of water being tested and the specific contaminants of concern. However, most monitoring procedures typically involve the following steps:

1. Sample collection: Water samples must be collected in a careful and sanitary manner to avoid contamination. The specific sampling method used will depend on the type of water being tested.
2. Sample analysis: Water samples must be analyzed using a variety of methods to identify and quantify contaminants. The specific analytical methods used will depend on the types of contaminants that are being tested for.
3. Data interpretation: The results of the water sample analysis must be interpreted to determine whether the water meets all applicable standards.

Advanced Technologies for Water Treatment: According to WRM (2020) Reports, Advanced water treatment technologies are emerging to help address the increasing challenges of water scarcity and pollution. These technologies can be used to remove a wide range of contaminants from water, including microorganisms, chemicals, and physical parameters. Some examples of advanced water treatment technologies include:

1. Membrane filtration: To filter impurities out of water, membrane filtration systems use membranes with microscopic pores. Bacteria, viruses, parasites, and heavy metals are just a few of the pollutants that can be eliminated via membrane filtration.
2. Reverse osmosis: Reverse osmosis is a kind of membrane filtration in which water is forced through a semipermeable membrane by atmospheric pressure. Salts and other dissolved materials can be taken out of water using a reverse osmosis process.
3. Ozone treatment: Ozone is a powerful oxidant that can be used to disinfect water and remove organic matter. Ozone treatment is also effective at eliminating certain chemicals, such as pesticides and herbicides.
4. Advanced oxidation processes: To produce very reactive free radicals, advanced oxidation processes (AOPs) combine energy and chemicals. Many pollutants, such as organic materials and specific compounds, can be eliminated by using free radicals.

Water can be treated using advanced water treatment technologies for a range of applications, including drinking water, wastewater treatment, and industrial water treatment. These technologies can aid in the provision of safe and clean water to all. Aside from the methods mentioned above, a number of other innovative water treatment technologies are being developed. For example, scientists are working on novel membranes that are more efficient and less expensive to run. They are also exploring novel ways to power water treatment operations with renewable energy.

Reverse Osmosis and Membrane Filtration: Membrane filtration and reverse osmosis are two modern water treatment technologies that may remove a variety of impurities from water, such as bacteria, chemicals, and physical characteristics. Membrane filtration removes pollutants from water by using membranes with microscopic pores. It can be used to eliminate a variety of contaminants such as bacteria, viruses, parasites, and heavy metals. Reverse osmosis is a membrane filtration method in which pressure forces water through a semipermeable membrane. It can remove salts and other dissolved substances from water WRM (2020). Both membrane filtration and reverse osmosis can treat water for a range of applications, including drinking water, wastewater treatment, and industrial water treatment.

Ultraviolet (UV) Disinfection: UV disinfection is a water treatment process that uses ultraviolet radiation to eliminate microorganisms such as bacteria, viruses, and parasites. UV light destroys the DNA of microbes, preventing them from reproducing. UV disinfection is a safe and effective method of water disinfection. It produces no hazardous chemicals or wastes. UV disinfection is also quite cheap to run and maintain. UV disinfection is used to treat water for a range of applications, including drinking water treatment, wastewater treatment, and industrial water treatment. It also cleans areas, including food contact surfaces and medical devices. Here are some of the most important advantages of UV disinfection: Security and efficiency.

1. Does not produce any harmful chemicals or byproducts,
2. Relatively inexpensive to operate and maintain
3. It can be used to disinfect water and surfaces

Water Quality Management Case Studies: Here are some water quality management case studies, WRM (2020):

Case Study 1: The Bay of Chesapeake

The Chesapeake Bay is the United States' largest estuary. It supports a diverse range of aquatic species but is also heavily contaminated by agricultural runoff, sewage, and industrial waste. The Chesapeake Bay Programme was established in 1983 to address the Bay's water quality issues. The program is a collaboration of the federal government, state and local governments, enterprises, and environmental organizations. The Chesapeake Bay Programme has made great strides towards bettering water quality. The amount of nitrogen entering the Bay, for example, has fallen by 25% since 1985. However, more work remains to be done.

Case Study 2: The Los Angeles River

The 51-mile-long Los Angeles River traverses the metropolitan area of Los Angeles. The river was channeled and covered in concrete in the early 20th century to prevent flooding and disease, although it was originally a lovely and natural river. A major push to revitalize the Los Angeles River has emerged in recent years. The main objectives are restoring the river to its more natural state and improving its usability as a public area. Improving water quality is a major obstacle to the revitalization of the Los Angeles River. The river is contaminated by sewage, industrial waste, and stormwater runoff, among other things. The non-profit Los Angeles River Revitalization Corporation (LARRC) is tasked with enhancing the river's water quality. The LARRC is attempting to restore natural wetlands and other habitats along the river and lessen pollution from stormwater runoff.

4. CHALLENGES, INNOVATIONS, AND FUTURE DIRECTIONS

Current Challenges in Water Demand and Quality Management: Water demand and quality management are facing a number of challenges, including:

1. Urbanisation and population growth: People are migrating to cities, and the world's population is expanding. Water supplies are under more and more strain as a result of this.
2. Climate change: Droughts and floods are among the more extreme weather occurrences brought on by climate change. As a result, water resource management and water quality maintenance are becoming more challenging.
3. Pollution: In many regions of the world, water pollution is a serious issue. Several things can produce pollution, such as sewage, industry, and agriculture.
4. Infrastructure requirements: Water infrastructure has to be updated because it is aging in many parts of the world. This presents a serious financial obstacle.

These problems are intricately linked and multifaceted. There isn't a solution that works for everyone. Nonetheless, a variety of actions can be taken to alleviate these issues, including:

1. Water conservation: Lowering the demand for water resources requires water conservation. Water conservation can be achieved in several ways, including replacing leaky faucets, water-efficient fixtures, and appliances, and reducing grass irrigation.
2. Pollution control: Pollution control is essential for protecting water quality. This can be done by implementing regulations and investing in treatment technologies.
3. Infrastructure investment: To satisfy the demands of an expanding population and prepare for climate change, investments in water infrastructure are crucial. This entails investing in brand-new pipelines, storage facilities, and water treatment plants.

Climate Change Adaptation: The process of adapting to the predicted or actual effects of climate change is known as climate change adaptation. WRM (2020), To minimize harm or seize opportunities, this may include modifying both natural and human systems. There are many different ways to adapt to climate change. Some examples include:

1. Building seawalls to protect coastal communities from sea level increase.
2. Developing drought-resistant crops.
3. Planting trees to reduce urban heat islands.
4. Improving early warning systems for extreme weather events.

Climate change adaptation is critical for mitigating the effects of climate change on people and ecosystems. It is also necessary for constructing a more resilient future. Here are some concrete examples of climate change adaptation (WQP 2021):

1. In the Netherlands, people have been building seawalls and dikes for centuries to protect themselves from flooding.
2. In Bangladesh, farmers are planting floating rice varieties that can withstand flooding.
3. In India, cities are planting trees and creating green spaces to reduce urban heat islands.
4. In the Caribbean, countries are developing early warning systems for hurricanes and other extreme weather events.

Socioeconomic and Political Challenges: Socioeconomic and political challenges are significant obstacles to effective water demand and quality management. These challenges include:

1. Poverty: Poverty is a significant obstacle to receiving safe drinking water and sanitation services. Poverty sometimes prevents people from paying for water services or investing in water-saving equipment and fixtures.
2. Inequality: Inequality can also contribute to water scarcity. Women and girls, for example, are more likely than men to lack access to safe water and sanitation facilities.
3. Conflict: Water conflict is becoming increasingly common in many world regions. Competition in water resources can lead to violence and instability.
4. Poor governance: Poor governance can make it difficult to manage water resources effectively. Corruption and a lack of openness can lead to water resource mismanagement and corruption.

These problems are complicated and interwoven. There is no such thing as a one-size-fits-all answer. However, several things may be done to solve these issues, including:

1. Investing in social programs: Investing in social programs to eliminate poverty and inequality will assist in enhancing everyone's access to safe drinking water and sanitation services.
2. Gender equality: Gender equality can help to guarantee that women and girls have equal access to water resources and that their needs are taken into account in water management decisions.
3. Promoting cooperation over conflict: Promoting cooperation over conflict is critical for long-term water management.
4. Improving governance and encouraging openness and accountability can help to improve water management and minimize corruption.

To address the socioeconomic and political difficulties of water demand and quality management, governments, corporations, and civil society must work together. (WQP 2021) We can work together to create a more just and equitable world in which everyone has access to clean water. Here are a few particular instances of how socioeconomic and political issues can affect water demand and quality management:

1. Millions of people in India live in poverty and lack access to safe drinking water. This is a major public health problem.
2. Water scarcity and conflict over water resources are major problems in the Middle East. This has led to instability and violence in the region.
3. In the United States, there is a growing gap between the rich and the poor. This gap is evident in access to water resources. Poor communities are more likely to live in areas with contaminated water and to pay higher water rates.

Innovations in Water Management Technologies

Several water management technology solutions are being developed and applied to solve the concerns of water scarcity and pollution. Among these innovations are:

1. Smart water meters: Smart water meters may collect real-time data on water usage. This information can detect leaks, decrease water waste, and boost water efficiency.
2. Desalination: The process of eliminating salt from saltwater to produce fresh water is known as desalination. Desalination technologies are becoming more efficient and cost-effective, making them a viable choice for addressing water needs in water-stressed areas.
3. Water recycling and reuse refers to preparing wastewater for reuse in non-potable uses such as industrial or agricultural irrigation. Water reuse and recycling can help to reduce the demand for freshwater resources.

4. Green infrastructure: Using natural and manmade systems to regulate stormwater runoff and improve water quality is referred to as green infrastructure. Green infrastructure can aid in the reduction of flooding, the filtering of pollutants from runoff, and the replenishment of groundwater supplies.

Here are some specific examples of how innovative water management technologies are being used around the world:

1. In Singapore, smart water meters are being used to identify leaks and reduce water waste.
2. In Saudi Arabia, desalination plants are being used to produce freshwater from seawater.
3. In California, recycled water is being used to irrigate crops and landscapes.
4. In New York City, green infrastructure is being used to reduce flooding and improve water quality.

Internet of Things (IoT) Applications in Water Management: The Internet of Things (IoT) is a network of physical items embedded with sensors, software, and other technologies that communicate with other devices and systems over the Internet. IoT applications in water management are numerous and diverse, including:

1. Smart water meters: Smart water meters can collect real-time data on water usage, which can be used to detect leaks, decrease water waste, and increase water efficiency.
2. Water quality monitoring: IoT sensors can be used to continuously monitor water quality parameters like pH, temperature, turbidity, and conductivity. This information can be utilized to detect and manage water quality issues early on.
3. Leak detection and prevention: IoT sensors can be used to detect leaks in water pipes and other infrastructure. This data can be used to repair leaks quickly and prevent further damage.
4. Irrigation management: IoT sensors can be used to monitor soil moisture and meteorological conditions in order to optimize irrigation schedules. This can assist to reduce water waste and increase crop yields.
5. Industrial water management: IoT sensors can be used to monitor and control water usage in industrial processes. This can help to minimize water consumption and enhance water efficiency.

Here are some concrete instances of how IoT is being used in global water management:

1. Intelligent water meters are being deployed in Singapore to detect leaks and reduce water waste. This has contributed to Singapore having one of the lowest rates of water use in the world.
2. In the United States, IoT sensors are being utilized to monitor the Great Lakes' water quality. This information is being utilized to identify and solve water contamination issues as they arise.
3. In Israel, IoT sensors are being used to detect leaks in water pipes. This has helped Israel to reduce water losses by over 20%.
4. IoT sensors are being utilized in India to monitor soil moisture and meteorological conditions in order to optimize irrigation plans. This has assisted Indian farmers in reducing water consumption and increasing crop yields.

Big Data Analytics for Predictive Water Management: Big data analytics can be used for predictive water management in a number of ways, including According to the report WRHP (2022):

1. Water demand prediction: Big data analytics can be used to forecast water consumption based on parameters such as population increase, economic development, and meteorological conditions. This data can be used to create water management strategies and guarantee that adequate water is available to fulfill demand.
2. Water quality prediction: Big data analytics can be used to predict water quality based on pollution sources, weather conditions, and land use. This data can be used to detect and solve water quality issues early on.
3. Water-related hazard prediction: Big data analytics can be used to anticipate water-related hazards like floods, droughts, and storm surges. This data can be utilized to create early warning systems and evacuate individuals from disaster areas.

Future Directions and Policy Recommendations: A variety of developments are anticipated to impact the future of water management, including the WRI (2020) report:

1. Climate change: Climate change is already having a substantial influence on the world's water resources. Climate change is anticipated to cause more extreme weather occurrences, such as droughts and floods, in the future. This will make water management more difficult.
2. Population growth: By 2050, the world's population is predicted to exceed 9 billion people. As the population grows, so will the need for water resources.

3. Urbanisation: The world is growing more urban. Over 60% of the world's population is expected to reside in cities by 2050. This urbanization will put further strain on urban water resources.

According to UNESCO (2017) reports, Water managers will need to develop innovative ways to water management to overcome these concerns. Some of the most important future directions in water management are as follows:

1. Integrated water management (IWM): IWM is a water management technique that considers all aspects of the water cycle, including surface water, groundwater, and wastewater. IWM also takes into account the social, economic, and environmental aspects of water management.
2. Water conservation: Water conservation is critical for minimizing water resource demand. Fixing leaking faucets, watering lawns less, and installing water-efficient appliances and fixtures are all examples of water-saving methods Hall, J. W., et al. (2014).
3. Water reuse: The process of treating wastewater so that it can be reused for non-potable applications such as irrigation or industrial use is known as water reuse. Water reuse can contribute to a reduction in the demand for freshwater resources.
4. Desalination: The process of eliminating salt from saltwater to produce freshwater is known as desalination. Desalination technologies are becoming more efficient and cost-effective, making them a viable choice for addressing water needs in water-stressed areas.

According to the United Nations Development Programme (2016) Report, in addition to these technical measures, it is also important to develop policy and institutional frameworks that support sustainable water management. Some of the key policy recommendations for the future of water management include:

1. Pricing water to reflect its actual value: Water is a valuable resource, and it should be priced accordingly. This will help to reduce water waste and encourage water conservation.
2. Investing in water infrastructure: It is important to invest in water infrastructure to meet the needs of a growing population and to adapt to climate change. This includes investing in things like water treatment plants, pipelines, and storage facilities.
3. Strengthening water governance: It is essential to enhance water governance to ensure that water resources are managed equitably and sustainably. This includes

things like improving transparency and accountability in water management decision-making.

Policy Reforms for Sustainable Water Management: World Resources Institute(WRI) (2020). There are a number of policy reforms that can be implemented to promote sustainable water management. Some of the key reforms include:

1. Investing in water infrastructure: It is important to invest in water infrastructure to meet the needs of a growing population and to adapt to climate change. This includes investing in things like water treatment plants, pipelines, and storage facilities.
2. Strengthening water governance: It is essential to enhance water governance to ensure that water resources are managed equitably and sustainably. This includes things like improving transparency and accountability in water management decision-making.
3. Promoting water conservation and efficiency: Governments can promote water conservation and efficiency through a variety of policy measures, such as public education campaigns, water-efficient standards for appliances and fixtures, and financial incentives for water conservation.
4. Pricing water to reflect its actual value: Water is a valuable resource, and it should be priced accordingly. This will help to reduce water waste and encourage water conservation United Nations Development Programme (UNDP) (2016)

In addition to these broad policy changes, a variety of specialized policy initiatives can be introduced to solve specific water management issues. In water-stressed areas, for example, governments may need to enact rules that limit water consumption for specific purposes such as agriculture or landscaping. Governments in flood-prone areas may need to invest in flood control infrastructure and establish early warning systems, UNESCO (2017). It is critical to recognize that there is no one-size-fits-all solution to water policy change. The particular policy adjustments required will differ depending on the context. However, the policy reforms described above are an excellent place to start when designing policies for sustainable water management, Bhargava, A., & Finger, S. (2017).

5. CONCLUSION

Summary of Key Findings: Water is an essential resource for life, and its sustainable management is critical for the well-being of humanity and the environment. However, water resources are facing a number of challenges, including climate change, population growth, and urbanization. To address these challenges, it is important to adopt new approaches to water management and implement effective policies. Key findings include:

1. Integrated water management (IWM) is an essential approach to managing water resources in a sustainable and equitable manner. IWM takes into account all aspects of the water cycle, including surface water, groundwater, and wastewater, as well as the social, economic, and environmental dimensions of water management.
2. Water conservation is critical for reducing the demand for water resources. Various water conservation measures, such as fixing leaky faucets, watering lawns less, and using water-efficient appliances and fixtures, can be implemented.
3. Water reuse can help to reduce the demand for freshwater resources by treating wastewater so that it can be reused for non-potable purposes, such as irrigation or industrial use.
4. Desalination is a viable option for meeting water needs in water-scarce regions. Desalination technologies are becoming more efficient and affordable, making them a more cost-effective option than in the past.

In addition to these technical measures, developing policy and institutional frameworks that support sustainable water management is also important. Key policy recommendations include:

1. Pricing water to reflect its true value: Water is a valuable resource and should be priced accordingly. This will help to reduce water waste and encourage water conservation.
2. Investing in water infrastructure: It is important to invest in water infrastructure to meet a growing population's needs and adapt to climate change. This includes investing in water treatment plants, pipelines, and storage facilities.
3. Strengthening water governance: It is important to strengthen water governance to ensure that water resources are managed equitably and sustainably. This includes things like improving transparency and accountability in water management decision-making.

Implications for Policy and Practice: The implications of the key findings and recommendations for policy and practice are significant. First, the findings emphasize the need for a comprehensive and integrated approach to water management, which includes considering all aspects of the water cycle, including surface water, groundwater, and wastewater, as well as the social, economic, and environmental dimensions of water management. However, before implementing these technologies, it is critical to carefully consider their environmental and social impacts. In terms of policy implications, the findings suggest that governments should:

1. Price water to reflect its true value, which will help to reduce water waste and encourage water conservation.
2. Invest in water infrastructure to meet a growing population's needs and adapt to climate change.
3. Strengthen water governance to ensure water security.
4. Protect water quality by developing and enforcing water quality regulations, investing in wastewater treatment infrastructure, and controlling pollution from agriculture and industry.
5. Adapt to climate change by developing water management strategies that are more resilient to drought, floods, and other extreme weather events.

REFERENCES

Alliance for Water Stewardship. (n.d.). https://a4ws.org/

Bhargava, A., & Finger, S. (2017). Internet of Things: Applications and challenges in smart cities: A case study of IBM smart city projects. In *Smart Cities* (pp. 61–84). Springer.

Big Data Analytics for Smart Water Management: A Review. (2019). *Water Resources Management*, *33*(11), 4101–4123.

Big Data Analytics for Water Demand Prediction: A Review. (2020). *Water Resources Management*, *34*(13), 4637–4659.

Big Data Analytics for Water Quality Prediction: A Review. (2021). *Water Resources Management*, *35*(15), 5591–5613.

Big Data Analytics for Water-Related Hazard Prediction: A Review. (2022). *Water Resources Management*, *36*(17), 6459–6481.

Falkenmark, M., & Rockström, J. (2006). The new blue and green water paradigm: Breaking new ground for water resources planning and management. *Journal of Water Resources Planning and Management*, *132*(3), 129–132. doi:10.1061/(ASCE)0733-9496(2006)132:3(129)

Gleick, P. H. (2003). Water use. *Annual Review of Environment and Resources*, *28*(1), 275–314. doi:10.1146/annurev.energy.28.040202.122849

Global Water Futures: Perspectives on the Future of Water Resources and Management. (2017). UNESCO.

Hall, J. W. (2014). Robust climate policies under uncertainty: A comparison of robust decision making and info-gap methods. *Risk Analysis*, *34*(4), 673–683. PMID:22519664

Ministry of Jal Shakti, Government of India. (2021). National *Water Policy*. Author.

Ministry of Urban Development, Government of India. (2015). *Atal Mission for Rejuvenation and Urban Transformation*. AMRUT.

Ministry of Urban Development, Government of India. (2015). *Smart Cities Mission*. Author.

Policy Reforms for Sustainable Water Management. (2020). World Resources Institute.

Sharma, A., & Malano, H. (2009). Evolution of urban water demand in a rapidly growing low-income semi-arid city. *Water Resources Management*, *23*(8), 1477–1494.

Smil, V. (2017). *Harvesting the Biosphere: What We Have Taken from Nature*. MIT Press.

The Future of Water Management: Challenges and Opportunities. (2020). World Resources Institute.

The Nature Conservancy. (n.d.). https://www.nature.org/en-us/what-we-do/our-insights/water-security/

United Nations. (2018). *Sustainable Development Goal 6 Synthesis Report on Water and Sanitation*. Author.

United Nations Water. (n.d.). https://www.unwater.org/

Water Management in the 21st Century: Challenges and Opportunities. (2016). United Nations Development Programme.

World Health Organization. (2017). *Guidelines for Drinking-Water Quality*. Author.

World Resources Institute. (2021). *Aqueduct Water Risk Atlas*. Author.

World Resources Institute. (n.d.). https://www.wri.org/water

Chapter 4
Prediction of Water Quality Using Machine Learning

Tran Thi Hong Ngoc
An Giang University. Vietnam

Phan Truong Khanh
An Giang University, Vietnam

Sabyasachi Pramanik
https://orcid.org/0000-0002-9431-8751
Haldia Institute of Technology, India

ABSTRACT

With the fast growth of aquatic data, machine learning is essential for data analysis, categorization, and prediction. Data-driven models using machine learning may effectively handle complicated nonlinear problems in water research, unlike conventional approaches. Machine learning models and findings have been used to build, monitor, simulate, evaluate, and optimize water treatment and management systems in water environment research. Machine learning may also enhance water quality, pollution control, and watershed ecosystem security. This chapter discusses how ML approaches were used to assess water quality in surface, ground, drinking, sewage, and ocean. The authors also suggest potential machine learning applications in aquatic situations.

DOI: 10.4018/979-8-3693-1396-1.ch004

1. INTRODUCTION

Wastewater carrying toxins from fast economic growth threatens natural water ecology. So, several water pollution management methods evolved. Water quality analysis and assessment have greatly enhanced water pollution control efficiency. The multivariate statistical approach, fuzzy inference, and water quality index (WQI) are among the various methods used to monitor and measure water quality globally. While many water quality metrics may be traced in accordance with regulations, the final findings may vary according to parameter selection. Taking into account that all water quality metrics is impractical due to cost, technical difficulty, and inability to account for variability. Recently, developments in machine learning have led academics to anticipate that huge volumes of data may be achieved and assessed to accomplish complex and large-scale water quality monitoring needs.

ML algorithms are used in artificial intelligence to examine data and find patterns to forecast future information. With its accuracy, flexibility, and extensibility, machine learning has become a popular data analysis and processing tool in several fields. Machine learning simplifies the finding of underlying mechanisms for complex nonlinear relational data. Recently, ML showed a huge potential as a tool in ecological science and engineering due to its versatility. In spite of the difficulty of ML for water quality measurement and assessment, much precise results are predicted.

Complex water kinds include drinking, wastewater, and groundwater, surface, marine, and fresh. These water kinds have varied qualities, making quality study difficult. Previous research suggests that machine learning may effectively handle these difficulties. In this study, we address the pros and cons of typical ML approaches and their implementations and performance in surface water, groundwater, drinking water, wastewater, and ocean water (Fig. 1).

ML is commonly utilized to find insights or have predictions from vast data from many contexts. Prior to using ML, data collecting, algorithm selection, model training, and validation are needed. Among these methods, selecting algorithm is the key aspect.

Machine learning has two primary classes: supervised and unsupervised. Labels in datasets distinguish these two kinds. Supervised learning predicts from labeled training datasets. Input and anticipated output values are included in each training instance. Supervised learning algorithms discover input-output correlations and create a predictive model to estimate the outcome from the I/P data. Supervised learning methods, such as LR, ANN, decision trees, SVM, Naive Bayes, KNN, and random forests are designed for data classification and regression.

In contrast, unsupervised learning handles data generally without labels, addressing pattern recognition problems using unlabeled training datasets. Unsupervised learning classifies training data depending on features, primarily via dimensionality reduction and grouping. But, the quantity and significance of categories are unknown. Thus, unsupervised learning is often employed for classification and association mining. PCA, K-means, and other unsupervised machine learning methods are popular. Reinforcement learning, which allows machines to generalize and solve unlearned problem is a different kind of ML method. In contrast to the other 2 ML classes, it is sometimes used in the aquatic aspect.

Figure 1. Water systems employ machine learning extensively. WWTP, wastewater treatment facility
Source: Zhu et al. (2022)

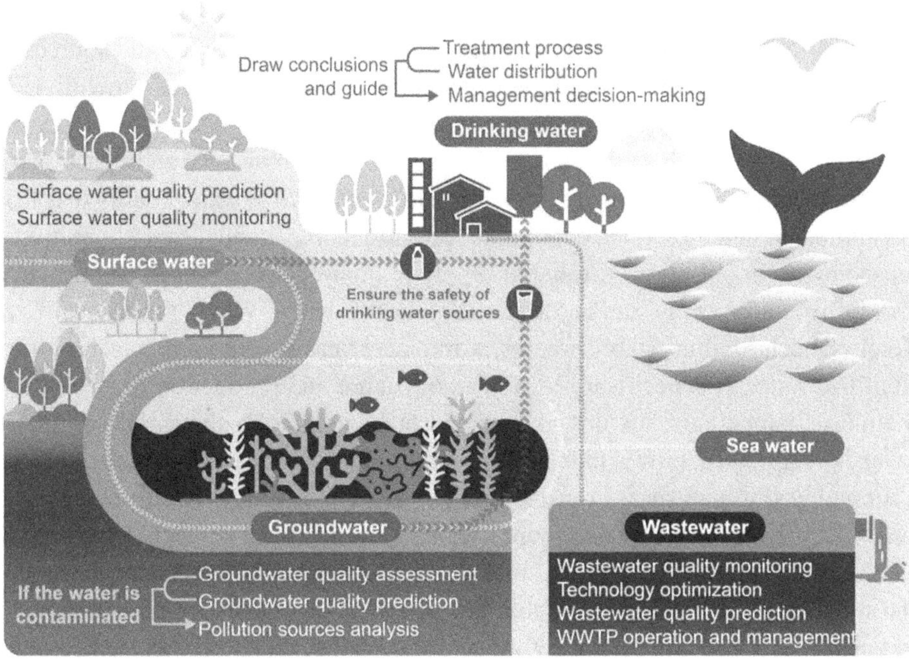

2. WHAT PURPOSES DOES WATER SERVE?

Water is a resource that is essential to life and may be used for a wide range of tasks, from everyday household tasks to supporting the operations of whole enterprises. Domestic Water Uses Cooking requires water to assist bring food to a boil, steam, or simmer. Water is a useful agent to assist wash clothing, dishes, and food items because of its special ability to generate solutions and emulsions. Water washing aids in the removal of pollutants. It is necessary for maintaining personal hygiene, cleaning teeth, and having showers. At home, it's used for gardening. Many household equipment, including air coolers, need water. Electricity is produced from water and utilized for household needs. Our bodies lose water as we breathe, perspire, and digest food. Drinking on a regular basis can help the body replace its water content and prevent dehydration and other health issues. It is advised to drink one to seven liters of water each day to prevent dehydration. Water Use in Agriculture Every year, agriculture uses a significant portion of the water supply. Large amounts of water are needed in agriculture in order to cultivate fresh food and support animals. It is used for crop chilling, frost management, fertilizer and pesticide application, and irrigation. Water use has to be managed properly in order to control agricultural production and output. Many water-saving techniques should be used by farmers in order to practice sustainable agriculture. In order to create oxygen via photosynthesis, plants need water and sunshine. Rainfall, groundwater from wells, and surface water from rivers, streams, open canals, ponds, reservoirs, and lakes are the sources of agricultural water. All conventional agricultural practices, such as growing rice, wheat, sugarcane, etc., need the usage of water. Illegal chemical dumping in water and the atmosphere, as well as widespread industrialization, may have an impact on the quality of water. In addition to affecting food crop quality, low water quality may lead to a number of illnesses. Food contamination and accompanying diseases are mostly caused by water pollution. A reliable supply of high-quality water is groundwater. Fish farms, dairy farms, and cattle farms all utilize water. Water's Industrial Uses Hotels, hostels, dining establishments, workplaces, and other commercial buildings all utilize water. It is necessary for the manufacture and production of steel, food, chemicals, paper, cars, textiles, dying, and other items. A substantial volume of water is required by power plants for cooling. Since water is a universal solvent, it is employed in industrial manufacturing to dissolve a variety of substances. It is less often used as a catalyst and more frequently utilized as a solvent. Additionally, a number of industrial operations make use of water vapor. Petroleum refineries and smelting plants both utilize water. Water is used in industry to fabricate, dilute, and process goods. There are easy ways to go about thanks to rivers, canals, seas, and oceans. When compared to land transportation without any obstructions, there won't be as much friction in the water. It provides economical options together with a cargo-transporting method that works. One source of hydroelectricity is water. To hold water that is later used to power turbines, dams are built across rivers and lakes. Worldwide, hydroelectricity is a renewable energy source that is in use. It has reduced reliance on fossil fuels as a source of energy. Gases, minerals, and oil are extracted with the help of water. It is necessary for a number of important mining tasks. Water is utilized in heat exchangers and steam turbines in addition to being a solvent. Water's Medicinal Properties Water is used in medicine for a variety of purposes, including in dentistry and hemodialysis. It is used to clean surgical instruments and supplies. It is necessary for hydrotherapy as well. Since some patients have compromised immune systems; water contamination by bacteria is a serious problem in the hospital context. Direct touch, ingestion, inadvertent contact, aspiration of water, and blood contact are all possible ways for the waterborne illness to spread. Hospitals and clinics should not utilize water unless it has undergone extensive treatment. In order to prevent contaminating water that the patient will consume, clinicians should take the necessary precautions.

3. METHODOLOGY (MACHINE LEARNING IN VARIOUS WATER APPLICATIONS)

Various researchers have utilized ML to solve water treatment and management system problems (Fig. 2), like real-time tracking, estimation, pollutant source estimating, concentration prediction, water resource allotment, and technology optimization.

3.1 Implementations in Surface Water

Civic and industry wastewater from human actions is the principal cause of urban water quality degradation. ML in surface water quality analysis is trendy. Several surface water quality estimation and assessment techniques exist. Various attempts were made to optimize ML models and improve estimation accuracy.

Data collecting is essential to machine learning model development. Additionally, aggregated and periodic water quality estimating findings may be benchmarked in water system management. Conventional environmental tracking approaches are used by agents. But, established approaches for in situ tracking have challenges. Remote sensing technologies provide real-time, huge-scale water quality tracking and show pollution movement and dispersion patterns that traditional approaches cannot identify. (Khanh, P. T. et al. 2023) saw that experiment-dependent ML let advanced optimization using real-time sensor data and satellite data, and the partial least squares (PLS), support vector regression, and DNN approaches were more accurate than conventional models. Since certain water quality variables, like pathogen concentrations, aren't optically active or need high-spatial-resolution hyperspectral data, remote sensing cannot directly detect them, but it may estimate those utilizing different measurable data. Wu et al. built a CNN model which can distinguish clean and dirty water from photos. Multiple comparison studies on a sea surface picture dataset confirmed the efficiency of this CNN. The benefit of CNN is that it uses the reflection picture as input without requiring feature engineering or parameter modification. Because of equipment or human error, few data will be missing, inaccurate, or destroyed, creating a sparse matrix and poor model application performance.

Figure 2. Various ML methods in water treatment and management systems. SVM, RF, ANN, Decision Tree, PCA, XGBoost, DO, and MP are used in the analysis.
Source: Zhu et al. (2022)

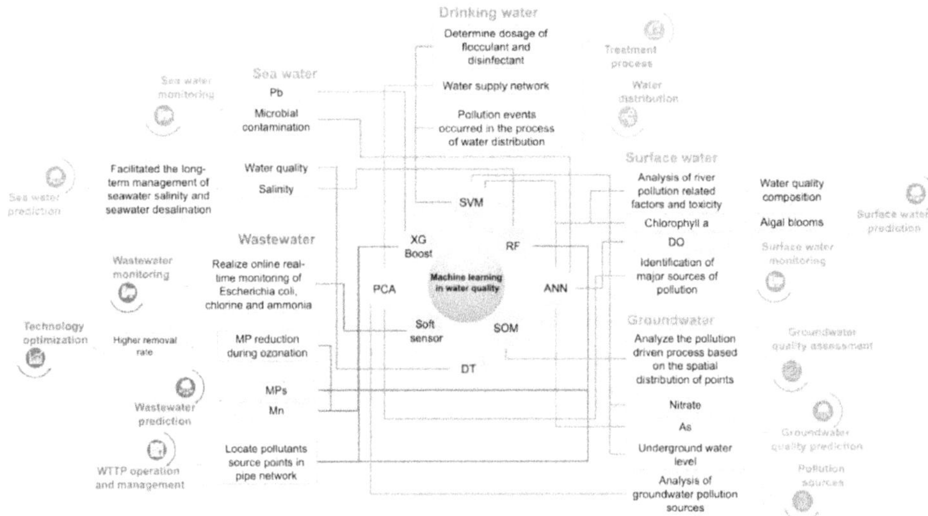

Data cleansing, another crucial ML stage, becomes necessary. Data cleaning may be done by not utilizing the raw data, utilizing averages or medians, or utilizing ML and matrix completion. (Ngọc, T. H. et al. 2023) used Deep Neural Network and deep matrix factorization to estimate organic O_2 requirement. The rationality and dependability of the technique were verified utilizing California Harbor waters as a case history. Data cleaning enhances machine learning model accuracy by improving data quality.

Machine learning prediction accuracy depends on model selection and training dataset quality. ANN and SVM predict water quality components well. SVM may outperform ANN in prediction and generalization. One explanation is because neural network model parameter tuning is unstable; therefore nonlinear disturbances greatly impair ANN accuracy. SVM minimizes generalization error better than ANN because it limits it rather than training error. Effective river management requires real-time monitoring of water quality or relying on other data while monitoring circumstances are insufficient due to complicated dynamic changes over time. Researchers have confirmed that LSTM (Long Short-Term Memory) and BWNN (Biased Wavelet Neural Network) networks can manage fluctuating and non-seasonal water-quality data. Traditional linear statistical models like the autoregressive integrated moving average (ARIMA) model may estimate time series. It is substandard to the BWNN model that is motivated by the self-adaption of the ANN and the time-frequency features of wavelet basis functions, and the Long Short Term Memory technique that masters straightway from time-series data. Long Short Term Memory and BWNN recognize nicely the nonlinear connection between variables and their projected variables, and utilize past data to forecast the future.

The characteristics used to train machine learning models affect prediction accuracy. Redundant variables increase model complexity and decrease inverse power and accuracy. A most important surface water quality measures is dissolved oxygen (DO), which reflects the water environment and its capacity to support water life. The linear polynomial neural network approach predicted Danube River DO concentration. The forecast accuracy was most affected by temperature, pH, BOD, and phosphorus content among 17 water quality factors. For prediction of DO accumulation in Columbia River, USA, among 5 variables chloride, NOX, cumulative dissolved solids, pH, and water temperature are closely associated with dissolved oxygen and may impact forecast accuracy. These findings align with (Roy, A. et al. 2023), who found that I/P parameters impact model estimation achievement. Furthermore to water parameters, eutrophication affects surface water quality forecast. (Ngoc, T. T. H. et al. 2023) discovered that nutrients, organic matter, and environmental factors generated algal blooms using the adaptive neuro-fuzzy inference system technique. (Jayasingh, R. et al. 2022) utilized meteorological and weekly water quality data to forecast chlorophyll-a concentrations in two U.S. locations. SVM and ANN showed equal accuracy. The incorporation of meteorological parameters greatly enhanced forecast accuracy. The machine learning model may also include regional hydrological and socioeconomic elements to provide complete regional water environment management.

ML model performance relies on its architecture, hence understanding algorithm logical structure is crucial to machine learning effectiveness. PNN's advantage over other neural network models in identifying crucial model parameters is that the quantity of hidden neurons and PNN layers is selected by data, preserving trial time. The Deep Neural Network model utilized to predict BOD has 19.20%–25.16% lower RMSE than conventional machine learning. Because there are DNN has numerous layers between its I/P and O/P, and it requires latest activation functions improve model convergence and minimize training difficulty compared to classic ANNs (sigmoid). Time series-based LSTM predicts water quality effectively. It has 3 information gates: input, forget, and output gate, cell state, and hidden state govern sequential information transmission, forgetting, and storage. BWNN can accomplish this aim, but needs enough input characteristics for accuracy.

3.2 Applications in Groundwater

Drinking water comes from groundwater. Thus, groundwater safety is vital to human well-being. ML has several applications in groundwater study, which includes quality evaluation and contamination estimation. Multivariate statistical analysis has become popular for groundwater quality assessments. PCA and cluster analysis are popular. Groundwater quality evaluation has been used SVM, DT, RF, and ANN algorithms. Studies on groundwater quality mostly examine ML approaches to determine viable solutions for certain challenges. In Tabriz City, Iran, (Pramanik, S. et al. 2021) compared 5 data mining techniques: normal DT, Random Forest, chi-square automatic interaction detector, and iterative dichotomizer to spot semiarid groundwater parameters and their impact on high-quality groundwater. (Pramanik, S. et al. 2022) calculated Seoul's urban groundwater quality spatial pattern using a self-organizing neural network and fuzzy c-means clustering. The researchers used a self-organizing map approach to classify groundwater samples into three categories by contamination degree and examined the pollution-directed technique dependent on geographic distribution. GIS methods were utilized to create groundwater quality maps to better detect contamination.

The intricate hydrogeology of groundwater contrasted predicting quality change trends is more challenging when dealing with surface water. Machine learning has been used to analyze regional data on a big scale and forecast water quality. PSO and SVM were used to assess and forecast groundwater WQI by (Samanta, D. et al. 2021) demonstrating its feasibility. A single groundwater contaminant, like nitrate and arsenic, may be estimated. (Pramanik, S et al. 2023) predicted groundwater nitrate constituent and distribution using an SVM. (Pramanik, S. et al. 2023) assessed groundwater nitrate contamination risk utilizing enhanced regression tree, multivariate discriminant analysis, and Support Vector Machine, and found that India's Khadar Plain is at higher risk of nitrate contamination. (Andrushia et al. 2023) utilized ML to forecast US groundwater nitrate levels and discovered that it can also predict national groundwater quality.

(Vidya Chellam, V. et al. 2023) predicted groundwater arsenic contamination in Ohio, and Japan utilizing an ANN. Additionally, groundwater levels may be forecast. In a study by (Praveenkumar, S. et al. 2023), DNN was shown to be the most accurate and efficient machine learning approach for predicting groundwater levels throughout the season, compared to ANFIS and SVM. (Pramanik, S. 2023) used ensemble modeling to estimate Indian city groundwater levels with 85% accuracy.

The investigation of contamination sources is crucial for groundwater safety. Studies employ PCA and clustering extensively. After PCA reduced dimensionality and K-means grouped data, (Veeraiah, V. et al. 2023) examined natural and anthropogenic hydro-geochemistry sources. (Pramanik, S. and Bandyopadhyay, S. 2023) used multivariate statistical analysis and Principal Component Analysis to discover ground-water quality parameters.

Groundwater quality and resources are frequently examined using a data mining decision tree. This algorithm can identify and describe connections between input and output variables using certain criteria. Performance and capacity to generalize rules to detect drinking-quality groundwater are benefits of RF. Since continuous data sets are better for groundwater law induction than discrete data sets, RF's greatest performance (accuracy of 97.10%) gives effective groundwater resource planning and management judgments. Current groundwater quality index prediction research uses the integrated model that may merge numerous weak learners into a lone strong learner and increase estimation performance. Integration is fantastic with boosting. To create models with lower variance, it is important to reduce over fitting when merging excellent models.

3.3 Drinking Water Application

In drinking water source operation, analysis, distribution, and decision taking, machine learning is frequently used. Drinking water is usually surface or groundwater. The examination and estimation of source water quality using ML may help reduce pollution early. (Bansal, R. et al. 2022) investigated multisensor-dependent Artificial Neural Network and Support Vector Machine approaches for dynamic water quality tracking in 2022. The models generate satisfactory identification rates for the two water groups. Support Vector Machine was more stable than ANN. (Dushyant, K. et al. 2022) used drinking water quality information from 4 Norwegian metropolitans to suggest adaptive frequency analysis. Their findings supported early drinking water quality threat cautioning, administration, and decision-taking. Moreover, (Gupta, A. et al. 2022) combined LSTM and Deep Neural Network to forecast time series data and created a water quality estimation approach which can accurately estimate water quality in the following 6 months. (K.aushik, D. et al. 2022) employed a Support Vector Machine to forecast pollution episodes under unknown circumstances using UV absorption. All four SVM datasets have good detection rates and low error rates. Most investigations have used chemical or physical measures, while microbiological factors, notably for E. coli, have seldom been studied. Machine learning can estimate drinking water plant coagulant and disinfectant levels. SVM algorithms are prominent in occlusion and disinfection construction because to their simple structure and resilience. (Bhattacharya, A. et al. 2021) suggested a predictive chemical dose control technique that was more successful than proportional-integral-derivative feedback control based on SVM-predicted residual free chlorine.

The necessity of drinking water has motivated researchers to concentrate on typical metropolitan water supply framework functioning, fault tracking, and catastrophe estimation. Due to the intricacy of water supply networks, drinking water therapy facilities which satisfy regulations can be re-polluted in transportation that may be monitored utilizing biological stability measures and purified. Cluster analysis may show water quality differences between networks. Furthermore, (Mandal,

A. et al. 2021) employed cluster analysis to determine how mixed water sources, including Al migration and seasonal fluctuations, affect aluminum remnants in urban drinking water supply frameworks. (Meslie, Y. et al. 2021) correctly casts water quality utilizing an RF technique. Burst water supply pipelines cause massive water loss and microbiological and chemical contamination. DL algorithms may forecast bursting locations with high uncertainty. (Pramanik, S. 2022) introduced a water leakage detection method using a random DT bagging classifier with shuffled frog-leaping optimization and minimal sensors at ideal locations in a WSN (water supply network). Pipe longevity matters in water-supply management. A sophisticated meta-learning framework dependent on a neural network by (Taviti Naidu, G. et al. 2023) demonstrated that residual chlorine affects pipeline service life. An SVM method may also forecast water distribution system contamination. (Chandan, R. R. et al. 2023) used PCA, the analytic hierarchy process, Random Forest and XGBoost frameworks to quantify catastrophe effects on water supply systems. However, real-time data collecting challenges restrict the practicality of this strategy.

Regional development and population growth are limited by water production capacity. (Mondal, D. et al. 2023) created a hybrid ANN-genetic algorithm weighted variables. The test findings indicate that this system effectively monitors water quality, having a strong correlation between projected and real values.

Machine learning can enhance wastewater treatment systems by analyzing historical data. (Ghosh, R. et al. 2023) used an SVM and adaptive evolutionary algorithm to mimic anaerobic, anOXic, and oXic conditions to reduce the anoXic tank capacity and conserve land area. Additionally, machine learning optimizes tertiary wastewater treatment methods including RO, NF, ozonize, and absorption. According to (Mall, P. K. et al. 2023) using an RF to anticipate micro pollutant decrease in ozonization led to improved removal efficiency. Using a high-resolution fluorescence excitation-emission matrix for machine learning may improve accuracy by computing the complicated nonlinear connection between statistical models for drinking water performance prediction.

The hybrid statistical model accurately predicts water production responses to different parameter variations, making it a valuable tool for changing water treatment systems rapidly and effectively. (Pandey, B. K. et al. 2023) suggested a time-series clustering-based autonomous tracking system for urban water administration and discovered significant water demand from 6 to 9 AM summertime owing to civic and public garden watering. (Pandey, B. K. et al. 2023) used a gated RNN and a 20-minute time step to forecast short-term water consumption for 15 and 24 hours. (Ahamad, S. et al. 2023) utilized DAN2 and KNN to anticipate every day, every week and every month water consumption in Iraq. DAN2 performed best of the three models. The every day, every week, and every month models predicted 97%,

98%, and 99% accurately. Machine learning can balance water-supply systems. Using accurate water demand predictions to allocate water resources is promising.

In conclusion, Artificial Neural Network and Support Vector Machine are often used in the area of drinking water, particularly in big quantities. The training phase's few seconds of computer time allows dynamic tracking frameworks to check drinking water grade and safeness in real time. Increased training strategies have significantly increased ANN recognition rates, notwithstanding their effectiveness in susceptible to noise.

3.4 Utilizations in Wastewater

ML is frequently utilized in water quality tracking and estimation, technological improvement, and WWTP operation and administration. Industrial and domestic wastewater includes contaminants, requiring water quality testing before therapy. Multiresolution analysis with Principal Component Analysis gave (Veeraiah, V. et al. 2023) a much sensitive method for tracking sewage measures at different scales than PCA. Big data collection, processing, and analysis depend on real-time internet tracking. A black box-based soft sensor was suggested for online, real-time E-monitoring. (Dhamodaran, S. et al. 2023) indicated that E. coli rose significantly after high rains, presumably owing to urban runoff resuspending sewer silt. Soft sensors and ANNs may reduce the expenditure and complexity of WWTP action and ailment and provide real-time chlorine and ammonia monitoring. The boosting-iterative predictor weighting-partial least squares approach and several sensors were used to create a water quality tracking system with a UV spectrometer and turbidimeter to measure COD and cumulative suspended solids. Boosting-IPW-PLS subdued factors irrelevant to water quality by allocating tiny weights and created a wastewater quality estimation framework.

(Dhamodaran, S. et al. 2023) employed Decision Tree to identify the specific sequence of MP removal by RO and NF, revealing particle dimension elimination, electrostatic repulsion, and adsorption as the primary separating processes. Additionally, XGBoost predicts MP removal efficiency in RO and NF. Two neural-network-based models by () helped practitioners choose the right adsorbent for a contaminant. According to the preceding situations, machine learning approaches may be extensively used in the future to treat wastewater with MPs and new contaminants.

Machine learning has improved water treatment analysis by providing predicted data. ANN solves difficult nonlinear environmental issues, including pollution elimination. (Veeraiah, V. et al. 2023) developed an Artificial Neural Network model to estimate COD and BOD levels in wastewater. Currently, water quality estimation algorithms focus on identifying specific pollutant levels. (Pandey, B. K.

et al. 2022) developed CatBoost to reliably anticipate tetracycline (TC) elimination using a photodegradation rate under varied realistic settings.

Metal–organic system (Pramanik, S. 2023) used 3 models (RF, SVM, and ANN) to estimate the elimination of 5 variables. RF provided the most accurate findings after verifying all models. Machine learning can predict biological signs. Bayes methods, such as naive and semi Naïve Bayes networks, were used to forecast pathogen elimination capability and reflect the relationship between pathogen minimization, operational circumstances, and tracking parameters. RF predicted wastewater Clostridiales and Bacteroidales by (Pramanik, S. and Bandyopadhyay, S. 2023). Comprehensive assessment and computation techniques for estimating excrement contamination sources have been developed using RF, reducing the spread of water-borne illnesses.

Many variables may impact the effluent quality of WWTPs. WWTP operation and maintenance might be difficult when expenses must be managed. Thus, machine learning may help WWTP managers save expenses and enhance operations. Toxic contaminants in sewage networks may disrupt WWTP functioning. XGBoost and RF identified contaminants and their source sites in a wastewater network to avoid such situations. Sensors for flow measurement are often put in sewage pipelines. However, contaminants rust, and excessive turbidity may cause measurement instability and erroneous results. Deep learning may increase sensor accuracy in numerous situations. (Anand, R. et al. 2022) utilized previous sensor settings to explain common failures. Once a defect was found, the model could adapt the process to maintain WWTP functioning.

3.5 Machine Learning in Aquatic Settings

Seawater contamination is threatening Earth's ecosystems. Machine learning can monitor marine pollution to solve these problems. The lead-prediction algorithm was trained utilizing past tracking data from Queensland's Simon stations using XGBoost. The researchers discovered the framework selected I/P parameters and predicted water quality effectively. (Choudhary, S. et al. 2022) presented an RF and UAV trash mapping program to track seaside plastic garbage. An ensemble ML system with a 2-layered learning framework predicted beach water coastal microbial contamination concentration. (Sinha, M. et al. 2021) used an LSTM-CNN framework to predict a single antibiotic resistance gene (ARG) in beach water to enhance accuracy. Using machine learning classification techniques, (Reepu, et al. 2023) recognized differentially indicated genes in dolphins revealed to seawater contaminants. Many researchers have also developed monitoring devices for algal blooms that might cause serious pollution. (Bansal, R. et al. 2021) trained the XGBoost framework using water type and algal bloom spectral properties to recognize and classify algal bloom algae. (Dutta, S. et al. 2021) used Mahalanobis distance-based hierarchical cluster analysis to assess the North Yellow and Bohai Seas' water quality. In conclusion, machine learning can detect seawater contaminants, quantify their quantity and distribution, and analyze marine creature state.

Protecting marine life requires monitoring seawater quality ecosystems. Many researchers assess seawater quality using machine learning. (Pramanik, S. and Raja, S. S. 2020) suggested a KNN-based near-shore water quality prediction model in 2022. BPNN, SVR, and Long Short Term Memory frameworks were used to create a water quality prediction technique that significantly enhanced accuracy. (Pramanik, S. et al. 2020) suggested a water quality estimation approach using an upgraded grey regression analysis algorithm and Long Short Term Memory based on multivariate correlation and time-series features. (Pramanik, S. 2023) used a geographic-based water quality assessment approach to evaluate geosynchronous ocean color image data and 1240 water quality sample stations in Zhe-jiang's coast.

Moreover, 80% of the earth's population will experience a freshwater catastrophe by 2040. In locations with severe water shortages, desalinated saltwater provides freshwater. However, seawater desalination still faces challenges such as poor efficiency and dependability.

Apostolopoulos et al. (2020) utilized a Convolutional Neural Network approach with transfer learning to categorize saline molecules in water at various concentrations to enhance water treatment facility saltwater treatment attainment. (Boussouf, S. et al. 2023) used regression and ML methods such logistic regression, RF, SVM, and LSTM to anticipate the Salton Sea's salinity and development tendency, enabling prolonged control of seawater saltness and salt removal.

The single water estimation framework was extensively explored, and the integrated model has emerged recently. Distinct models have distinct methods for different input characteristics, resulting in varying predicted performance. The aggregation model of (Guardieiro, V. et al. 2023) favors classifier selection. Upon entering fresh data, the most suitable prediction model is chosen before generating predictions. This method selects models using input characteristics, while (Aydin, H.E. et al. 2023) XGBoost approach can cover I/P features and pick 7–10 out of 25 to integrate with Artificial Neural Network and various implementation techniques without losing information during model training. A potential modeling algorithm is XGBoost. XGBoost is a quick, accurate feature selection technique that relies on sample size.

4. CONCLUSION

Because it can forecast water gradations, improve water resource management, manage water resource deficiencies, and more, machine learning is frequently utilized to handle water environment issues. There are also hurdles in employing machine learning to assess water quality, including the need for vast volumes of high-quality data. Accurate data collection in water treatment and administration systems is challenging due to expenditure or technical constraints. Due to the complexity of actual water treatment and management systems, present algorithms are limited to specific systems, limiting the widespread use of ML techniques. Practical use of ML approaches needs researchers to have professional expertise.

The following factors may help overcome the aforesaid challenges:

(1) In future research and engineering, explore developing sophisticated sensors, such as soft sensors, for water quality monitoring to acquire correct data for ML algorithms.
(2) Improve the feasibility and reliability of techniques and design more general techniques and models for water treatment and management.
(3) Train diverse personnel to create sophisticated ML methods and use them in technical procedures.

REFERENCES

Ahamad, S., Veeraiah, V., Ramesh, J. V. N., Rajadevi, R., Reeja, S. R., Pramanik, S., & Gupta, A. (2023). *Deep Learning based Cancer Detection Technique. In Thrust Technologies' Effect on Image Processing*. IGI Global.

Anand, R., Singh, J., Pandey, D. K., Pandey, B., Nassa, V. K., & Pramanik, S. (2022). Modern Technique for Interactive Communication in LEACH-Based Ad Hoc Wireless Sensor Network. In M. M. Ghonge, S. Pramanik, & A. D. Potgantwar (Eds.), *Software Defined Networking for Ad Hoc Networks*. Springer. doi:10.1007/978-3-030-91149-2_3

Andrushia, A. D., Neebha, T. M., Patricia, A. T., Sagayam, K. M., & Pramanik, S. (2023). Capsule Network based Disease Classification for VitisVinifera Leaves. *Neural Computing & Applications*. Advance online publication. doi:10.1007/s00521-023-09058-y

Apostolopoulos, I. D., & Mpesiana, T. A. (2020). Covid-19: Automatic detection from X-ray images utilizing transfer learning with convolutional neural networks. *Physical and Engineering Sciences in Medicine*, *43*(2), 635–640. doi:10.1007/s13246-020-00865-4 PMID:32524445

Aydin, H. E., & Iban, M. C. (2023). Predicting and analyzing flood susceptibility using boosting-based ensemble machine learning algorithms with SHapley Additive exPlanations. *Natural Hazards, 116*(3), 2957–2991. doi:10.1007/s11069-022-05793-y

Bansal, R., Jenipher, B., Nisha, V., Makhan, R., Dilip, K., Pramanik, S., Roy, S., & Gupta, A. (2022). Big Data Architecture for Network Security. In Cyber Security and Network Security. Wiley. doi:10.1002/9781119812555.ch11

Bansal, R., Obaid, A. J., Gupta, A., Singh, R., & Pramanik, S. (2021). Impact of Big Data on Digital Transformation in 5G Era. *2nd International Conference on Physics and Applied Sciences (ICPAS 2021).* 10.1088/1742-6596/1963/1/012170

Bhattacharya, A., Ghosal, A., Obaid, A. J., Krit, S., Shukla, V. K., Mandal, K., & Pramanik, S. (2021). Unsupervised Summarization Approach with Computational Statistics of Microblog Data. In D. Samanta, R. R. Althar, S. Pramanik, & S. Dutta (Eds.), *Methodologies and Applications of Computational Statistics for Machine Learning* (pp. 23–37). IGI Global. doi:10.4018/978-1-7998-7701-1.ch002

Boussouf, S., Fernández, T., & Hart, A. B. (2023). Landslide susceptibility mapping using maximum entropy (MaxEnt) and geographically weighted logistic regression (GWLR) models in the Río Aguas catchment (Almería, SE Spain). *Natural Hazards, 117*(1), 207–235. doi:10.1007/s11069-023-05857-7

Chandan, R. R., Soni, S., Raj, A., Veeraiah, V., Dhabliya, D., Pramanik, S., & Gupta, A. (2023). Genetic Algorithm and Machine Learning. In Advanced Bioinspiration Methods for Healthcare Standards, Policies, and Reform. IGI Global. doi:10.4018/978-1-6684-5656-9

Choudhary, S., Narayan, V., Faiz, M., & Pramanik, S. (2022). Fuzzy Approach-Based Stable Energy-Efficient AODV Routing Protocol in Mobile Ad hoc Networks. In M. M. Ghonge, S. Pramanik, & A. D. Potgantwar (Eds.), *Software Defined Networking for Ad Hoc Networks.* Springer. doi:10.1007/978-3-030-91149-2_6

Dhamodaran, S., Ahamad, S., Ramesh, J. V. N., Muthugurunathan, G., Manikandan, K., Pramanik, S., & Pandey, D. (2023). Food Quality Assessment using Image Processing Technique. In Thrust Technologies' Effect on Image Processing. IGI Global.

Dhamodaran, S., Ahamad, S., Ramesh, J. V. N., Sathappan, S., Namdev, A., Kanse, R. R., & Pramanik, S. (2023). *Fire Detection System Utilizing an Aggregate Technique in UAV and Cloud Computing. In Thrust Technologies' Effect on Image Processing.* IGI Global.

Dushyant, K., Muskan, G., Gupta, A., & Pramanik, S. (2022). Utilizing Machine Learning and Deep Learning in Cyber security: An Innovative Approach. In M. M. Ghonge, S. Pramanik, R. Mangrulkar, & D. N. Le (Eds.), *Cyber security and Digital Forensics*. Wiley. doi:10.1002/9781119795667.ch12

Dutta, S., Pramanik, S., & Bandyopadhyay, S. K. (2021). Prediction of Weight Gain during COVID-19 for Avoiding Complication in Health. *International Journal of Medical Science and Current Research*, *4*(3), 1042–1052.

Ghosh, R., Bhunia, R., Pramanik, S., Mohanty, S., & Patnaik, P. K. (2023). Smart City Healthcare System for Survival Forecast for Cardiac Attack Situations using Machine Learning Techniques. In *Data-Driven Mathematical Modeling in Smart Cities*. IGI Global. doi:10.4018/978-1-6684-6408-3.ch019

Guardieiro, V., Raimundo, M. M., & Poco, J. (2023). Enforcing fairness using ensemble of diverse Pareto-optimal models. *Data Mining and Knowledge Discovery*, *37*(5), 1930–1958. doi:10.1007/s10618-023-00922-y

Gupta, A., Verma, A., & Pramanik, S. (2022). Security Aspects in Advanced Image Processing Techniques for COVID-19. In S. Pramanik, A. Sharma, S. Bhatia, & D. N. Le (Eds.), *An Interdisciplinary Approach to Modern Network Security*. CRC Press.

Jayasingh, R., & Kumar, J. (2022). Speckle noise removal by SORAMA segmentation in Digital Image Processing to facilitate precise robotic surgery. *International Journal of Reliable and Quality E-Healthcare*, *11*(1), 1–19. Advance online publication. doi:10.4018/IJRQEH.295083

Kaushik, D., Garg, M. A., Gupta, A., & Pramanik, S. (2022). Application of Machine Learning and Deep Learning in Cyber security: An Innovative Approach. In Cybersecurity and Digital Forensics: Challenges and Future Trends. Wiley. doi:10.1002/9781119795667.ch12

Khanh, P. T., Ngọc, T. H., & Pramanik, S. (2023). Future of Smart Agriculture Techniques and Applications. In A. Khang & I. G. I. Global (Eds.), *Advanced Technologies and AI-Equipped IoT Applications in High Tech Agriculture*. doi:10.4018/978-1-6684-9231-4.ch021

Mall, P. K., Pramanik, S., Srivastava, S., Faiz, M., Sriramulu, S., & Kumar, M. N. (2023). FuzztNet-Based Modelling Smart Traffic System in Smart Cities Using Deep Learning Models. In *Data-Driven Mathematical Modeling in Smart Cities*. IGI Global. doi:10.4018/978-1-6684-6408-3.ch005

Mandal, A., Dutta, S., & Pramanik, S. (2021). Machine Intelligence of Pi from Geometrical Figures with Variable Parameters using SCILab. In D. Samanta, R. R. Althar, S. Pramanik, & S. Dutta (Eds.), *Methodologies and Applications of Computational Statistics for Machine Learning* (pp. 38–63). IGI Global. doi:10.4018/978-1-7998-7701-1.ch003

Meslie, Y., Enbeyle, W., Pandey, B. K., Pramanik, S., Pandey, D., Dadeech, P., Belay, A., & Saini, A. (2021). Machine Intelligence-based Trend Analysis of COVID-19 for Total Daily Confirmed Cases in Asia and Africa. In D. Samanta, R. R. Althar, S. Pramanik, & S. Dutta (Eds.), *Methodologies and Applications of Computational Statistics for Machine Learning* (pp. 164–185). IGI Global. doi:10.4018/978-1-7998-7701-1.ch009

Mondal, D., Ratnaparkhi, A., Deshpande, A., Deshpande, V., Kshirsagar, A. P., & Pramanik, S. (2023). Applications, Modern Trends and Challenges of Multiscale Modelling in Smart Cities. In *Data-Driven Mathematical Modeling in Smart Cities*. IGI Global. doi:10.4018/978-1-6684-6408-3.ch001

Ngọc, T. H., Khanh, P. T., & Pramanik, S. (2023). Smart Agriculture using a Soil Monitoring System. In A. Khang *Advanced Technologies and AI-Equipped IoT Applications in High Tech Agriculture*. doi:10.4018/978-1-6684-9231-4.ch011

NgocT. T. H.PramanikS.KhanhP. T. (2023). *The Relationship between Gender and Climate Change in Vietnam*. The Seybold Report. DOI doi:10.17605/OSF.IO/KJBPT

Pandey, B. K., Pandey, D., Nassa, V. K., George, A. S., Pramanik, S., & Dadheech, P. (2023). Applications for the Text Extraction Method of Complex Degraded Images. In The Impact of Thrust Technologies on Image Processing. Nova Publishers.

Pandey, B. K., Pandey, D., Nassa, V. K., Hameed, A. S., George, A. S., Dadheech, P., & Pramanik, S. (2023). A Review of Various Text Extraction Algorithms for Images. In *The Impact of Thrust Technologies on Image Processing*. Nova Publishers. doi:10.52305/ATJL4552

Pandey, B. K., Pandey, D., Wairya, S., Agarwal, G., Dadeech, P., Dogiwal, S. R., & Pramanik, S. (2022). Application of Integrated Steganography and Image Compressing Techniques for Confidential Information Transmission. In Cyber Security and Network Security. doi:10.1002/9781119812555.ch8

Pramanik, S. (2022). Carpooling Solutions using Machine Learning Tools. In *Handbook of Research on Evolving Designs and Innovation in ICT and Intelligent Systems for Real-World Applications*. IGI Global. doi:10.4018/978-1-7998-9795-8.ch002

Pramanik, S. (2022). An Effective Secured Privacy-Protecting Data Aggregation Method in IoT. In *Achieving Full Realization and Mitigating the Challenges of the Internet of Things*. doi:10.4018/978-1-7998-9312-7.ch008

Pramanik, S. (2023). Intelligent Farming Utilizing a Soil Tracking Device. In *Convergence of Cloud Computing, AI and Agricultural Science*. doi:10.4018/979-8-3693-0200-2.ch009

Pramanik, S. (2023). *Intelligent Farming, Utilizing a Soil Tracking Device, Convergence of Cloud Computing, AI and Agricultural Science*. IGI Global. doi:10.4018/979-8-3693-0200-2.ch009

Pramanik, S., & Bandyopadhyay, S. (2023). Identifying Disease and Diagnosis in Females using Machine Learning. In I. G. I. John Wang (Ed.), *Encyclopedia of Data Science and Machine Learning*. Global. doi:10.4018/978-1-7998-9220-5.ch187

Pramanik, S., & Bandyopadhyay, S. (2023). Analysis of Big Data. In *Encyclopedia of Data Science and Machine Learning*. Global. doi:10.4018/978-1-7998-9220-5.ch006

Pramanik, S., Galety, M. G., Samanta, D., & Joseph, N. P. (2022). Data Mining Approaches for Decision Support Systems. *3rd International Conference on Emerging Technologies in Data Mining and Information Security*.

Pramanik, S., Joardar, S., Jena, O. P., & Obaid, A. J. (2023). An Analysis of the Operations and Confrontations of Using Green IT in Sustainable Farming. *Al-Kadhum 2nd International Conference on Modern Applications of Information and Communication Technology, AIP Conference Proceedings, 2591*. 10.1063/5.0119513

Pramanik, S., Obaid, A. J., N M, & Bandyopadhyay, S.K. (2023). Applications of Big Data in Clinical Applications. *Al-Kadhum 2nd International Conference on Modern Applications of Information and Communication Technology, AIP Conference Proceedings, 2591*. 10.1063/5.0119414

Pramanik, S., Sagayam, K. M., & Jena, O. P. (2021). *Machine Learning Frameworks in Cancer Detection*. ICCSRE. doi:10.1051/e3sconf/202129701073

Pramanik, S., Singh, R. P., & Ghosh, R. (2020). Application of Bi-orthogonal Wavelet Transform and Genetic Algorithm in Image Steganography. *Multimedia Tools and Applications*, *79*(25-26), 17463–17482. Advance online publication. doi:10.1007/s11042-020-08676-1

Pramanik, S., & Suresh Raja, S. (2020). A Secured Image Steganography using Genetic Algorithm. *Advances in Mathematics: Scientific Journal*, *9*(7), 4533–4541. doi:10.37418/amsj.9.7.22

Praveenkumar, S., Veeraiah, V., Pramanik, S., Basha, S. M., Lira Neto, A. V., De Albuquerque, V. H. C., & Gupta, A. (2023). *Prediction of Patients' Incurable Diseases Utilizing Deep Learning Approaches, ICICC 2023*. Springer. doi:10.1007/978-981-99-3315-0_4

Reepu, S. Kumar, Chaudhary, M. G., Gupta, K. G., Pramanik, S., & Gupta, A. (2023). Information Security and Privacy in IoT. In Handbook of Research in Advancements in AI and IoT Convergence Technologies. IGI Global.

Roy, A., & Pramanik, S. (2023). A Review of the Hydrogen Fuel Path to Emission Reduction in the Surface Transport Industry. *International Journal of Hydrogen Energy*.

Samanta, D., Dutta, S., Galety, M. G., & Pramanik, S. (2021). A Novel Approach for Web Mining Taxonomy for High-Performance Computing. *The 4th International Conference of Computer Science and Renewable Energies (ICCSRE'2021)*. 10.1051/e3sconf/202129701073

Sinha, M., Chacko, E., Makhija, P., & Pramanik, S. (2021). Energy Efficient Smart Cities with Green IoT. In C. Chakrabarty (Ed.), *Green Technological Innovation for Sustainable Smart Societies: Post Pandemic Era*. Springer. doi:10.1007/978-3-030-73295-0_16

Taviti Naidu, G., Ganesh, K. V. B., Vidya Chellam, V., Praveenkumar, S., Dhabliya, D., Pramanik, S., & Gupta, A. (2023). Technological Innovation Driven by Big Data. In *Advanced Bioinspiration Methods for Healthcare Standards, Policies, and Reform*. IGI Global. doi:10.4018/978-1-6684-5656-9

Veeraiah, V., Shiju, D. J., Ramesh, J. V. N., Ganesh, K. R., Pramanik, S., & Pandey, D. (2023). *Healthcare Cloud Services in Image Processing. In Thrust Technologies' Effect on Image Processing*. IGI Global.

Veeraiah, V., Shiju, D. J., Ramesh, J. V. N., Ganesh Kumar, R., Pramanik, S., Gupta, A., & Pandey, D. (2023). *Healthcare Cloud Services in Image Processing. In Thrust Technologies' Effect on Image Processing*. IGI Global.

Veeraiah, V., Talukdar, V., Manikandan, K., Talukdar, S. B., Solavande, V. D., Pramanik, S., & Gupta, A. (2023). Machine Learning Frameworks in Carpooling. In Handbook of Research on AI and Machine Learning Applications in Customer Support and Analytics. IGI Global. doi:10.4018/978-1-6684-7105-0.ch009

VidyaChellam, V., Veeraiah, V., Khanna, A., Sheikh, T. H., Pramanik, S., & Dhabliya, D. (2023). *A Machine Vision-based Approach for Tuberculosis Identification in Chest X-Rays Images of Patients, ICICC 2023*. Springer. doi:10.1007/978-981-99-3315-0_3

Zhu, M., Wang, J., Yang, X., Zhang, Y., Zhang, L., Ren, H., Wu, B., & Ye, L. (2022). A review of the application of machine learning in water quality evaluation. *Eco-Environment & Health, 1*(2), 107-116. doi:10.1016/j.eehl.2022.06.001

Chapter 5
Nowcasting Various Forms of Precipitation Using Improvised Random Forest Classifier

Ashritha Pilly
https://orcid.org/0009-0004-9685-4832
Osmania, India

C. Kishor Kumar Reddy
Stanley College of Engineering and Technology for Women, India

ABSTRACT

Weather forecasting is the utilization of science and technology to foresee the conditions of the atmosphere for a given location and time. Weather forecasting is high priority since it helps to settle future climate changes and provides information on critical weather conditions. As the weather has a great impact on various aspects of human life, aquatic life, aviation industry, and others, efforts have been made for decades to improve the efficiency of weather forecasting to ensure a better life and to reduce economic loss, but the result is not more precise than expected. The present research focuses on improving the efficiency of weather forecasting, focusing on various forms of precipitation such as rain, snow, hailstorms, and snowflakes by making use of historical numerical weather datasets across the globe. The efficiency in terms of performance measures has been compared with existing models.

DOI: 10.4018/979-8-3693-1396-1.ch005

1. INTRODUCTION

Objectives:

Research Problem: Analyzing on Nowcasting various forms of Precipitation using different Methods.

Significance: Helps to determine future climate changes.

Objectives: This research aims to achieve maximum percentage of performance measures and to find best model to predict the weather .

In the most recent century, weather forecasting has been one of the world's most experimentally and technologically frustrating problems. Because of the abrupt changes that occur, environmental change has long sought a good amount of thought. There are various limitations on better weather forecasting, thus it ends up being reliable for accurately predicting the weather right now. In meteorology, weather forecasting plays a crucial role. Making an accurate prediction is one of the major frustrations facing meteorologists anywhere in the world. Because they are used to protect property and human life, weather warnings are crucial. Farming depends on forecasts that take into account the temperature, outlook, humidity, and wind.

As this technology can conceal a lot of climatic data and improve itself the more it is used for more precise forecasts, the display of machine learning-based replication will substantially support weather models. With the huge volumes of raw data meteorologists collect daily, weather forecasting and machine learning are a perfect match. Few models can analyze past forecasts and observations, compare them to weather conditions noticed in reality and helps to eliminate errors in new forecasts. Machine learning helps to provide the best-in-class forecasts and hyper local now casts (Minseok, 2023).

Estimates can be used to plan activities around these times and to prepare for as well as survive them because heavy rain, wind cold, and snow significantly reduce outdoor activities. Without accurate weather forecasts people might find themselves in dangerous conditions for which they were unprepared and suffer hurt or worse. Among other issues, learning weather representation from a sizable amount of weather datasets is a concern in weather forecasting.

Precipitation is the term for the different ways that water falls from the air. They are all created by troposphere clouds, which are located between 8 and 16 kilometers (4 and 11 miles) above sea level. Any type of water particle that descends to the earth's surface from the upper atmosphere is known as precipitation. The drop is moved to the ground by gravity and surface frictional drag. A stormy wake is left behind after a particle leaves the cloud, encouraging quicker and more frequent drops. The crystallized ice may reach the ground as ice pellets or snow, or it may melt and change into raindrops before it reaches the earth's surface depending on the atmospheric circumstances.

The six types of precipitation are as follows:

1. Rain: Rain is one type of precipitation that falls as water droplets onto the surface of the earth. The size of these drops exceeds 0.5mm. When bigger water droplets fall to the earth's surface, they break apart into smaller droplets.
2. Drizzle: Although drizzle is a very light form of rain, unlike droplets, drizzle droplets fall to the ground. Drizzle moves with the air currents because the water droplets are so small and near to one another. When compared to drizzle, raindrops are larger, measuring less than 0.5 mm in diameter and moving at a speed of 1 mm per hour.
3. Snow: Ice crystals made of snow have a density of 0.1g/cc on average. Higher elevations and colder climates both have snow. These ice crystals can, however, somewhat melt if the temperature is above freezing (Soroosh, 2018). These slightly melted crystals soon evaporate as a result of proportionately warm temperatures. This evaporation aids in the cooling of the ice crystals and causes them to adhere to the surface of the earth as snow.
4. Sleet: When rain falls while the air is below freezing, sleet condenses into frozen raindrops. Ice pellets are another name for these. Sleet typically manifests as a tiny, translucent ball of ice which resemble hailstones but are smaller in size than hailstones, which are more common.
5. Glaze: When rain or drizzle comes into contact with a cloud and the earth's surface at 0 degrees Celsius, the glaze is typically generated. The droplets have time to spread out into a sheet even though they promptly freeze. The ice coating layer is firmer, thicker, and seems more transparent. Glaze is also known as Black-Ice.
6. Hail: Hailstones are large, irregularly shaped ice balls that fall from the sky during thunderstorms. The size of hailstones exceeds 8mm. Solid hailstones form when they precipitate. Most hail is composed of water ice, which ranges in size from 0.2 to 6.

It's important to note that weather prediction is a complex science, and meteorologists often use a combination of these methods to create accurate forecasts. Additionally, advancements in technology and data analysis techniques continue to improve weather prediction capabilities over time. This paper exhibits a classifier approach utilizing "Improvised Random Forest Classifier" (Method in Machine Learning) for weather forecasting. At present system, state of weather is divided into some attributes like as Precipitation Type, Visibility (km) Loud Cover, Pressure (millibars),Temperature, Apparent Temperature, Wind Speed (km/hr), Humidity, Wind Bearing (degrees). The class label "Weather Forecasting" will be predicted by the system using these attributes. Two fundamental operations training and testing will be implemented in the system. Research and organizational changes have aided in understanding, halting, and reducing the effects of climate change, as well as incorporating proof evidence into decision-making and climate policy, for more than 50 years. Climate change research has used a variety of methods, including previous studies that demonstrated global temperature rises, the use of ice cores that include 800,000 years of continuous Earth climate records, and climate modeling using computers. a table containing fictitious information for 19 instances of extreme weather variations (Ying-Nong, 2022).

Table 1. Statistics on Drastic weather changes over the period 2020 to 2023.

S.no	Year	Location	Drastic Weather Change
1	2020	New York City	Blizzard with record Snowfall
2	2020	Los Angeles	Severe Wildfires
3	2020	Chicago	Polar Vortex &Extreme Cold
4	2020	Miami	Category 5 Hurricane
5	2020	Seattle	Unpredicted Heat wave
6	2021	Sydney	Heavy Flooding and Landslides
7	2021	Tokyo	Typhoon and Storm Surge
8	2021	London	Hailstorm with Golf Ball-sized Hail
9	2021	Mumbai	Intense Monsoon Rains and Flooding
10	2021	Rio de Janeiro	Heat wave and Drought
11	2022	Beijing	Sandstorm and Dust Cloud
12	2022	Moscow	Record-breaking Cold Snap
13	2022	Buenos Aires	Tornado Outbreak
14	2022	Vancouver	Heavy Snow and Avalanche
15	2023	Mexico City	Intense Rain and Urban Flooding
16	2023	Cairo	Severe Sandstorm
17	2023	New Delhi	Extreme Heat and Heat wave
18	2023	Johannesburg	Hailstorm and Damaging Winds

Weather forecasting is an essential tool for various industries and activities, and its accuracy has improved significantly over the years. However, there can still be losses due to weather forecasting errors or limitations. Some of the potential losses include - Agriculture: Farmers heavily depends on accurate weather forecasts to aim their planting, irrigation, and harvesting schedules. Incorrect forecasts can lead to crop damage due to unexpected weather events like frost, hail, drought, or heavy rainfall.

- Transportation: Weather conditions can impact air, land, and sea transportation. Incorrect weather forecasts may result in flight cancellations, delays, or rerouting of ships, causing disruptions to travel plans and logistics.
- Tourism: The tourism industry depends on weather forecasts to attract visitors and plan activities accordingly. Inaccurate forecasts can lead to disappointed tourists and potential revenue losses for hotels, restaurants, and local businesses.
- Energy: Weather plays a crucial role in energy demand and supply. Incorrect forecasts can lead to suboptimal energy production planning, affecting electricity generation from renewable sources like wind and solar, as well as impacting heating and cooling demand.
- Retail: Weather forecasts influence consumer behavior. For example, unanticipated severe weather might lead to reduced foot traffic in retail stores, affecting sales.
- Outdoor events: Weather conditions are vital for outdoor events such as concerts, festivals, and sports events (Kang, 2023). Poor weather forecasting can result in event cancellations or postponements, leading to financial losses for organizers and disappointment for attendees.
- Construction: Weather forecasts are crucial for construction planning. Unexpected severe weather can cause delays and safety concerns, impacting construction timelines and budgets.
- Insurance: Inaccurate weather forecasts can lead to insurance claims for damages that were not accurately predicted or accounted for, potentially affecting insurance company's bottom line.

It's primary to note that while weather forecasting has come a long way and continues to improve; it is still a complex and challenging task. Meteorologists use sophisticated models and data to make predictions, but some inherent uncertainties in weather systems can lead to occasional inaccuracies. As technology and data collection methods advance, the aim is to minimize these losses and provide more reliable forecasts to various industries and the public. The loss due to weather forecasting can have significant economic, social, and safety implications. Inaccurate weather predictions can lead to economic losses in various industries, disruption of services, and potential harm to human life and safety. Resource management and planning may be adversely affected, and public trust in meteorological services can be eroded if forecasting errors become frequent.

Despite the advancements in technology and meteorological understanding, weather forecasting remains a challenging and uncertain science. However, efforts to improve forecasting accuracy continue, with the hope of minimizing the negative impacts of inaccurate predictions.

To mitigate the effects of forecasting errors, it is crucial for individuals, businesses, and governments to remain vigilant, stay informed about weather updates, and take proactive measures when possible. Additionally, ongoing research and investment in weather prediction technologies will contribute to more reliable forecasts and reduce the potential losses associated with unpredictable weather events (Ying-Nong, 2022).

The following are the Chapter main contributions:

1. Assembling of Dataset from various sources.
2. Preprocessing the collected data.
3. Creating training and test sets from the dataset.
4. Evaluate the model using performance measure.
5. Deploy it to perform predictions on new, unseen data.

The remaining chapters are structured as follows: Section 2 Relevant work typically refers to the work experience or professional background that is directly related to weather forecasting. Section 3 Proposed Methodology referred to as a research methodology which is a detailed outline or plan that describes the methods and procedures a researcher intends to conduct their study or research project on weather forecasting. Section 4 Results and Discussion presents the raw data and findings obtained from the research and the Discussion section follows the Results section and is where the researchers interpret and explain the significance of the results obtained by comparing different models like Random Forest, CNN, RNN, Logistic Regression and may more. Section 5 Conclusion is the final section of a research paper which serves as a summary of the key points discussed in the study and provides a clear and concise statement of the overall findings and their implications. Section 6 References also known as works cited is a list of all the sources cited or referenced in a research paper or scholarly work.

2. RELEVANT WORK

Minseok Seo (2023) proposed an "Intermediate and Future Frame Prediction of Geostationary Satellite Imagery" with warp and refine network using GK2A (GEO-KOMPSAT-2A) satellite dataset by Total Variation-L1 algorithm with an Accuracy of 69% and includes Time step parameter τ to 0.25,weight parameter λ to 0.15 using images and it's analyses. Yuhang Jiang et al. (2022) proposed a "CSIP-Net: Convolutional Satellite Image Prediction Network for Meteorological Satellite Infrared Observation Imaging" using advanced geosynchronous radiation imagers and it's observed values which is executed with FY-4A Dataset by recurrent neural networks (RNN), convolutional neural networks (CNN), deep learning algorithm with Accuracy of 79.08%.

Ying-Nong Seo (2022) proposed "Rainfall Forecast Using Machine Learning with High Spatiotemporal Satellite Imagery Every 10 Minutes" with IMERGE Dataset using Graphs and scaling for study by machine learning algorithms, LSTM Model, RF Algorithm for better approach to gain maximum Accuracy of 83%. Huiyun Ma Seo (Huiyun, 2023) proposed a new system for detecting the weather named "A Novel ST-ViBe Algorithm for Satellite Fog Detection at Dawn and Dusk" with both Satellite Datasets & Validation Datasets using Detection Algorithm & ST-ViBe algorithm to demonstrate Accuracy-78.97%, Precision, and specificity.

Tian Wang (2023) proposed "Multi-Model Ensemble Forecasts of Surface Air Temperatures in Henan Province" using TIGGE datasets by observing Bar graphs with Deep Neural Network, Feed-Forward Neural Network, The Convolutional Neural Network, Experimental Design, Verification Metrics, Feed forward Neural Network which resulted an Accuracy of 89.088% . Waytehad Rose (2021) put forword an " Application of Deep Learning Architectures for Satellite Image Time Series Prediction" using GK2A (GEO-KOMPSAT-2A) satellite dataset by executing models like The Multi-Layer Perceptron (MLP), The Convolutional Neural Networks (CNN), Deep learning for an Accuracy of 85.0% and an Recall of 82.7%.

Walid Abdelrahman Seo (Walid, 2023) presented "High-Precision AI-Enabled Flood Prediction Integrating Local Sensor Data and 3rd Party Weather Forecast" using Model Evaluation dataset by tracking all the sensors on weather with a multi-linear regression model, LSTM neural networks, multiple artificial neural network methods with an Accuracy of 91.04% and sub datasets for training 72%, validation 18% and testing 10%. Mónica Estébanez-Camarena (2023) proposed "The Potential of Deep Learning for Satellite Rainfall Detection over .Data-Scarce Regions, the West African Savanna" with validation and test datasets of Contingency tables type using deep neural networks, PERSIANN-CCS, IMERG algorithms with Accuracy of 76%

Zeyu Zhou (2023) proposed "A Novel Hybrid Intelligent SOPDEL Model with Comprehensive Data Preprocessing for Long-Time-Series Climate Prediction" using M6–M9 Models, SOPDEL model by mapping pattern, pie charts data types using ELM neural networks, machine learning, optimization algorithm, SO–DBN–LST-ME algorithm, LSTM-based deep neural networks with 86.46%. Soroosh (2018) presented "Short-Term Precipitation Forecast Based on the PERSIANN System and LSTM Recurrent Neural Networks" using test datasets and it's illustrations by PERSIANN, ANN-based precipitation retrieval algorithm, LSTM algorithm with an Accuracy of 84.4%.

Kang Chen (2023) proposed "Pushing the skill full global medium-range weather forecast beyond 10 days lead" using ERA5 (Hersbach 2020) with Linear graph by executing data-driven methods, AI-based method, Accuracy 10.75 days (ACC of z500 > 60%). Hemaja Patoju (2023) proposed "Weather forecasting using digital image processing" with HYTA dataset by analyzing Clusters, Performance Graph Plot, Error Histogram Plot using supervised learning: decision trees, random forests, logistic regression, and linear regression. K-means, the Apriori method for association rule learning issues, Principal Component Analysis, Singular Value Decomposition, independent Component Analysis, semi-supervised learning, and reinforcement learning with an accuracy of 82.9% are examples of unsupervised learning.

Liangliang (2020) proposed "Combining Optical, Fluorescence, Thermal Satellite, and Environmental Data to Predict County-Level Maize Yield in China Using Machine Learning Approaches" using Model Evaluation dataset with an Accuracy of 89.08%. Chris Funk (2015) proposed "The climate hazards infrared precipitation with stations a new environmental record for monitoring extremes" using infrared Precipitation with Stations (CHIRPS) dataset, Cen Trends version 1.0 datasets executing - The ARIMA model, with mean bias errors (MBE) <15%,Regression estimates half (10 percent) of the decline and an Accuracy of 82.77% was obtained.

3. PROPOSED METHODOLOGY

Using Improvised Random forests Classifier for research papers is a popular and effective approach in various domains, particularly in the area of data analysis, and predictive modeling. Random forests are an ensemble learning method that merges multiple decision trees to increase the overall accuracy and robustness of the predictions. Popular machine learning algorithm Random Forest uses the supervised learning method. Regression and Classification issues in machine learning can both be pre-owned. It is found on the idea of ensemble learning, which is the act of combining different classifiers to address a complicated issue and improve model performance. According to what its name implies, "Random Forest is a classifier that contains a number of decision trees on various subsets of the given dataset and takes the average to improve the predictive accuracy of that dataset".

Instead of relying on just one decision tree, random forest gathers forecasts from all decision trees and predicts the outcome based on the majority of votes. Greater accuracy is obtained and the overfitting issue is controlled when there are more Decision trees in the Random forest. It is possible that certain decision trees can predict the right output while others cannot since the random forest combines numerous decision trees to forecast the dataset's class. But when all the decision trees are combined, they forecast the right result. In light of this, the following two predictions for an improved Random forest classifier: The feature variable in the dataset needs to contain some real values in order for the classifier to forecast accurate outcomes rather than results.

Figure 1. Improvised random forest classifier (IRFC) model for weather forecasting

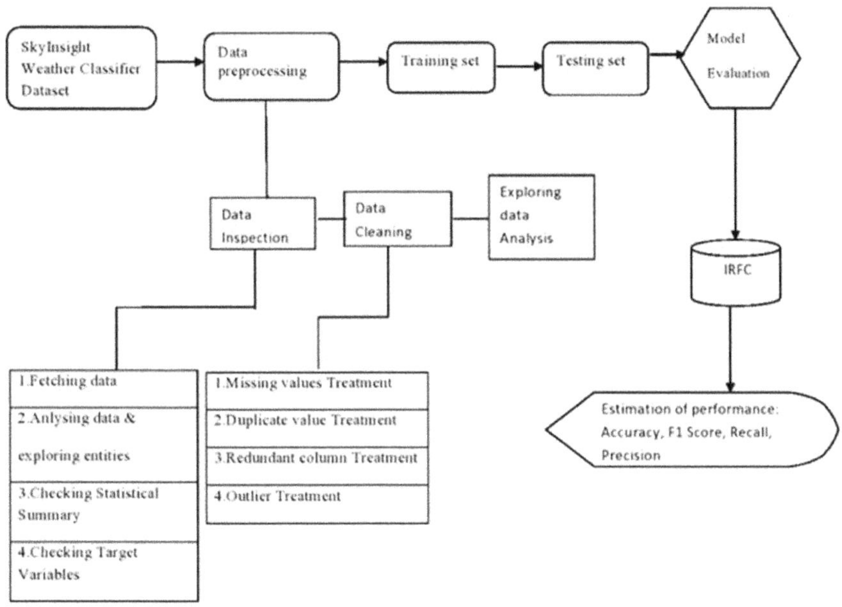

Each tree's predictions must be closely related. It has the ability to carry out both classification and regression tasks. It handles large datasets with high structuralism effectively. It highlights the model's accuracy and ignores the overfitting problem (Huiyun, 2023). In this study, historical weather data, including atmospheric conditions, temperature, humidity, wind speed, and other relevant parameters, were collected for a specific region. Creating training and testing sets from the dataset. The Random Forest algorithm was employed to build the weather forecasting model using the training set. The hyper parameters of the model were tuned using grid search and cross-validation techniques to optimize its performance. Subsequently, the model was evaluated using the validation set to prevent overfitting. Finally, the model's generalization capability was tested using the independent testing set.

3.1 Dataset Used

The "SkyInsight Weather Classifier" dataset is a well known dataset used for weather forecasting research and analysis. It contains data of weather change occurs, Summary (Partly cloudy, mostly cloudy, overcast), Precipitation Type (rain, snow, null), temperature, humidity, wind bearing in degrees, wind speed computed in km/hr, Loud cover, Pressure calculated in millibars with 4,822 records and 12 attributes in it. The aim is to predict the weather condition.

Table 2. Dataset description

Apparent Temperature (F)	Humidity (g/kg)	Wind Speed (km/hr)	Visibility (km)	Pressure (millibars)	Precipitation Type	Summary
7.388889	0.89	14.1197	15.8263	1015.13	Rain	Partly Cloudy
7.227777	0.86	14.2646	15.8263	1015.63	Rain	Partly Cloudy
6.977778	0.83	11.0446	15.8263	1016.51	Rain	Mostly Cloudy
13.77222	0.6	17.0982	15.8263	1007.89	Rain	Over cast
14.25556	0.59	16.5025	14.9569	1007.89	Rain	Over cast
13.57778	0.9	20.9622	7.6153	1004.08	Rain	Over cast
10.91111	0.86	22.3951	2.6565	1004.61	Rain	Foggy
12.49444	0.91	3.9445	9.982	1015.51	Rain	Clear
3.277778	0.75	15.939	0	1027.62	Null	Clear
15	1	0.3542	1.9642	1016.04	Rain	Foggy
5.855556	0.83	13.9748	0	1025.85	Null	Clear
15.19444	0.8	18.7243	0	1022.49	Rain	Drizzle
11.51111	0.89	7.7602	6.8586	1013.66	Rain	Light Rain
12.68333	0.82	9.9498	9.982	1018.28	Rain	Drizzle
5.477778	0.93	17.2109	3.059	1008.2	Rain	Light Rain
13.91667	0.96	13.2503	8.05	998.15	Rain	Over cast
11.08889	0.93	21.5418	10.0464	1000.35	Rain	Over cast

3.2 Data Preprocessing

Information preprocessing in weather forecasting refers to the initial steps taken to clean, format, and organize raw weather data before it can be used for analysis and modeling. Weather forecasting relies heavily on data collected from various sources, such numerical models. However, this data may come in different formats, may contain errors or missing values, and could be collected at irregular intervals. Information preprocessing aims to overcome these challenges and ensure the data is suitable for further analysis and forecasting.

The main steps involved in information preprocessing for weather forecasting include:

- Data Collection: Gathering data from various sources, such as weather stations, satellites, and radar systems.

- Data Cleaning: Removing any inaccuracies, errors, or outliers present in the data. This process involves identifying and correcting faulty measurements, dealing with missing data, and smoothing out noisy signals.
- Data Integration: Combining data from numerous sources into a unified format or database, ensuring consistent units, and aligning timestamps.
- Data Transformation: Converting the data into a format suitable for analysis and modeling. This may involve resampling or interpolating data to ensure uniform time intervals and spatial resolutions.
- Feature Extraction: Extracting relevant features from the raw data that are crucial for weather prediction (Zeyu, 2023). For example, features like temperature, humidity, wind speed, and precipitation are commonly used in weather forecasting models.
- Data Aggregation: Reducing the volume of data by summarizing or aggregating information over specific time periods or spatial regions.

This can help make the data more manageable and computationally efficient. Quality Control: Applying additional checks to ensure the overall quality and accuracy of the preprocessed data. Quality control measures can help identify any remaining inconsistencies or errors that might affect the forecast.

After completing these preprocessing steps, the data is ready for use in weather forecasting models, such as numerical weather prediction models, statistical models, or machine learning algorithms. Proper information preprocessing is essential to ensure that the forecast models produce reliable and accurate predictions, which are crucial for public safety and decision-making in various sectors like agriculture, transportation, and emergency response (Wang, 2023).

3.3 Training Set

In weather forecasting, the training set plays a crucial role in developing and improving predictive models. The training set is a subset of historical weather data that is used to teach or train the forecasting model. Essentially, the model “learns” from this historical data to identify patterns, relationships, and trends that can be used to make predictions about future weather conditions. Here, for the given dataset 80% is considered as the training dataset i.e. 964 records from the data set (Rose, 2021).

3.4 Testing Set

After a forecasting model has been trained using the training set, its performance and accuracy are evaluated using the testing set in weather forecasting. A portion of historical weather data that the model has not seen before during training makes up the testing set. It serves as an independent dataset to assess how well the model can generalize its predictions to new, unseen data. Here for the given dataset 20% is considered as the testing dataset i.e. 3858 records from the data set [6].

3.5 Proposed Model: Random Forest

It is a crucial step Machine learning to sustain the accuracy and to make better expectations for the given information. The proposed model is "Random forest" to proceed the further research and to show the best accuracy values with least loss%.

3.6 Model Evaluation

Model evaluation is the process of estimating the performance and effectiveness of a predictive or machine learning model using various metrics and systems. It involves measuring the performance of a trained model on a testing dataset. The models include Accuracy, Recall, F1 Score, Precision, error rate. It helps to avoid false prediction and pretend the precise one.

Algorithm for Improvised Random Forest Classifier [IRFC]

Step 1: Collect the data from various sources .
Step 2: Preprocess the collected data.
Step 3: Divide the data into both training set and testing set.
Step 4: Evaluate the proposed IRFC model with existing approaches.
Step 5: Compare the model with other models.
Step 6: Conclude the model with highest Measures.

4. RESULTS AND DISCUSSION

The results section of a weather forecasting report typically presents the outcome of the forecasting process. This includes the predictions made by the weather forecasting models and the actual weather observations, which serve as a basis for evaluating the accuracy of the forecast and the In the discussion section, the weather forecasting results are interpreted, and the implications of the findings are explored. Key points to include in the discussion are: Forecast Accuracy, Weather Events, Improvement Strategies Forecast Value and many more. The specific content of the results and discussion section will depend on the nature of the weather forecasting study, the data available, and the specific objectives of the analysis. Additionally, including visual aids such as graphs, maps, and charts can help illustrate the results and enhance the understanding of the discussion (Walid, 2023).

Accuracy is analyzed by comparing different models and implementing them. In general Accuracy can be a common evaluation metric used to measure the performance of a classification model. It represents the proportion of correctly predicted cases (or data points) out of the total number of instances/cases in the dataset. Accuracy is especially useful when the dataset is well-balanced, meaning that the classes have roughly equal representation. Performance Metrics terms - True Positive (TP): The TP, which is also known as sensitivity or recall, is used to calculate the proportion of real positives that are actually recognized. A True Negative (TN) is an outcome for which the model correctly predicts the negative class. False Positive (FP) is a measurement of a test's accuracy. Technically speaking, the likelihood of incorrectly rejecting the null hypothesis is the definition of the false positive rate. False Negative (FN): Also known as the miss rate, this statistic measures the likelihood that a true positive would go unnoticed by the test.

Accuracy = (Number of Correct Predictions) / (Total Number of Predictions)

$$\text{Accuracy} = \frac{TP + TN}{TP + TN + FP + FN} \tag{1}$$

Table 3 shows comparison between different models for accuracy which are obtained as follows: The model proposed so called IRFC with the highest Accuracy of 94.1% and then Total Variation-L1 algorithm [TV-L1] at 69%,Convolutional Neural Networks [CNN] at 79.08%,Random forest with different optimizers [RF] model with 83%, ST-Vibe model [ST-Vibe] at 78.97%, Feed forward Neural Network [FNN] at 89.088%, Multi linear regression [MLR] at 91.04%, Integrated Multi-satellite Retrievals [IMERGE] model at 76%, Optimization Algorithm[OA] at 86.46%, long short term memory under RNN [LSTM]at 84.4%, Artificial Intelligence [AI] at 60%, Reinforcement learning [RF] at 82.9%, Model Evaluation[ME] at 89.08% and Autoregressive Integrated Moving Average [ARIMA] at 82.77% .

Precision is defined as a performance metric used in the field of machine learning and statistics to rate the accuracy of a model's positive predictions. It is a measure of how many of the predicted positive instances are actually true positives, relative to the total number of predicted positive instances. Precision represents the ability of the model to avoid false positives. In other words, it measures the proportion of positive predictions made by the model that are correct.

Table 3. Comparing accuracy with various models

Model Name	Accuracy (%)
AI	60
TV –L1	69
IMERGE	76
ST-V	78.9
CNN	79.08
ARIMA	82.77
RL	82.9
RF	83
LSTM	84.4
OA	86.46
ME	89.08
FNN	89.088
MLR	91.04
IRFC	94

Figure 2. Comparison of accuracy with various models graphically

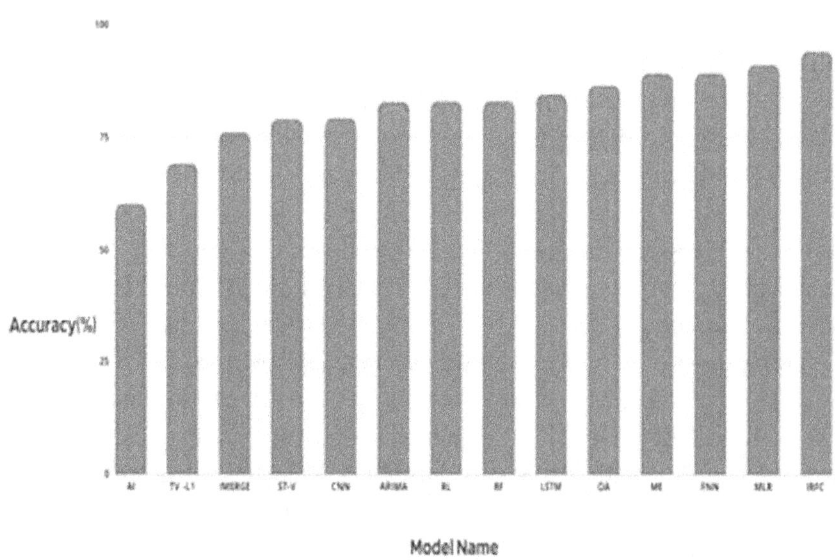

$$\mathbf{Precision} = \frac{\mathbf{TP}}{\mathbf{TP} + \mathbf{FP}} \tag{2}$$

Table 4 shows comparison between different models for precision which are obtained as follows:: The proposed model IRFC with the highest precision of 93.81% and remaining are as –Gaussian NB [GNB] at79.2%, Logistic Regression [LR] at 83%,Support vector Machine [SVM] at 82.3%, SGD Classifier[SGD C] at 82.1%,Decision Tree [DT] at 89%, K Nearest Neighbor [KNN] at 90.5% and Extra Tree Classifier [ETC] With 93% .

Table 4. Comparing precision with various models.

Model Name	Precision (%)
G NB	79.2
SGD	82.1
SVM	82.3
LA	83
DT	89
KNN	90.5
ETC	93
BC	93.9
IRFC	94.1

Figure 3. Comparing of precision with various models graphically

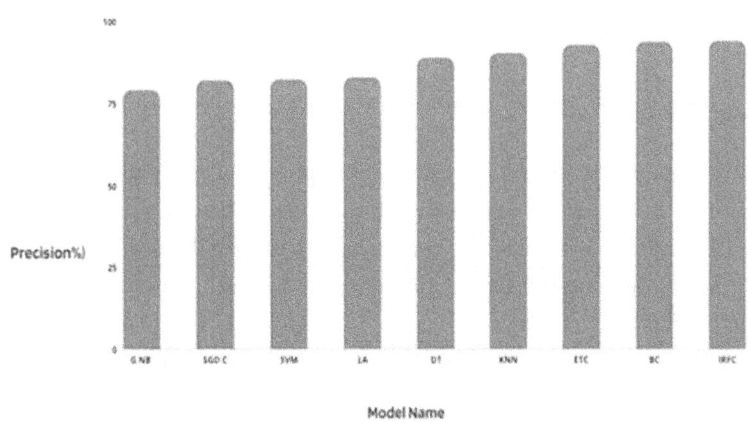

Recall is another performance metric used in the field of statistics, particularly in the context of binary classification problems (Liangliang, 2020). Recall also referred as sensitivity, hit rate, or true positive. It measures the skill of a model to exactly identify all positive occurrences from the total number of actual positive occurrences in the dataset. In binary classification, recall is calculated using the following formula:

$$\mathbf{Recall} = \frac{\mathbf{TP}}{\mathbf{TP} + \mathbf{FN}} \tag{3}$$

Table 5 shows comparison between different models for recall /sensitivity which are obtained as follows: The proposed model IRFC with the highest precision of 94.1% and remaining are as –Gaussian NB [GNB] at 79.5%, Logistic Regression [LR] at 83.4%,Support vector Machine [SVM] at 82.5%, SGD Classifier [SGD C] at 82.3%,Decision Tree [DT] at 88.9%, KNearest Neighbor [KNN] at 90.5% and Extra Tree Classifier [ETC] with95% Bagging classifier [BC] at 93.9% .

Table 5. Comparing recall with various models

Model Name	Recall (%)
G NB	79.5
SGD C	82.3
SVM	82.5
LA	83.4
DT	88.9
KNN	90.5
ETC	93.7
BC	93.9
IRFC	94.1

A performance metric that is frequently applied in binary classification issues, where there are two classes: positive and negative, is the F1 score. It is a balanced indicator of a model's accuracy when it comes to unbalanced class distributions because it is the harmonic mean of precision and recall. The following formula is used to determine the F1 score:

$$\textbf{F1 Score} = \frac{2 * (\textbf{Precision} * \textbf{Recall})}{(\textbf{Precision} + \textbf{Recall})} \quad (4)$$

Figure 4. Comparison of recall with various models graphically

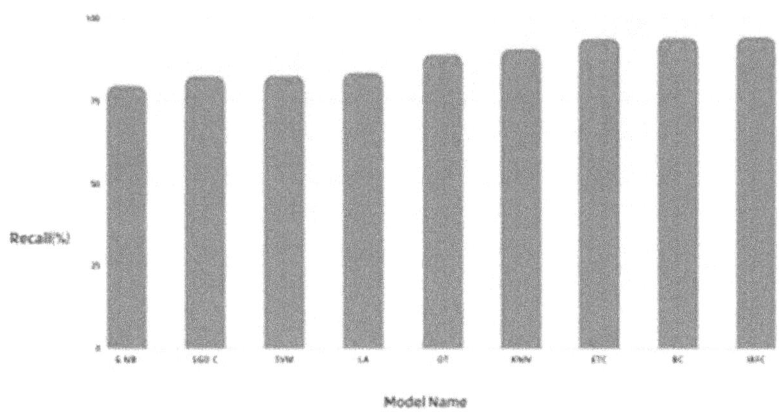

Table 6 shows comparison between different models for F1 Score which are obtained as follows: The proposed model IRFC with the highest precision of 94.0% and remaining are as –Gaussian NB[GNB] at 79.2%, Logistic Regression [LR] at 83.0%,Support vector Machine [SVM]at 82.4%, SGD Classifier [SGD C] at 82.2%,Decision Tree [DT] at 89%,K Nearest Neighbor [KNN] at 90.5% and Extra Tree Classifier [ETC] with 92% .

Table 6. Comparing F1-score with various models

Model Name	F1-Score (%)
G NB	79.2
SGD C	82.2
SVM	82.4
LA	83.0
DT	89.0
KNN	90.5
ETC	92.0
IRFC	94.0

Figure 5. Comparison of F1-score with various models graphically

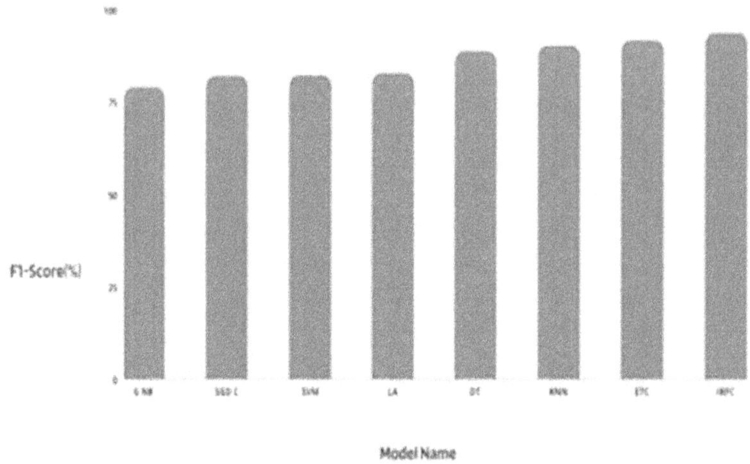

Loss in accuracy refers to a decrease in the accuracy of a model that is predicted when it is applied to new or unseen data. In the context of machine learning and predictive modeling, accuracy is a commonly used evaluation metric that measures the segment of correctly predicted instances over the total number of instances. There are several factors that can lead to a loss in accuracy: Overfitting, Limited or Biased Training Data, Changing Data Distribution, Noise in Data, Feature Engineering, Hyper parameter Tuning and many more (Hemaja, 2023).

The Loss in Accuracy score is calculated using the following formula:

Loss in Accuracy% = **Total Accuracy**(100)% - **Obtained Accuracy**% (5)

Table 7 shows comparison between different models for Loss in Accuracy which are obtained as follows: The proposed model IRFC with the least Loss in Accuracy of 5.9% and then Total Variation-L1 algorithm [TV-L1] at 31%, Convolutional Neural Networks [CNN] at 20.92%,Random forest with different optimizers [RF] model with 17%, ST-Vibe model [ST-Vibe] at 21.03%,Feed forward Neural Network [FNN] at 10.912%, Multi linear regression [MLR] at 8.96%, Integrated Multi-satellite Retrievals [IMERGE] model at 24%, Optimization Algorithm[OA] at 13.54%, long short term memory under RNN [LSTM]at 15.6%, Artificial Intelligence [AI] at 40%, Reinforcement learning [RL] at 17.1%, Model Evaluation [ME] at 10.92% and Autoregressive Integrated Moving Average [ARIMA] at 17.23% (Monica, 2023).

Table 7. Comparing loss in accuracy with various models

Model Name	Loss in Accuracy %
AI	40
TV –L1	31
IMERGE	24
ST-V	21.03
CNN	20.92
ARIMA	17.23
RL	17.1
RF	17
LSTM	15.6
OA	13.54
ME	10.92
FNN	10.912
MLR	8.96
IRFC	5.9

Figure 6. Comparison of loss in accuracy with various models graphically

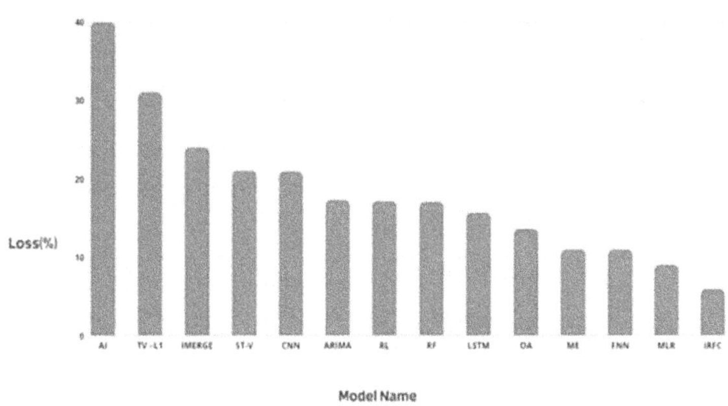

5. CONCLUSION

In conclusion, our research demonstrates the considerable potential of the Random Forest algorithm in weather forecasting. By achieving improved accuracy and generalization, this machine learning approach presents a valuable alternative to traditional meteorological models. The findings of this study encourage further exploration and refinement of machine learning techniques for weather prediction, ultimately leading to more reliable and advanced forecasting systems with broad societal impacts. The suggested model improves the performance of accuracy, Precision, F1 Score, and Recall. Furthermore, the interpretability of the Improvised Random Forest Classifier model allows meteorologists to gain insights into the most influential meteorological variables, aiding in a better understanding of weather dynamics and potentially leading to advancements in meteorological research.

Overall, our research highlights the potential of Random Forest in revolutionizing weather forecasting methodologies. As the field of machine learning continues to advance, we can anticipate more sophisticated and accurate weather prediction systems, ultimately benefiting various sectors of society, including agriculture, transportation, disaster management, and more. With ongoing research and collaboration between meteorologists and data scientists, Random Forest and other machine learning approaches are likely to play a pivotal role in shaping the future of weather forecasting, improving our ability to prepare for and respond to weather-related challenges. Hence proposed "Improvised Random Forest Classifier" method obtained the performance measures as Accuracy measure as 94.0%, precision measure as 94.1%, Recall measure as 94.1%, F1 Score measure as 94.0%.

REFERENCES

Anisha, Reddy, Nguyen, Kumar, & Hanafiah. (2022). Intelligent Systems and Machine Learning for Industry: Advancements, Challenges and Practices. CRC Press, Taylor & Francis.

Hemaja, Charan, & Charan. (2023). *Weather forecasting using digital image processing*. Academic Press.

Jiang, Cheng, & Gao. (2022). CSIP-Net: Convolutional Satellite Image Prediction Network for Meteorological Satellite Infrared Observation Imaging. *Atmosphere, 14*(1), 2. DOI: doi:10.3390/atmos14010025

Kishor Kumar Reddy, Anisha, Khan, Hanafiah, P., & R. (2024). Sustainability in Industry 5.0: Theory and Applications. CRC Press, Taylor & Francis.

Kishor Kumar Reddy, Anisha, Marlia Mohd Hanafah, Pragathi, Ramana Murthy, & Madana Mohana. (2023). *An intelligent optimized cyclone intensity prediction framework using satellite images*. Springer Earth Science Informatics. https://doi.org/10.1007/s12145-023-00983-z

Ma, Liu, Jiang, Jiang, Feng, & Hu. (2023). *A Novel ST-ViBe Algorithm for Satellite Fog Detection at Dawn and Dusk*. https://doi.org/10.3390/rs15092331

Moskolaï, Abdou, & Dipanda, & Kolyang. (2021). Application of Deep Learning Architectures for Satellite Image Time Series Prediction. *Remote Sensing*, *13*(23), 4822. doi:10.3390/rs13234822

Seo, Peterson, Landsfeld, Pedreros, Verdin, Shukla, Husak, Rowland, Harrison, Hoell, & Michaelsen. (2015). *The climate hazards infrared precipitation with stations a new nvironmental record for monitoring extremes.* Academic Press.

Seo, Han, Gong, Bai, Ling, Luo, Chen, Ma, Zhang, Su, Ci, Li, Yang, & Ouyang. (2023). *Pushing the skillfull global medium-range weather forecast beyond 10 days lead.* ariXiv:2304.02948 [cs.AI]

Seo, L., Zhang, Z., & Luo, Y. (2020). Combining Optical, Fluorescence, Thermal Satellite, and Environmental Data to Predict County-Level Maize Yield in China Using Machine Learning Approaches. *Remote Sensing*, *12*(1), 21. doi:10.3390/rs12010021

SeoM.ChoiY.RyH.ParkH.BaeH.LeeH.SeoW. (2023). Intermediate and Future Frame Prediction of Geostationary Satellite Imagery. *arXiv:2303.04405.*

Seo, M. E.-C., & Taormina, R. (1922). The Potential of Deep Learning for Satellite Rainfall Detection over. Data-Scarce Regions, the West African Savanna. *Remote Sensing*, *15*(7). https://doi.org/10.3390/rs15071922

Seo, Z. Z., Tang, W., Li, M., Cao, W., & Yuan, Z. (2023). A Novel Hybrid Intelligent SOPDEL Model with Comprehensive Data Preprocessing for Long-Time-Series Climate Prediction. *Remote Sensing*, *15*(7), 1951. doi:10.3390/rs15071951

Sorooshian, Asanjan, Yang, Hsu, Lin, & Peng. (2018). *Short-Term Precipitation Forecast Based on the PERSIANN System and LSTM Recurrent Neural Networks.* AGU. https://doi.org/ doi:10.1029/2018JD028375

Wang, Q., & Abdelrahman, W. (2023). High-Precision AI-Enabled Flood Prediction Integrating Local Sensor Data and 3rd Party Weather Forecast. *Sensors (Basel), 23*(6), 3065.

Wang, Zhang, Zhi, & Ji. (2023). Multi-Model Ensemble Forecasts of Surface Air Temperatures in Henan Province. *Atmosphere, 14*(3), 520. https://doi.org/. doi:10.3390/atmos14030520

Ying-Nong Seo, F. (2022). Rainfall Forecast Using Machine Learning with High Spatiotemporal Satellite Imagery Every 10 Minutes. *Remote Sensing*, *14*(23), 5950. doi:10.3390/rs14235950

Chapter 6
Mapping of Glacial Lakes and Glacial Lake Outburst Flood in Kinnaur District, Himachal Pradesh Using Remote Sensing and GIS

Poonam Vishwas
Delhi Technological University, India

K. C. Tiwari
Delhi Technological University, India

Gopinadh Rongali
National Center for Medium Range Weather Forecasting (NCMRWF), MoES, Noida, India

Rubeena Vohra
Bharati Vidyapeeth's College of Engineering, India

ABSTRACT

Glacial lake outburst floods (GLOFs) are increasingly threatening the sustainability and infrastructure of high-mountain regions globally. This study specifically addresses the GLOFs risk in the Kinnaur district of Himachal Pradesh. Using a comprehensive analytical framework, the authors explore factors affecting GLOF initiation, map potential downstream impact zones, and assess societal vulnerability to these climate-induced calamities. For precise analysis, they utilize Landsat images (1989-2018) and Cartosat DEM (30 m) data to map glacial lakes. ArcGIS 10.8 generates geospatial characteristics, while the maximum likelihood classification (MLC) algorithm is

DOI: 10.4018/979-8-3693-1396-1.ch006

employed for land use and land cover (LULC) mapping. Indices like NDSI, NDWI, and NDPI contribute additional insights. Identifying 39 glacial lakes, 10 deemed hazardous due to size (>0.1 km2), the study underscores the specific threat of cross-border GLOFs from China's Upper Satluj River Basin to Eastern Himachal Pradesh. In a prospective deglaciated scenario, it is envisaged that a substantial escalation in glacial lake outburst flood (GLOF) hazard levels will prevail across the majority of administrative units. Given the increasing encroachment of glacial lakes towards precipitous headwalls, the potential for glacial lake outburst flood initiation is significantly amplified. This elevates the risk to surrounding areas, as the displacement of ice and rock can initiate substantial GLOFs. In light of this prognosis, an essential initial action for adaptation in these circumstances involves strengthening resilience and implementing capacity-building measures designed to reduce the risk of GLOFs.

INTRODUCTION

The Indian Himalayan Region (IHR) faces significant and complex challenges in addressing the adverse impacts of climate change. This region, similar to numerous other mountainous areas worldwide, exhibits heightened vulnerability to the consequences of global climate shifts, impacting both its natural environment and man-made environment. The recession of mountain glaciers and the emergence of large glacial lakes stand out as prominent and evolving outcomes of climate change (Bolch et al. 2008). The risks associated with glacial lake outburst floods (GLOFs) are on the rise, evolving in tandem with changes in the stability of the surrounding rock and ice walls (Clague and Evans, 1992; 2000; Mool 2005). With the expansion of residential, tourism, and particularly hydroelectric developments into alpine valleys, coupled with a growing human dependence on limited hydrological resources, there is an urgent need to enhance the integration of strategies for mitigating the risk of glacial lake outburst floods (GLOFs) into climate adaptation planning (Quincey et al. 2007).

Glacial lake outburst floods (GLOFs) represent the sudden discharge of a water reservoir that has developed beneath, on the side of, in front of, within, or on the surface of a glacier. Dams associated with these events can consist of ice, moraine, or bedrock. Given the anticipated warming and ongoing reduction of alpine glaciers (Reynolds 2000; Benn et al. 2000), the focus has shifted from monitoring and evaluating existing GLOF risks to predicting the emergence of potentially dangerous lakes in the future. These lakes are expected to primarily form within depressions on the exposed glacier bed (Frey et al. 2010; Ukita et al. 2011) and will be impeded by either bedrock or moraine barriers, contingent on various factors such as glacier morphology, erosional characteristics, and the rate of glacier retreat (Watanabe et al. 1994). Nonetheless, due to the fact that the majority of glacial lake outburst floods (GLOFs) have taken place in remote and scarcely inhabited regions, the limited availability or quality of data can result in substantial misinterpretations, especially concerning GLOFs that occurred prior to the introduction of satellite remote sensing technology (Sakai et al. 2000).

Applying satellite remote sensing, particularly utilizing optical imagery such as Landsat data with archives dating back to 1972, proves to be a highly effective modern approach for monitoring and researching historical glacial lake outburst floods (GLOFs) in remote and difficult-to-access high-altitude areas. Geomorphic indicators, including features like V-shaped depressions, extensive debris accumulations, and paleo river channels, have been employed in conjunction with remote sensing information to detect and ascertain past occurrences of outburst events (Thompson et al. 2010). However, it's important to note that these characteristics may also emerge as a result of other cryo-hydrologic events apart from glacial lake outburst floods (GLOFs). Therefore, caution is warranted when solely relying on satellite data for the detection of GLOF events (Yao et al. 2010). Confirming GLOF incidents robustly involves integrating evidence from historical records, field surveys, and long-term satellite observations, including high-resolution image archives like Google Earth. This approach enhances understanding of hydrogeomorphic processes triggered by GLOF events in the Himalayan region (Molnia 2009).

Numerous scholars have utilized remote sensing to map and monitor glacial lakes and associated Glacial Lake Outburst Floods (GLOFs). Allen et al. (2021) has investigated the Chorabari GLOF event in 2013, which occurred in the Kedarnath region of Uttarakhand. Using remote sensing and GIS they characterised the GLOF event and its impact on the region. Another noteworthy research has been done by Sattar et al. (2019) where they investigated the potential risks associated with GLOFs in the Gilgit-Baltistan region of Pakistan. Majeed et al. (2021) analyzed the recession of Gya Glacier and the 2014 glacial lake outburst flood in the Trans-Himalayan region of Ladakh, India with the aid of cutting edge remote sensing and GIS tools. In a very recent study, Zaidi et al. (2020) investigated the impact of climate change on the formation of glacial lakes and the potential risks associated with glacial lake outburst floods (GLOFs) in the Himalayan region. In another significant study, Gilany and Iqbal (2007) discussed the potential impacts of GLOFs on the socio-economic conditions of the Himalayan region.

The present study, therefore, represents an effort to employ satellite data for the mapping and ongoing monitoring of glacier lakes situated in the Kinnaur district of Himachal Pradesh. In light of the previously discussed insights in the literature, the current research aims to make a valuable contribution to existing knowledge on GlOF's in Himalaya. This is achieved through a set of defined objectives: (1) mapping the status of glaciers in the Kinnaur District spanning from 1998 to 2018, (2) creating an inventory of glacial lakes and assessing their spatial and temporal changes, (3) identifying glacial lakes with potential hazards, and (4) to validate the findings of the study in the field. In addition, our analysis extends to the assessment of future GLOF threats, aligning them with other climate-related risks such as fluvial floods and droughts. Policymakers rely on and seek the best available knowledge regarding expected changes over the coming decades to the next century, and this comprehensive study aims to meet these informational needs and demands.

Figure 1. Location of the study area

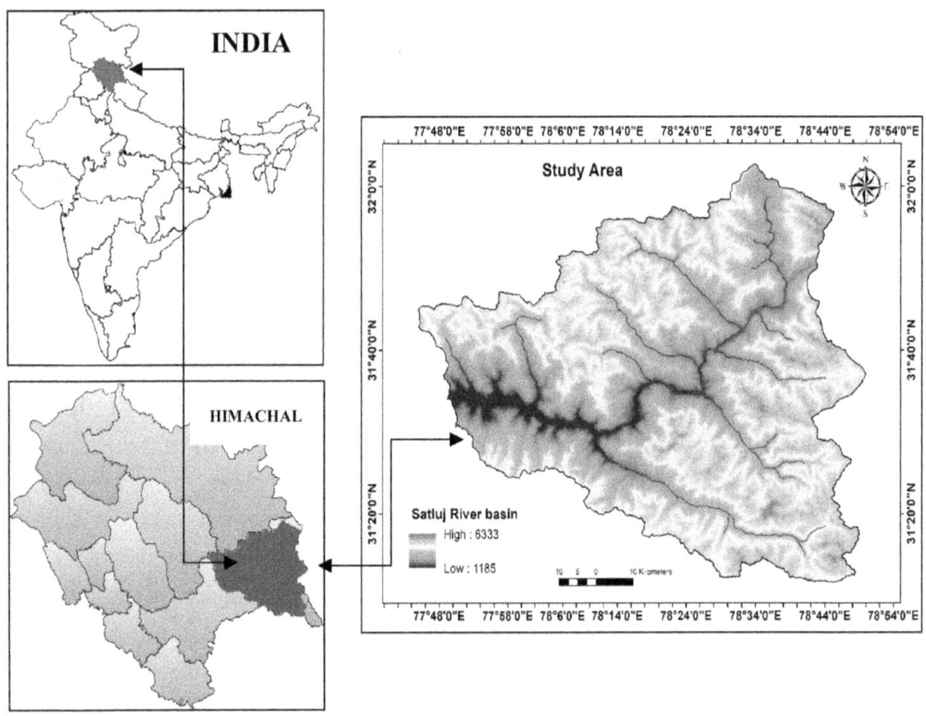

THE STUDY AREA

Considering the framework outlined above, this research focuses on the Kinnaur district (Latitude: 31.9167° N to 32.1667° N, Longitude: 78.1667° E to 79.2333°) within Himachal Pradesh as the designated study area for this research work. Situated in the north-eastern extremity of Himachal Pradesh, Kinnaur shares its border with Tibet to the east and is approximately 235 km from Shimla, the state capital. The topography of the study area is distinguished by the presence of three significant mountain ranges: the Zanskar, Himalayas, and the valleys of Sutlej, Spiti, and Baspa, accompanied by their respective tributaries. These elevated terrains are adorned with rich timber resources, thriving orchards, pastoral lands, and hamlets dispersed throughout the region. The study area boasts a varied climate, spanning alpine to temperate zones, shaped by its Himalayan setting. Winters are cold with substantial snowfall, while summers are brief and pleasant. Monsoons bring essential rainfall but also pose landslide risks. The region showcases diverse vegetation, from sub-alpine and alpine flora in higher elevations to coniferous and mixed deciduous forests at lower altitudes. Kinnaur is renowned for its apple orchards and hosts a range of wildlife, including Himalayan black bears, leopards, and snow leopards. It's a critical habitat for rare species like the Western Tragopan and the Himalayan Musk Deer, and a seasonal haven for migratory birds.

DATASET

In this research study, high-resolution satellite imagery, along with a digital elevation model, served as the primary remote sensing datasets. High-resolution satellite imagery, specifically Landsat 8 OLI and Landsat 4-5 TM data for the years, 1989, 1998, 2008, 2018, served as integral components in this study. We meticulously chose images with a cloud cover not exceeding 10% and minimal snow cover, specifically focusing on the period between October and November. This selection was made after a comprehensive evaluation of all accessible datasets. We obtained a total of 24 Landsat images from the USGS website (https://earthexplorer.usgs.gov). These images were sourced for the months of October (comprising 12 scenes) and November (comprising 12 scenes) for the years 1989, 1998, 2008, 2018 respectively. Additionally, we utilized the Cartosat Digital Elevation Model (30m) to acquire topographic data as well as drainage network. For the purpose of manual validation and analysis, Google Earth images were also incorporated in our study. The study heavily relies on the field photographs, videos and discharge data collected during the field work in the vicinity of Pangi and Kashag Gramang, Himachal Pradesh, from October 17th to October 26th, 2021.

METHODOLOGY

Database Generation

The process of mapping glaciers holds significant importance within the framework of assessing glacial lake hazards. This significance stems from the critical role glaciers play in relation to the position, slope, and surface area of lakes, all of which are integral factors in evaluating the potential for glacial lake hazards. Notably, remote sensing emerges as the primary and, in many cases, the sole feasible approach for conducting the mapping and ongoing monitoring of these glaciers. We primarily utilized Landsat 8 OLI and Landsat 4-5 TM data for spatial and temporal analysis of snow cover and glacier lakes. Subsequently, we performed essential image pre-processing for the satellite scenes, involving both Geometric and Radiometric correction, using ERDAS IMAGINE software. To encompass the entire study area, we created a mosaic of all the images. For relief and drainage analysis of the study area, we integrated Cartosat DEM.

Glacier and Glacier Lake Mapping

To map the glacial cover and glacial lakes we employed Landsat (TM and OLI) imageries for the years 1989, 1998, 2008, and 2018, respectively to prepare land use and land cover maps, applying the maximum likelihood supervised technique. Following this, we employed these images to calculate indices such as NDSI, NDWI, and NDPI, which were subsequently utilized for our analysis and result findings. NDWI (Normalized Difference Water Index) values typically fall within the range of 0.60 to 0.85 for standard lake surfaces, as indicated by Huggel et al. (2002). For the present study area, an index value between 0.65 and 0.90 has proven to be suitable for the purpose of mapping glacial lakes. NDWI has been calculated as under:

Normalized Difference Water Index (NDWI)

$$\text{NDWI} = \frac{GREEN - NIR}{GREEN + NIR}$$

Where, NIR is Near Infrared.

To get the most accurate results when determining snow cover, it's widely recognized that using a combination of techniques, like setting thresholds for NDSI images and carefully outlining snow areas manually, works best. NDSI is a commonly employed technique for satellite-based snow cover mapping (Paul 2004). It relies on distinguishing the varying reflectance properties of snow, with higher reflectance in the visible (green) region and lower reflectance in the short-wave infrared region (Hall et al. 1995; Hall et al. 2002; Kulkarni et al. 2005; Kulkarni et al. 2006). Remarkably, NDSI is capable of accurately delineating and mapping snow even in areas affected by mountain shadows (Kulkarni et al. 2005). NSDI for the present study has been calculated using the following formule:

Normalized Difference Snow Index (NDSI)

$$\text{NDSI} = \frac{Green - SWIR}{Green + SWIR}$$

Where, SWIR = Short Wave Infra-Red

Normalised Difference Pond Index (NDPI)

$$\text{NDPI} = \frac{Green - SWIR}{Green + SWIR}$$

Where, SWIR = Short Wave Infra-Red.

We utilized a threshold value of 0.4 to delineate snow-covered regions from the NDVI and NDWI data. Additionally, we selected specific samples from these datasets to delineate the lakes within the study area. Subsequently, we calculated the area of these lakes and, based on the computed areas, estimated the volume of the glacial lakes. Extensive manual editing and delineation procedures were applied by utilizing Google Earth imagery to create the definitive lake maps. It's worth noting that certain lakes, particularly those concealed under ice or not discernible in the imagery, have been excluded from the inventory. We employed a labelling system known as the "binary tree format" to assign labels to glacial lakes. This system initiates at the basin's outlet and utilizes a "left, root, right" rule to sequentially number the lakes according to their positions in relation to the associated streams.

Identification of Potentially Dangerous Lakes

The Himachal Himalayan region currently lacks precise estimates for the volume of its glacial lakes based on their water surface areas. However, there are available estimations for glacial lakes in the Swiss Alps, as documented by Huggel et al. (2002). Given the absence of specific volumetric data for glacial lakes in the Himachal Himalaya, it is deemed scientifically reasonable to apply the relationships and methodologies developed for estimating lake volumes in the Swiss Alps to make analogous estimations for the lakes in this region. The empirical relation provided by Huggel et al. (2002) is written as:

The lake volume; $V = 0.104\ A^{1.42}$ (1)

Where, V designates lake volume in m^3 and A is the area of the lake in m^2.

It's crucial to emphasize that increased lake volumes are indicative of higher potential hazards. To ensure the accuracy of our identified glacier lakes, we conducted further validation by cross-referencing them with Google Earth images.

Figure 2. Methodological framework of the study

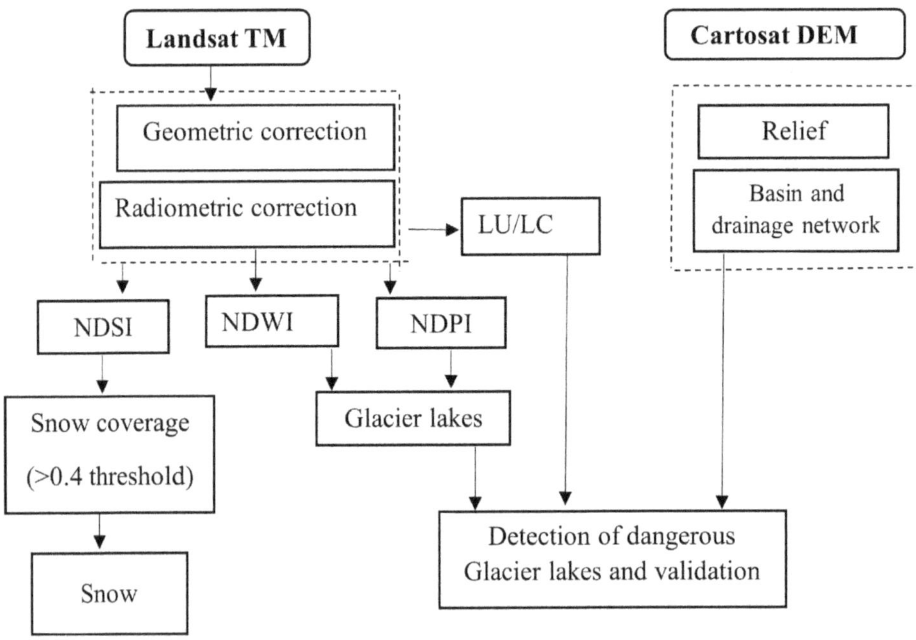

For selecting potentially dangerous lakes, following parameters have been used.

1: Lakes that are smaller than 0.1 square kilometres are seen as safe because they contain relatively small amounts of water. This assessment is based on the empirical relationship between lake area and volume, as described by Huggel's formulae.

2: Lakes near or attached to parent glaciers are considered more hazardous because they can grow larger. In contrast, isolated lakes usually remain stagnant, without expanding, and often release their water gradually rather than bursting. To assess the lakes' positions in relation to glaciers, we conducted overlay analysis.

3: We also looked at the steepness of the slopes around the lakes, as this can affect the water flowing into them. We estimated whether the terrain was thick or thin by visually analysing high-resolution satellite images on Google Earth and calculating distances.

Identification of the Most Vulnerable Lake

For the final step of the study, we identified the most vulnerable lake. Conventionally, volume is commonly regarded as the foremost criterion, and the largest lake is typically designated as such. However, in this particular study, the process of identifying the most vulnerable lake took into account four primary factors:

1) Area of the lake;
2) distance from the outlet of the basin;
3) growth in lake area since 1990 (change detection);
4) Elevation of the lake.

Field Validation

To validate the research outcomes and gather additional data regarding various physical parameters of the lake, an extensive field expedition was conducted in the vicinity of Pangi and Kashag Gramang, Himachal Pradesh, from October 17th to October 26th, 2021. The field data has been collected in the form of lake photographs, videos and discharge data.

RESULTS AND DISCUSSION

In the this study, four distinct land use and land cover maps were created, each corresponding to the years 1989, 1998, 2008, and 2018, respectively. The land use and land cover analysis aimed to investigate alterations in the land use pattern and identify glacier lakes. To accomplish this, we employed the maximum likelihood supervised technique within the ERDAS Imagine software. Subsequently, we categorized the findings into six distinct classes, which include glacier, forest, scrubland, wasteland, waterbody, and agriculture. The majority of the region was glaciated in the land cover maps for the years 1989, 1998, 2008, and 2018. As depicted in the Figure 3, our observations revealed that snow cover was less in 1989 compared to 1998, and there was also a reduced area of wasteland coverage. However, over the subsequent decade, we noted a consistent decrease in snow coverage and an expansion of wasteland. Additionally, the river channel appeared wider in the 2018 LULC map.

Figure 3. Land use and land cover area pattern of the study area

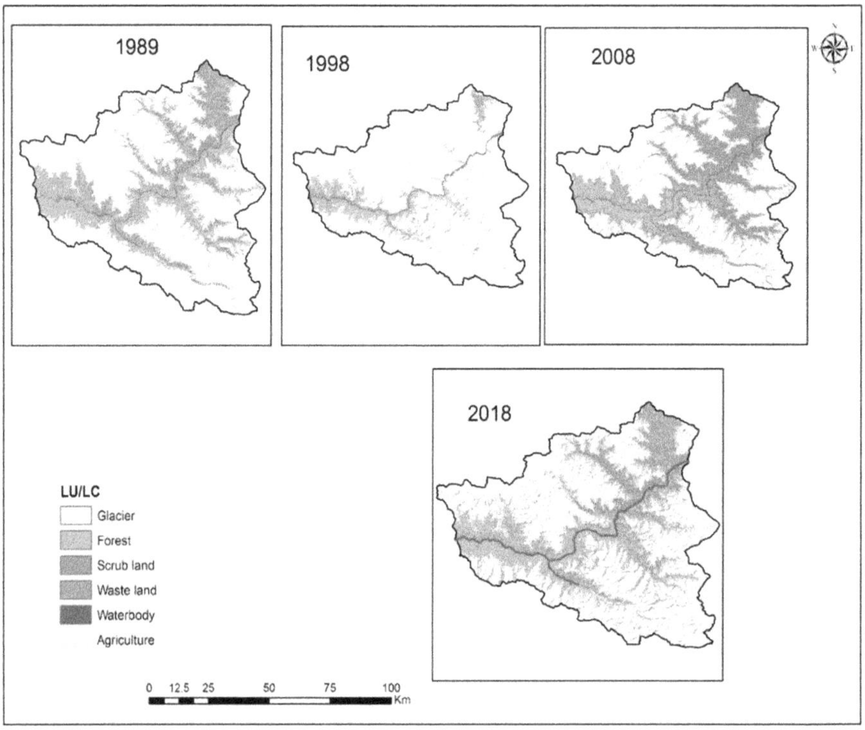

The NSDI maps created for the study area corroborate these findings, yielding similar results. These maps provide further validation of the observed trends and patterns within the study area. In accordance with our research methodology, we utilized the Normalized Difference Snow Index (NDSI) to map glaciers comprised of clean ice. It is worth noting that our observations indicated that a substantial number of Himalayan glaciers are characterized by a layer of debris covering their surfaces. Despite this debris cover, manual delineation remained a viable strategy due to the perceptible variations in illumination resulting from the distinctive shapes of these glaciers. To facilitate glacier mapping, we applied a threshold value of 0.4, categorizing all values exceeding this threshold as glacier areas, while those falling below it were classified as non-snow cover regions. The outcomes of the Normalized Difference Snow Index (NDSI) analysis are depicted in Figure 4. It is noteworthy that our observations reveal an augmented extent of snow coverage during the year 1998, followed by a discernible reduction in snow coverage during the year 2018.

For the assessment of glacial lakes, we applied the Normalized Difference Water Index (NDWI), which is expressed as (NIR - Blue) / (NIR + Blue), and also made use of the Normalized Difference Pond Index (NDPI). These indices served as valuable tools for the detection and analysis of alterations in the nature of the observed glacial lakes. To do so, we combined digital satellite data with a Digital Elevation Model (DEM) of the region to improve the accuracy of our glacier and glacial lake inventory. Additionally, we conducted ground truthing by cross-referencing our findings with Google Earth to correct any errors in pinpointing the exact locations of glaciers and glacial lakes. We extracted pixel samples from both established glacial lakes and other bodies of water within the range of -1 to +1 on NDWI and NDPI images.

Figure 4. NDSI Map of the study area

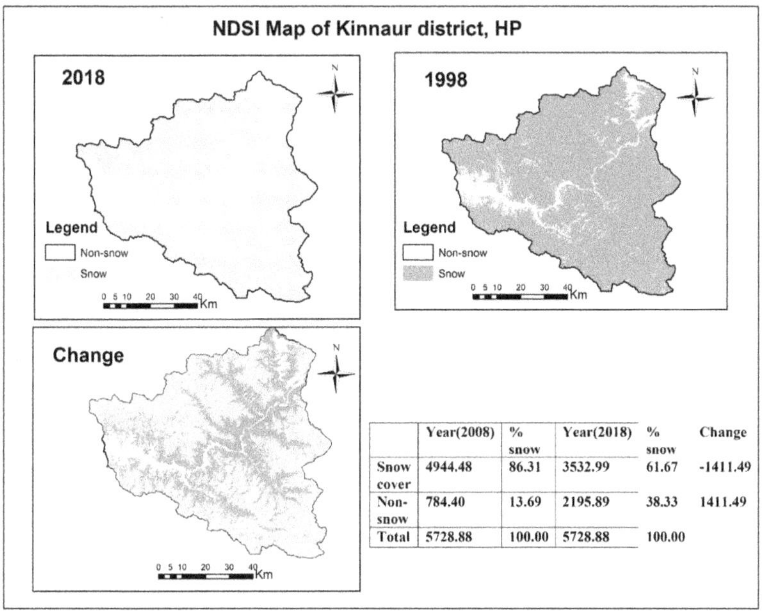

	Year(2008)	% snow	Year(2018)	% snow	Change
Snow cover	4944.48	86.31	3532.99	61.67	-1411.49
Non-snow	784.40	13.69	2195.89	38.33	1411.49
Total	5728.88	100.00	5728.88	100.00	

These samples were then used to compute the average values for the water bodies. The mean values we computed were used as a reference point to distinguish glacial lake features from other features. We applied a grayscale thresholding technique, which essentially divides the input into two categories. One category includes pixels with values below a predetermined gray level defined by the analyst, and the other category includes pixels with values above this specified threshold. The computed threshold value, which is -0.69, was used to distinguish glacial lake areas from other features by reclassifying the image in ArcGIS software. Through this process, we identified a total of 39 glacial lakes within Kinnaur district, spanning altitudes from 1,200 to 6,300 meters. The combined area of these lakes within the district measures approximately 4.02 square kilometres.

Figure 5. NDWI and NDPI maps of the study area for the years 1998 and 2018

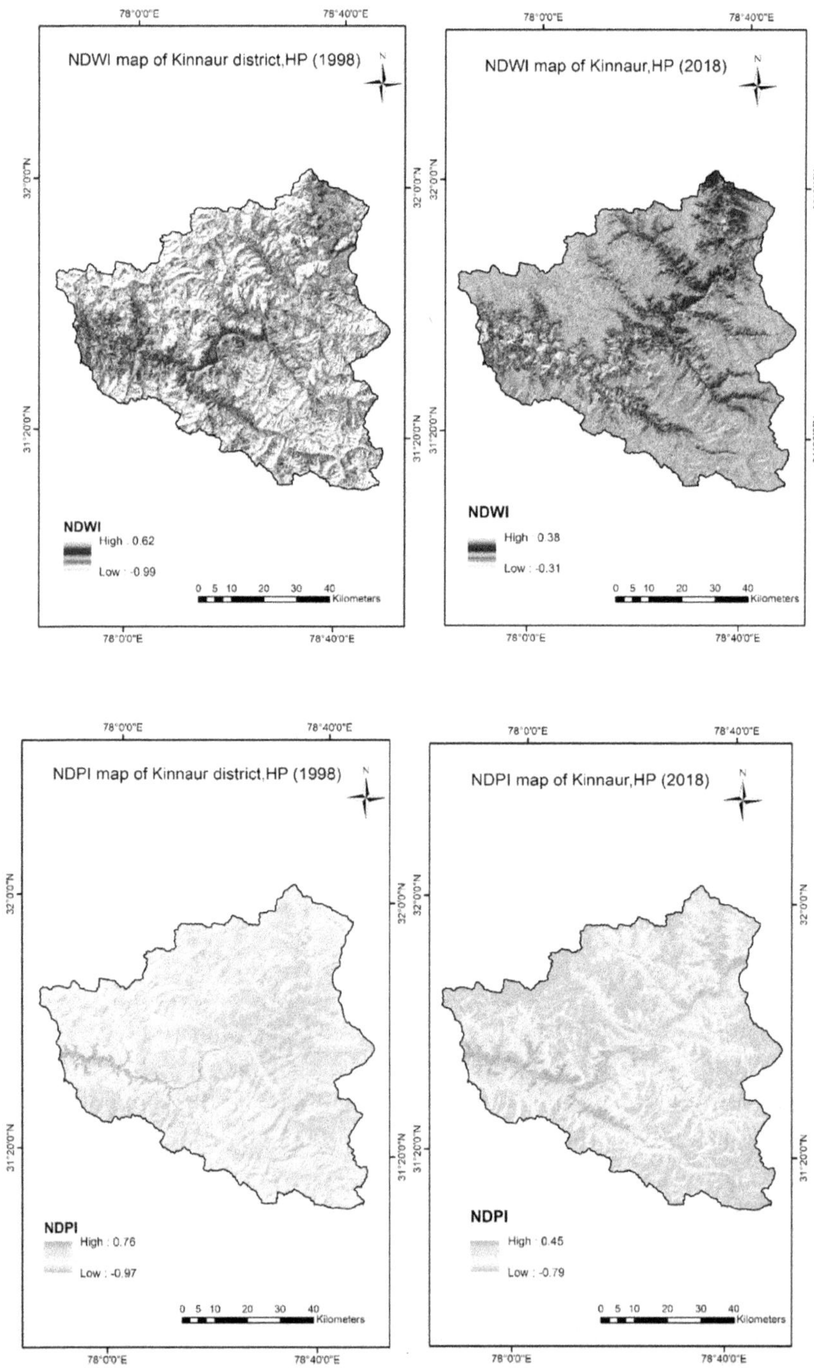

Figure 6. Location of glacial lakes in the study area

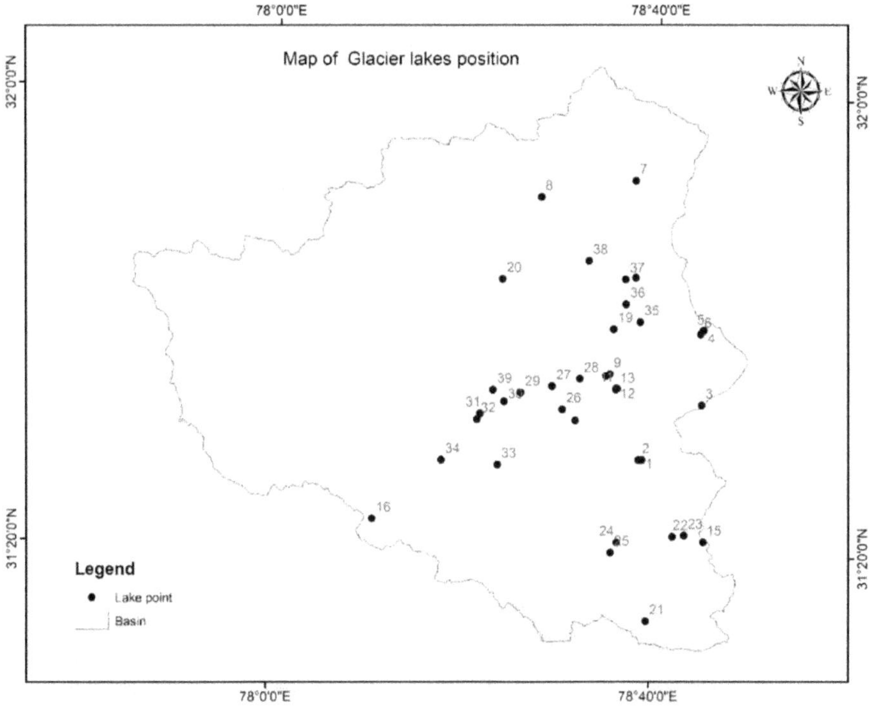

Table 1. Area, volume and distance from outlet of the selected potentially dangerous lakes

Lake no.	Area	Volume	Distance From Outlet
9	0.76	0.07	65.23
17	0.20	0.01	60.32
21	0.11	0.0045	78.21
24	0.13	0.0057	39.91
26	0.21	0.113	87.25
34	0.10	0.0039	73.32
31	0.11	0.0045	34.74
33	0.10	0.0039	46.32
36	0.33	0.0215	76.67
37	0.29	0.017	53.67

Figure 7. Area and volume of the identified 10 potentially dangerous lakes

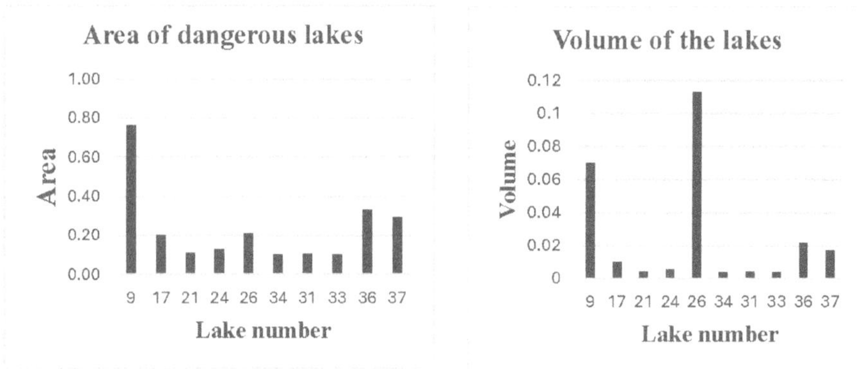

Figure 8. Validation of glacier lake were done through high resolution Google Earth images

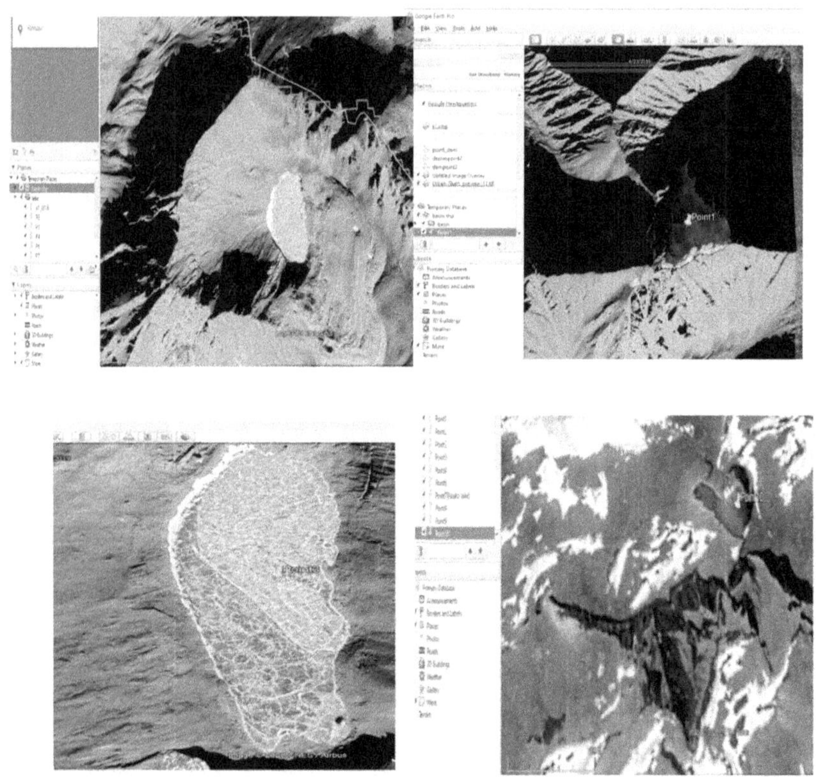

Out of a total of 39 glacial lakes identified in the Kinnaur district of Himachal Pradesh, 10 lakes have been classified as dangerous due to their substantial area exceeding 0.1 square kilometers. Table 1 displays a comprehensive list of potentially hazardous lakes, complete with labels identifying each lake, as well as crucial data regarding their surface area, volume, and their respective distances from the outlet.

Figure 9. Field photographs of the potentially most dangerous lake at Kashag Gramag, Himachal Pradesh

Lake as appears on satellite Image	Way to lake
Actual lake	The extent of water in Lake at 4276 m
GPS Survey at HPPCL intake	Team members near one of the lake

This Glacial lake is situated at an altitude of 4276 m (Lat/ Long: 31°37.472′ N, 78°17.463′E) and is possibly one of the highest potentially dangerous lake at this altitude. The maximum surface area of the lake as measured from the satellite image, acquired on 16 September 2018, is approximately 0.2396 sq.km^2. The identified lake is known as Mukim-Chikkim by the locals. The water coming from the lake and flowing through a stream is used by the Himachal Pradesh Power Corporation Limited (HPPCL) for hydro-electricity power generation. The intake for the HPPCL is situated 17 kms downstream from the lake. For the reaching the lake there is a motorable road up to the HPPCL intake, after which the difficult terrain starts.

CONCLUSION

The utilization of visual and digital image analysis within a Geographic Information System (GIS) proves instrumental in investigating glacial lakes and Glacial Lake Outburst Floods (GLOFs).The outcomes derived from an extensive satellite image analysis as well as field surveys have shed significant light on the genesis of the glacial lakes in the study area. Within the Kinnaur district of Himachal Pradesh, a total of 39 glacial lakes were identified, with 10 of them being categorized as hazardous lakes due to their area exceeding 0.1 square kilometers. The formation and growth of glacial lakes were closely tied to changes in the climate, particularly an increase in temperature and a decrease in precipitation. These changes led to the retreat of glaciers in the Kinnaur district. The growth rate of glacial lakes varied inversely with their proximity to the parent glacier. Closer lakes had higher growth rates because they primarily received water from the melting glacier. Due to the ongoing enlargement of lakes and the retreat of glaciers, the potential threat of GLOF is steadily increasing in the Kinnaur district. These observed trends underscore the critical importance of monitoring and assessing glacial lakes, especially those situated at higher altitudes, utilizing advanced technologies such as remote sensing and Geographic Information Systems (GIS). The practical application of this study lies in the informed management and mitigation of glacial lake outburst flood (GLOF) risks in the Kinnaur district of Himachal Pradesh. The findings from comprehensive research underscore the need to enhance strategies, particularly in light of the escalating threats in the Kinnaur district. The study recommends implementing continuous monitoring through advanced remote sensing and GIS techniques is essential for tracking the response of glacial lakes to climatic changes in the Kinnaur district. The findings from this comprehensive research underscore the need to enhance strategies, particularly in light of the escalating threats in the Kinnaur district. Future research directions may involve continuous spatio-temporal monitoring of glacial lakes, employing advanced remote sensing and GIS techniques as well as field based validation to analyse their response to climatic variations. Developing predictive models for GLOF hazards, coupled with assessing socio-economic and environmental vulnerabilities, is crucial for informed risk mitigation in the high Himalayan region.

ACKNOWLEDGMENT

The authors are grateful to the NASA for making the Landsat and ASTER GDEM datasets freely available under the umbrella of USGS web server. The present work was financially supported by the National Mission on Himalayan Studies (NMHS), MoES & CC, India under the Himalayan Research Fellowship (HRF) program.

Disclosure Statement: No potential conflict of interest was reported by the author(s).

Compliance With Ethical Standards: The author declares that no copyright norms have been violated, and the study has no conflict of interest.

REFERENCES

Aggarwal, A., Jain, S. K., Lohani, A. K., & Jain, N. (2013). Glacial lake outburst flood risk assessment using combined approaches of remote sensing, GIS and dam break modelling. *Geomatics, Natural Hazards & Risk.*

Allen, S., Sattar, A., King, O., Zhang, G., Bhattacharya, A., Yao, T., & Bolch, T. (2021). Glacial lake outburst flood hazard under current and future conditions: First insights from a transboundary Himalayan basin. *Nat Hazards Earth Syst Sci Discuss,* 1–30.

Allen, S. K., Linsbauer, A., Randhawa, S. S., Huggel, C., Rana, P., & Kumari, A. (2016). Glacial lake outburst flood risk in Himachal Pradesh, India: An integrative and anticipatory approach considering current and future threats. *Natural Hazards, 84*(3), 1741–1763. doi:10.1007/s11069-016-2511-x

Allen, S. K., Rastner, P., Arora, M., Huggel, C., & Stoffel, M. (2015). Lake outburst and debris flow disaster at Kedarnath, June 2013: Hydrometeorological triggering and topographic predisposition. *Landslides*, 1–13. doi:10.1007/s10346-015-0584-3

Arendt, A., Bliss, A., Bolch, T., Cogley, J. G., Gardner, A. S., Hagen, J.-O., Hock, R., Huss, M., Kaser, G., & Kienholz, C. (2015). *Randolph glacier inventory – a dataset of global glacier outlines. Version 5.0*. Global Land Ice Measurements from Space. Available from: https://www.glims.org/RGI/00çrgi50çTechnicalNote.pdf

Bolch, T., Buchroithner, M. F., Peters, J., Baessler, M., & Bajracharya, S. (2008). Identification of glacier motion andpotentially dangerous glacial lakes in the Mt. Everest region, Nepal using spaceborne imagery. *Natural Hazards and Earth System Sciences*, *8*(6), 1329–1340. doi:10.5194/nhess-8-1329-2008

Clague, J. J., & Evans, S. G. (1992). A self-arresting moraine dam failure, St. Elias Mountains, British Columbia. Current Research, Part A. Geological Survey of Canada Paper, 92(1A), 185–188.

Clague, J. J., & Evans, S. G. (2000). A review of catastrophic drainage of moraine-dammed lakes in British Columbia. *Quaternary Science Reviews*, *19*(17-18), 1763–1783. doi:10.1016/S0277-3791(00)00090-1

Frey, H., Haeberli, W., Linsbauer, A., Huggel, C., & Paul, F. (2010). A multi-level strategy for anticipating futureglacier lake formation and associated hazard potentials. *Natural Hazards and Earth System Sciences*, *10*(2), 339–352. doi:10.5194/nhess-10-339-2010

Gilany, N., & Iqbal, J. (2020). Geospatial analysis and simulation of glacial lake outburst flood hazard in Shyok Basin of Pakistan. *Environmental Earth Sciences*, *79*(6), 1–11. doi:10.1007/s12665-020-8867-y

Hall, D. K., Riggs, G. A., & Salomonson, V. V. (1995). Development V.V. Salomonson, I. Appel Remote Sensing of Environment 89 (2004) 351–360 359 of methods for mapping global snow cover using moderate resolution imaging spectroradiometer data. *Remote Sensing of Environment*, *54*(2), 127–140. doi:10.1016/0034-4257(95)00137-P

Hall, D. K., Riggs, G. A., Salomonson, V. V., DeGirolamo, N. E., Bayr, K. J., & Jin, J. M. (2002). MODIS Snow-cover products. *Remote Sensing of Environment*, *83*(1-2), 181–194. doi:10.1016/S0034-4257(02)00095-0

Huggel, C., Kaab, A., Haeberli, W., Teysseire, P., & Paul, F. (2002). Remote sensing based assessment of hazards from glacier lake outbursts: A case study in the Swiss Alps. *Canadian Geotechnical Journal*, *39*(2), 316–330. doi:10.1139/t01-099

Kulkarni, A. V., Rathore, B. P., Mahajan, S., & Mathur, P. (2005). Alarming retreat of Parbati Glacier, Beas basin, Himachal Pradesh. *Current Science*, *88*(11), 1844–1850.

Kulkarni, A. V., Rathore, B. P., Randhawa, S. S., Sood, R. K., & Kaul, M. (2005). Glacier Atlas Chenab Basin (H. P.). RSAM/SAC/RSA/RESIPA/MWRG-GLI/ SN22.

Kulkarni, A. V., Singh, S. K., Mathur, P., & Mishra, V. D. (2006). Algorithm to monitor snow cover using AWiFS data of RESOURCESAT for the Himalayan region. *International Journal of Remote Sensing, 27*(12), 2449–2457. doi:10.1080/01431160500497820

Majeed, U., Rashid, I., Sattar, A., Allen, S., Stoffel, M., Nüsser, M., & Schmidt, S. (2021). Recession of Gya Glacier and the 2014 glacial lake outburst flood in the Trans-Himalayan region of Ladakh, India. *The Science of the Total Environment, 756*, 144008. doi:10.1016/j.scitotenv.2020.144008 PMID:33293089

Molnia, B. F. (2009). Inventorying and monitoring the recent behavior of Afghanistan's glaciers, debris-covered glaciers, supraglacial lakes, and the potential for catastrophic flooding (jȯ̈kulhlaups). In European Geosciences Union General Assembly 2009, Geophysical Research Abstracts (vol. 11). Academic Press.

Mool, P. K. (2005). *Monitoring of glaciers and glacial lakes from 1970 s to 2000 in Poiqu Basin, Tibet Autonomous Region, PR China.* Unpublished report prepared by ICIMOD, Kathmandu, Nepal, and CAREERI, Gansu, China.

Paul, F., Huggel, C., & Kääb, A. (2004). Combining satellite multispectral image data and a digital elevation model for mapping debris-covered glaciers. *Remote Sensing of Environment, 89*(4), 510–518. doi:10.1016/j.rse.2003.11.007

Quincey, D. J., Richardson, S. D., Luckman, A., Lucas, R. M., Reynolds, J. M., Hambrey, M. J., & Glasser, N. F. (2007). Early recognition of glacial lake hazards in the Himalaya usingremote sensing datasets. *Global and Planetary Change, 56*(1–2), 137–152. doi:10.1016/j.gloplacha.2006.07.013

Reynolds, J.M. (2000). On the formation of supraglacial lakes on debris-covered glaciers. *IAHS Publ, 264*, 153–161.

Sakai, A., Chikita, K., & Yamada, T. (2000). Expansion of a moraine-dammed glacial lake, Tsho Rolpa, inRolwaling Himal, Nepal Himalaya. *Limnology and Oceanography, 45*(6), 1401–1408. doi:10.4319/lo.2000.45.6.1401

Sanjay, Lohani, Singh, Chaudhary, & Thakural. (2012). Glacial lakes and glacial lake outburst flood in a Himalayan basin using remote sensing and GIS. *Natural Hazards.*

Sattar, A., Goswami, A., & Kulkarni, A. V. (2019). Application of 1D and 2D hydrodynamic modeling to study glacial lake outburst flood (GLOF) and its impact on a hydropower station in Central Himalaya. *Natural Hazards*, *97*(2), 535–553. doi:10.1007/s11069-019-03657-6

Thompson, S., Benn, D., Luckman, A., & Kulessa, B. (2010). The evolution of supraglacial lakes, NgozumpaGlacier, Khumbu Himalaya, Nepal. *Geophysical Research Abstracts*, *12*, EGU2010–11396-1.

Ukita, J., Narama, C., Tadono, T., Yamanokuchi, T., Tomiyama, N., Kawamoto, S., Abe, C., Uda, T., Yabuki, H., Fujita, K., & Nishimura, K. (2011). Glacial lake inventory of Bhutan using ALOS data: Part 1, methods andpreliminary results. *Annals of Glaciology*, *52*(58), 2011. doi:10.3189/172756411797252293

Watanabe, T., Ives, J. D., & Hammond, J. E. (1994). Rapid growth of a glacial lake in Khumbu Himal, Nepal:prospects for a catastrophic flood. *Mountain Research and Development*, *14*(4), 329–340. doi:10.2307/3673729

Yao, T., Li, Z. G., Yang, W., Guo, X. J., Zhu, L. P., Kang, S. C., Wu, Y. H., & Yu, W. S. (2010). Glacial distribution and mass balance in the Yarlung Zangbo River and its influence onlakes. *Chinese Science Bulletin*, *55*(20), 2072–2078. doi:10.1007/s11434-010-3213-5

Zaidi, A. Z., Yasmeen, Z., & Siddiqui, M. D. (2013) glacial lake outburst flood (GLOF) Risk mapping in Hunza River Basin (Pakistan) using geospatial techniques. In *2013 6th International Conference on Recent Advances in Space Technologies (RAST)*. IEEE.

Chapter 7
Land Surface Temperature and Its Lapse Rate Estimation Using Landsat-8 TIRS Data in Beas River Basin, India and Computed Differences With MODIS-Terra

Gopinadh Rongali
National Centre for Medium Range Weather Forecasting, MoES, Noida, India

Ashok K. Keshari
Indian Institute of Technology Delhi, New Delhi, India

Ashwani K. Gosain
Indian Institute of Technology Delhi, New Delhi, India

R. Khosa
Indian Institute of Technology Delhi, New Delhi, India

Ashish Kumar
https://orcid.org/0000-0003-3466-8845
School of Computer Science Engineering and Technology, Bennett University, Greater Noida, India

DOI: 10.4018/979-8-3693-1396-1.ch007

ABSTRACT

In the study of snow melt runoff, the temperature lapse rate (TLR) is an essential variable. For the Beas River Basin in the Western Himalayas, it has been approximated in the current study. In this work, the split-window (SW) technique for TLR estimate has been used to recover the land surface temperature (LST). LST in the study area has a negative correlation with elevation values, and the trend shows that LST and elevation have an inverse relationship, according to data from the United States Geological Survey's (USGS) advanced spaceborne thermal emission and reflection radiometer (ASTER) and global digital elevation model (GDEM). The TLRs for the Beas River Basin region vary from 0.71°C/100 m to 0.87°C/100 m during the time period of 18 April 2013 to 27 June 2015. The findings were calculated using lapse rates calculated from maps produced by the moderate resolution image spectroradiometer (MODIS). There is excellent agreement between the MODIS-Terra data and the air temperature and LST from Landsat-8. The modelling of snow and glacier melt flow in the Himalayan region will benefit from the current work.

1. INTRODUCTION

An important consideration in research on snow and glacier melt runoff is the land surface temperature (LST). Typically, it is described as the ground's surface temperature. Recently advanced remote sensing technology is responsible for providing a substitute to mapping LST on a large scale. Many studies have tried to recover this information from satellite data (Jiang, 2013; Tang and Li, 2014; Jiménez-muñoz et al., 2014; Rajeshwari and Mani, 2014; Sattari and Hashim, 2014). These strategies are based on a split-window (SW) technique for estimating LST from the Landsat-8 satellite, which was jointly developed by the National Aeronautics and Space Administration NASA and the United States Geological Survey (USGS). The Landsat Data Continuity Mission satellite carried out the mission (LDCM). When the Landsat-8 satellite was sent into orbit in February 2013, it brought with it two sensors, one of which was a thermal infrared sensor (TIRS) that can detect heat. According to the underlying principle for LST remote sensing, the total amount of radiative radiation released by the ground's surface rises sharply with surface temperature. A ground object's energy emission's spectral distribution changes with ground temperature as well. Sensors that operate in the thermal infrared (TIR) region of the electromagnetic spectrum (8-12 m) at the atmospheric window may remotely detect the thermal energy of the real ambient temperature of the ground surface. The brightness temperature (TB) refers to the temperature of the ground at the satellite level.

Several studies have been conducted to estimate LST using TIR radiation emitted from surfaces using an SW algorithm (Becker and Li, 1990; Ulivieri et al., 1994; Coll and Caselles, 1997; Sobrino and Raissouni, 2000; Jain et al., 2008). To account for the different transmittances of the atmosphere, the SW technique uses two neighbouring TIR channels, centred at 11 m and 12 m, to get surface temperatures from Landsat-8. Based on the differential infrared absorption, the SW LST approach corrects atmospheric influences. The accuracy of the SW algorithm depends on the magnitude of the difference between the emissivities of the surface in the two bands (Becker, 1987; Ruescas et al., 2016).

A crucial factor in determining how much water melts from snow and glaciers are the temperature lapse rate (TLR). The air temperature data available for each elevation zone is spatially interpolated based on the TLR method defined as a rate of decrease of temperature with elevation (Singh, 1991). However, even in highly industrialised nations, there are few stations that measure air temperature, and they are sometimes hundreds of kilometres apart. Spatial coverage is even worse in developing countries where measurements are very few or non-existent (Lakshmi et al., 2001). Snow cover temperature has been mainly observed in polar regions by using the advanced very high-resolution radiometer (AVHRR) sensor (Key and Haefliger, 1992; Key et al., 1997; Stroeve and Steffen, 1998). In India, a constant value of TLR calculated from air temperature data has been used in snow melt studies in different catchments with different kinds of relief (Thapa, 1980; Bagchi, 1981; Agarwal et al., 1983; Jeyram et al., 1983; Seth, 1983). However, many researchers (Jain, 2001; Haritashya, 2005) have suggested using actual TLR data for better snow melt runoff estimation. Singh (1991) conducted a sensitivity analysis of TLR in the Beas basin and found that a difference of 1°C in TLR influences the snow melt runoff computation by 28%–37%. Saraf et al. (2005) has reported a negative linear correlation between LST and elevation. They constructed a digital elevation model (DEM) using nighttime thermal images captured by the National Oceanic and Atmospheric Administration's advanced very high-resolution radiometer (NOAA-AVHRR).

Temperature is the greatest indicator of the heat transfer mechanisms involved in snow melting. The kind of precipitation is also influenced by the temperature that prevails throughout the precipitation period (rain or snow). The temperature index approach of calculating snowmelt is the foundation of many snowmelt models. The majority of snowmelt simulation models were created with the idea of dividing basins into elevation bands in accordance with basin topography. Additionally, because the air temperature is monitored manually, there is a chance for inaccuracy. LST maps, on the other hand, are continuous datasets produced from satellite data. If an adequate radiometric calibration and atmospheric adjustment can be accomplished, the sensors aboard satellites might provide data that are objective since they automatically capture the key factors.

From the LST, we have calculated the TLR for this work. Towards this end, we have investigated the practicability of utilising Landsat-8 data for tracking snowmelt processes. We have been used the SW method formulated by Sobrino et al. (1996) for LST determination from Landsat-8 data in this study. The SW algorithm does take emissivity parameters into account. In particular, a uniform emissivity may be employed since snow covers the majority of the research region. Hence the emissivity values determined by Sobrino et al. (1996). In particular, a uniform emissivity may be employed since snow covers the majority of the research region. Sobrino et al. (1996) algorithm is chosen above others because of this. The findings from the Landsat-8 LST images have been computed using the MODIS-Terra LST maps for the relevant dates.

Figure 1. The Beas River Basin, USGS GDEM extends to Pandoh Dam, with three weather stations

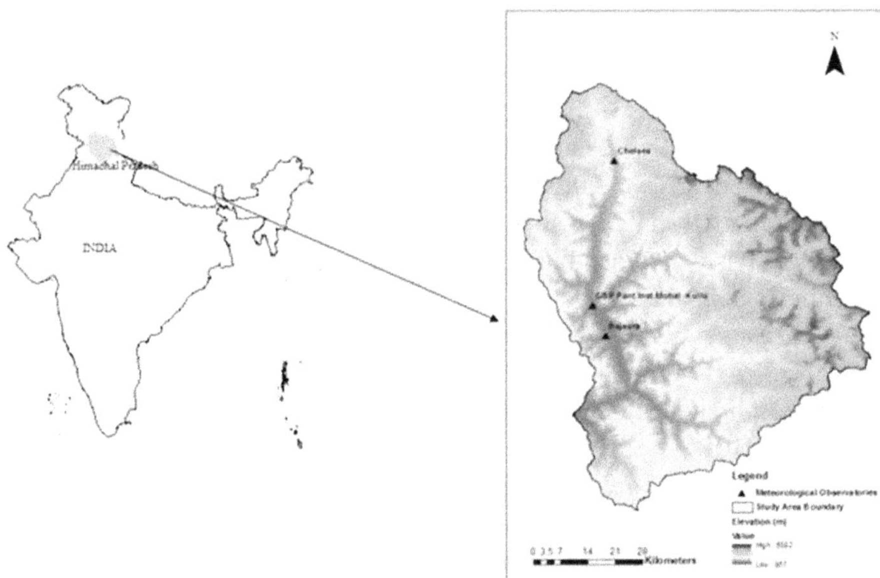

2. THE STUDY AREA

The Himalayas were studied in detail, along with the Beas River basin and the area upstream to the Pandoh Dam. From Rothang Pass, at an altitude of 3900 m, the Beas River runs approximately north to south all the way to Larji. At Larji, it turns right and continues flowing in the same way until it reaches the Pandoh dam, where it meets the Indus River as a significant tributary. There are around 5,600 square kilometres of forest inside the Beas river's catchment region up to the Pandoh dam. Catchment area elevations range from 857 m to 6582 m. About 3500 metres is the average elevation of the basin. The Beas River is 116 kilometres long from its source to Pandoh. The glaciers that feed the Parbati and Sainjkhad rivers are among its tributaries. The river valley is flanked by mountain ranges to the north and to the east. During the winter months, snow covers a large section of the river. The Beas River receives most of its water from melting snow throughout the summer.

3. DATA USED

The primary data sets used in the present study include the following:

(a) USGS Earth Explorer (https://earthexplorer.usgs.gov/) has been used to download Landsat-8 images of the study area. The satellite passes have occurred between 10:00 and 10:25 a.m. during Indian standard time. On these six dates (April 18, June 5, 2013, April 21, June 8, 2014, and June 27, 2015) throughout the day, six images were captured.

(b) Downloaded from the USGS Earth Explorer website (https://earthexplorer.usgs.gov/) is a 30 m resolution DEM of the study area taken from the GDEM of the Beas River basin produced using data from the advanced spaceborne thermal emission and reflection radiometer (ASTER).

(c) The National Snow and Ice Data Center's (NSIDC) website (https://nsidc.org/) has been used for MODIS-Terra LST maps (MOD11A1) for the relevant dates.

(d) Air temperature data has been downloaded from the website of space applications centre (SAC) meteorological and oceanographic satellite data archival center (MOSDAC) (https://www.mosdac.gov.in/)from three ground stations (Bajaura, Chelsea, and GB Pant Institute Mohal-Kullu) located within the Beas river basin (Figure 1).

In Table 1, we show the specifics of the data set utilised in this investigation.

Table 1. Specifics on the several data sets that contributed to this research

Type of Data	Details of Data	Source of Data
For the Beas River region, ASTER GDEM	Resolution of 30 metres	Source: US Geological Survey Earth Explorer, accessed at the URL https://earthexplorer.usgs.gov/
Landsat-8 images	Images obtained on 18 April 2013 to 27 June 2015 withspatial resolution of 30 m	Downloaded from the USGS earth explorer https://earthexplorer.usgs.gov/
MODIS LST MOD11A1	Images obtained on 18 April 2013 to 27 June 2015 withspatial resolution of 1 km	Downloaded from the NSIDC https://nsidc.org/
Air temperature readings from the Chelsea, GB Pant Inst; Mohal-Kullu; and Bajaura stations	AWS & IMD daily average air temperature data of Beas River basin	Downloaded from theSAC MOSDAC https://www.mosdac.gov.in/

4. METHODOLOGY

4.1 Database Creation

The DEM of the research region was derived from the USGS global DEM through the Beas river basin (Tile No. 0N31E077). Each of the six Landsat-8 images has been calibrated before being registered. The DEM was then used as the master image in an image registration process, with the other Landsat-8 images acting as slaves. In this case, we use a resampling strategy based on the closest neighbours and a transformation strategy based on the second order. Geo-referencing precision (root mean squared error inside a pixel) was increased by selecting over 30 ground control sites strategically placed throughout the research region.

Only one of the 12 layers that make up the MODIS daily LST product (MODIS LST day 1km) was utilised for this analysis. MODIS's raw LST data is supplied in a tree structure (HDF-EOS). HDF, or the High-Density File Format, is a multi-object file format created specifically for the transfer of scientific data across different computer systems and networks. Through the use of the ERDAS Imagine programme, this images originally stored in the HDF-EOS format has been converted to the ".img" format. A sinusoidal projection is used for the latest edition of MODIS MOD11A1 data. To convert from the sinusoidal projection to geographic coordinates and WGS84 datum, a re-projection is performed. As time went on, the USGS-DEM was used to geo-reference the MODIS LST images. After applying a scale factor of 0.02 to each of the LST images, the temperature in Kelvin was calculated by subtracting 273.15 from the resulting number. Detailed information on the MODIS LST product and methods used for MODIS LST/E product creation are given in the MODIS LST products users' guide (Wan, 2006).

Figure 2. A process flow diagram for calculating LST, Aster GDEM, and TLR using Landsat-8 TIR data

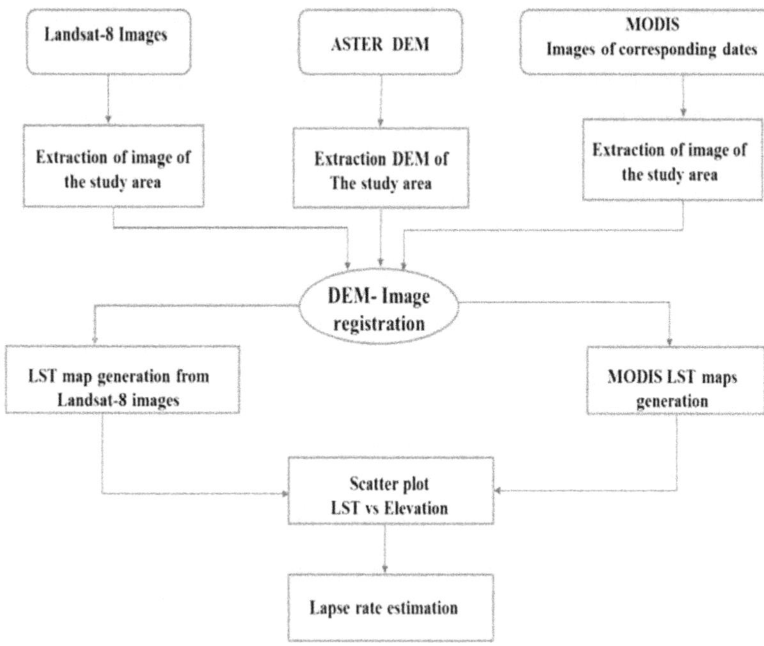

4.2 Assessment of LST

Several methods for estimating the surface temperature emerged as more and more thermal data became accessible. Converting satellite radiance to surface TB is the first issue that has to be addressed. Landsat-8 thermal channels need calibration for TB recovery. The Landsat 8 (L8) data users' manual explains how to calibrate the thermal channels in great detail. In the current investigation, the Landsat-8 thermal channels were calibrated using the method described in the user handbook. We have been used the SW algorithm proposed by (Sobrino et al., 1996) has been used to generate LST maps for the six selected Landsat-8 images. Two TIRS bands are used to calculate an estimate of the LST based on the mean and difference in the emissivity of the land surface (LSE). The equation is:

$$LST = TB_{10} + C_1(TB_{10}-TB_{11}) + C_2(TB_{10}-TB_{11})^2 + C_0 + (C_3+C_4W)(1-m) + (C_5+C_6W)\Delta m \quad (1)$$

where,

LST= land surface temperature (K)

C_0 to C_6 = split-window coefficient values (Table 2) (Sobrino et al., 1996; Sobrino and Li, 2003; Zhao et al., 2009; Skokovic et al., 2014)

TB_{10} and TB_{11} = brightness temperature of band10 and band11 (K)

m= mean LSE of TIRS bands

W= atmospheric water vapor content

Δm= difference in LSE

Table 2. SW coefficient values

Constant	C_0	C_1	C_2	C_3	C_4	C_5	C_6
Value	-0.26	1.37	0.18	54.30	-2.23	-129.20	16.40

4.2.1 The Top of Atmospheric Spectral Radiance

Multiplying each TIRS band by its associated multiplicative rescaling factor (0.000342) and then adding an additive rescaling factor (0.1) yields the value of the top of atmosphere (TOA) spectral radiance function. Here we show how to use the SW method to calculate the spectral radiation at the top-of-atmosphere (TOA) for TIRS bands 10 and 11 and OLI sensor bands 2 through 5. The original picture is converted to a spectral radiance image by means of this technique. In order to complete the work, we use the procedure of equation 2 inside the raster calculator in ArcGIS 10.3.

$$L\lambda = \left(\frac{L_{max}-L_{min}}{DN_{max}}\right) \times Band + L_{min} \quad (2)$$

where,

$L\lambda$ = upper atmosphere spectral radiance (in watts) / (m2×$srad$×μm)

Lm_{ax} = absolute peak intensity in each band's spectrum

L_{min} = absolute minimum of each band's spectral radiation

DN_{max} = $Qcal_{max}$ - $Qcal_{min}$ = difference of maximum and minimum calibration of the sensor

$L\lambda$ = 0.000342×band 10 + 0.1 (band 10_radiance) and 0.000342×band 11 + 0.1 (band 11_radiance)

4.2.2 Brightness Temperature

The term "terrestrial microwave radiation" refers to electromagnetic waves that propagate upward into the upper layers of Earth's atmosphere. TIRS thermal band DN values have been calibrated and may now be converted to TB. The spectral radiance at the TOA is required for a region's TB. Estimates of TB in TIRS Bands 10 and 11 were calculated using the following method. Thermosphere-Bound Radiation is the name given to the electromagnetic radiation that escapes into space from the uppermost layers of Earth's atmosphere. The thermal calibration of TIRS has been completed by converting the thermal DN values of the raw thermal bands into TOA spectral radiance using the TB equation presented in equation 3.

$$TB = K2/(\ln(K1/L\lambda+1)) - 273.15 \quad (3)$$

where,

$K1$ and $K2$ = metadata image file thermal constant for bands

L = layer of Spectral Radiation at the Top of the Atmosphere

TB = 1321.01/ln (774.89/band 10_radiance+1) – 273.15 for band 10_*TB* and 1201.14/ln (480.89/band 11_radiance+1) – 273.15 for band 11_*TB*

Figure 3. Process flow diagram for thermal calibration

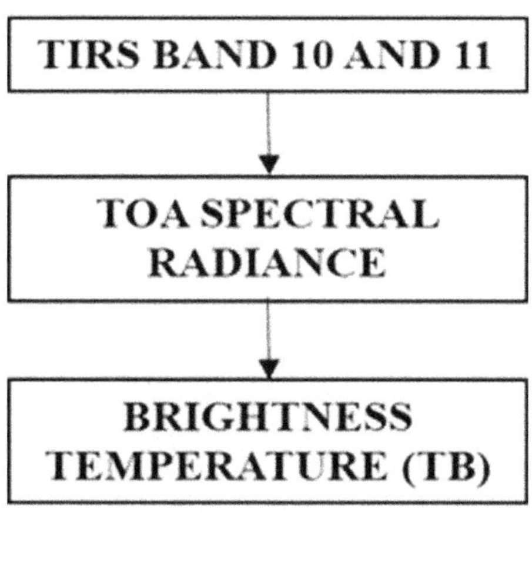

4.2.3 NDVI Threshold

Normalized difference vegetation index (NDVI) is determined using ERDAS Imagine after layer stacking data from four different OLI sensor bands (2, 3, 4, and 5). NDVI measured an output value of -0.26 to 0.61. The fraction of vegetation cover (FVC) may be calculated by first reclassifying the NDVI image into soil and vegetation to get NDVIs and NDVIv. After LSE has been generated for both TIR bands, the mean and difference LSE are determined. Following the layer stacking of bands (2, 3, 4, and 5), we used the technique shown in equation 4 to estimate NDVI using optical bands from the OLI sensor.

$$NDVI = \frac{Band_5 - Band_4}{Band_5 + Band_4} = \frac{NIR - RED}{NIR + RED} \tag{4}$$

Range: -1< $NDVI$ <+1

4.2.4 Fractional Vegetation Cover

Using equation 5, we have calculated an estimated FVC for an image based on the NDVI image that was acquired in the previous phase. An estimate of the percentage of land that is covered by vegetation is known as the fractional vegetation cover (FVC). The method of FVC is shown in the flowchart seen in Figure 4. In order to calculate LSE, the SW method makes use of FVC. ArcGIS 10.3 was used to reclassify the NDVI layer into soil and vegetation, calculate NDVI for soil and vegetation, and perform the procedure of the FVC of equation 6.

Figure 4. FVC flow diagram

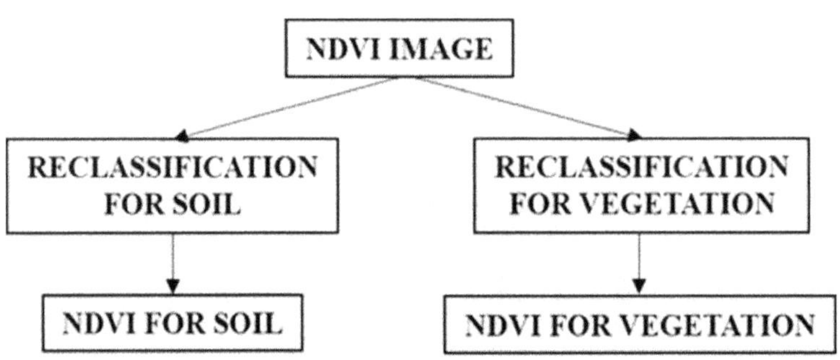

$$FVC = \frac{NDVI - NDVI(Soil)}{NDVI(Vegetation) - NDVI(Soil)} \quad (5)$$

$$FVC = \frac{NDVI - 0.15}{0.48 - 0.15} \quad (6)$$

4.2.5 Land Surface Emissivity

The region's LSE must be determined before the LST can be determined. The NDVI threshold approach is used to estimate LSE. Using the technique shown in equation 7, we have approximated the LSE using the FVC layer produced in the previous phase. LSE is a method for gauging the intrinsic properties of Earth's surface. Specifically, it evaluates how efficiently it transforms thermal or heat energy into radiant radiation. Bands 10 and 11 soil and vegetation emissivity are required for LSE estimate, and are listed in Table 3. Therefore, the LSE for bands 10 and 11 is calculated independently.

$$LSE = \varepsilon s \times (1 - FVC) + \varepsilon v \times FVC \quad (7)$$

where,

$\varepsilon s_{=}$ emissivity for soil

$\varepsilon v_{=}$ emissivity for vegetation

FVC = fractional vegetation cover

LSE= 0.971 × (1–*FVC*) + 0.987 × *FVC* (band10_*LSE*) and 0.977(1–*FVC*) + 0.989 × *FVC* (band11_*LSE*).

Table 3. Emissivity values

Emissivity	Band10	Band11
εs	0.971	0.977
εv	0.987	0.989

(**Source:** Sobrino et al., 1996; Sobrino et al., 2003; Zhao et al., 2009; Skokovic et al., 2014)

4.2.6 Mean and Difference of LSE

As stated in equations 8 and 9, we have combined the LSEs for bands 10 and 11 generated from the step-through mean and the difference between them.

$$\text{Mean of } LSE = m = \frac{LSE_{10} + LSE_{11}}{2} \tag{8}$$

$$\text{Difference of } LSE = \Delta m = LSE_{10} - LSE_{11} \tag{9}$$

4.2.7 Land Surface Temperature

Using the SW approach in equation 10, we have calculated an estimate for the LST. The LST is then expressed as a temperature in kelvin.

$$LST = TB_{10} + C_1(TB_{10} - TB_{11}) + C_2(TB_{10} - TB_{11})^2 + C_0 + (C_3 + C_4W)(1 - m) + (C_5 + C_6W)\Delta m \tag{10}$$

$$LST = T_{10} + 1.378(TB_{10} - TB_{11}) + 0.183(TB_{10} - TB_{11})^2 - 0.268 + (54.300 - 2.238(0.013))(1 - MeanLSE) + (-129.200 + 16.400(0.013))\,(Difference\ of\ LSE)$$

where,

TB_{10} and TB_{11} = Brightness temperature for bands 10 and 11
$C_0 - C_9$= Coefficients of the split window
m = Mean *LSE*
Δm= Difference of *LSE*
W = Atmospheric water-vapour content

Table 4. NDVI for soil and vegetation

NDVI for soil	0.15
NDVI for vegetation	0.48

To some extent, the emissivity of the surface is taken into account by the SW algorithm. Since much of the research area is blanketed in snow and the data used in this inquiry have a spatial resolution of 30 m, we have made the conservative assumption that the reflectance of snow is constant. Therefore, the algorithm designed by Sobrino et al. (1996) is used for the LST retrieval from Landsat-8 images. There has been a statistical comparison between the DEM and the LST maps from Landsat-8 and MODIS to establish a link between altitude and LST (Figures 6 and 7). The lapse rate for the study area over six dates is calculated using the regression equations.

5. RESULTS AND DISCUSSION

To some extent, the emissivity of the surface is taken into account by the SW algorithm. The 30 m spatial resolution of the data utilised in this investigation, together with the fact that snow covers most of the study region, leads to a reasonable assumption that snow's reflectivity is constant over the entire study area. There has been a statistical comparison between the DEM and the LST maps from Landsat-8 and MODIS to establish a link between altitude and LST (Figures 6 and 7). The lapse rate of the study region at six different times has been calculated using the regression equations. We compared the LST produced from the Landsat-8 and MODIS data to the air temperature observed at three distinct places since there are no on-the-ground techniques for determining LST. Figure 8 displays the consistency between Landsat-8 and MODIS observations of the relationships between air temperature and LST. The air temperature is usually greater than the LST. This study would be very useful for various studies in the Himalayan region because of worldwide interest in snow melt runoff as evident from several studies carried out by various researcher (Thapa, 1980; Bagchi, 1981; Agarwal et al., 1983; Jeyram et al., 1983; Seth, 1983; Jain, 2001; Haritashya, 2005; Tak and Keshari, 2020).

There is an inverse linear relationship between LST and altitude, as shown in the current research. Map depicting the day-to-day variations in LST on April 24, 2015, based on Landsat-8 data is shown in Figure 5(b). By contrasting the elevation map with the LST maps, we can see that the temperature in the study region follows the same general pattern as the topography. From the Landsat-8 and MODIS data for the chosen stations, the LSTs have been approximated for various dates using the lapse rate values and the linear regression formulae. Figure 8 displays the comparison between these LST values and their related air temperature values. Comparing LST and air temperatures (Figure 8) reveals a little discrepancy.

LST derived from Landsat-8 and MODIS datasets, as well as air temperature, are shown in a scatter format in Figures 6 and 7, respectively. As they demonstrate, there is a linear relationship between height and a temperature drop. The following may be said about the regression line equations based on the graphs:

$$Y = -aX + b \tag{11}$$

where X and Y are two numbers—the altitude and the temperature, respectively. With a negative slope (-a), the scatter plots of temperature and LST values show an inverse relationship with one another. For the scatter plots, we have also calculated the values of the coefficient of determination (R^2) (Figures 6 and 7).

Figure 5. Landsat-8 image from April 24, 2015 and the ASTER GDEM (a) and LST map obtained from it (b)

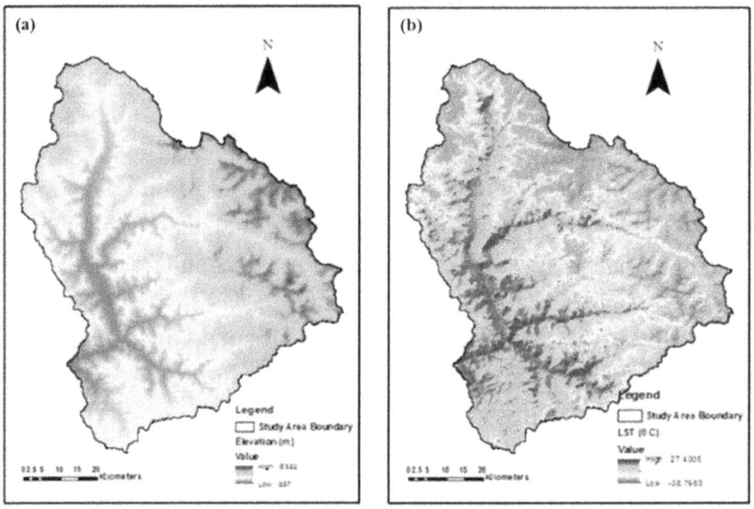

Figure 6. Temperature-altitude relationships (ASTER GDEM vs. LST map obtained from Landsat-8 images) are shown as scatter plots for a total of six time points

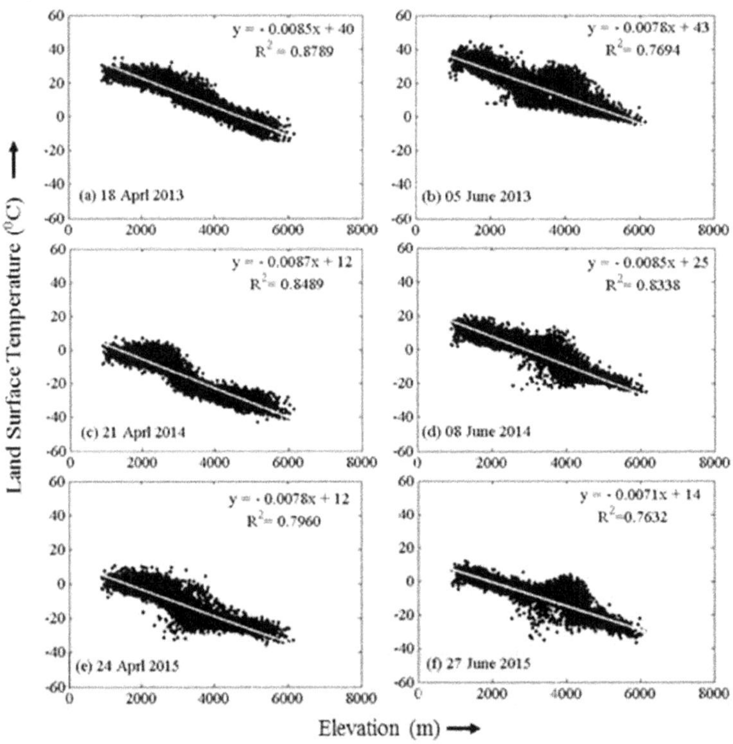

Figure 7. Temperature-altitude relationships (ASTER GDEM vs. LST map obtained from MODIS images) are shown as scatter plots for a total of six time points

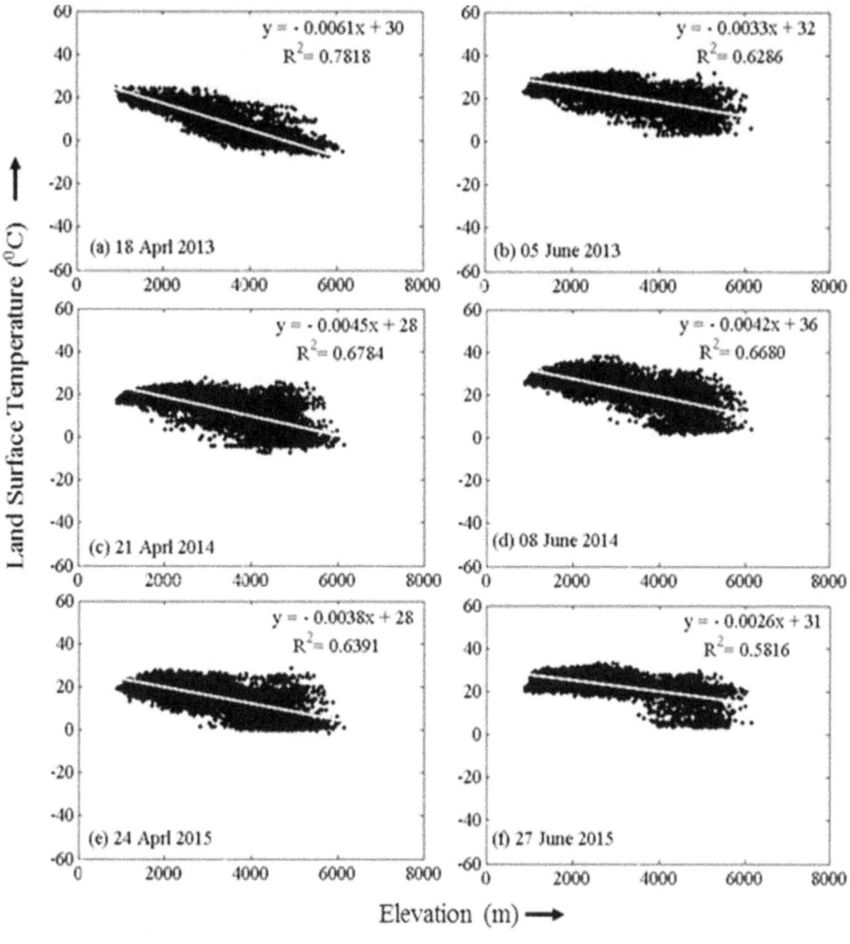

R^2 is a number between 0 and 1 that indicates how much of the original variance the connection explains. For there to be no correlation between the two variables, the R^2 value must be 0. LST readings from Landsat-8 images have been shown to have an R^2 of between 0.7632 and 0.8789, whereas MODIS LST maps have R^2 values ranging from 0.5816 to 0.7818. This proves the inverse relationship between LST and elevation. When compared to Landsat-8 LST data, MODIS LST data does not provide a satisfactory match with elevation values. Figures 6 and 7 show that the lapse rate is proportional to the equation's slope. In the research region, the TLR was calculated to be between 0.71°C/100 m and 0.87°C/100 m for the Landsat-8 LST data and between 0.26°C/100 m and 0.61°C/100 m for the MODIS LST values.

TLR changes with both latitude and time of year. According to (Bolstad et al., 1998), TLR estimated from the air temperature data shows slightly seasonal trends and averaged 4°C/km to 10ºC/km when the maximum temperature is considered. Minimum TLRs average between -3°C and 5.7°C, however this range is affected by the time of year. As the images used to calculate LST are from various times of year, a deviation of up to 0.15 °C/100 m (or 1.5 °C/km) is tolerable.

Figure 8. Difference between air temperature data, LST values from Landsat-8, and LST values from MODIS for six distinct days at three separate sites

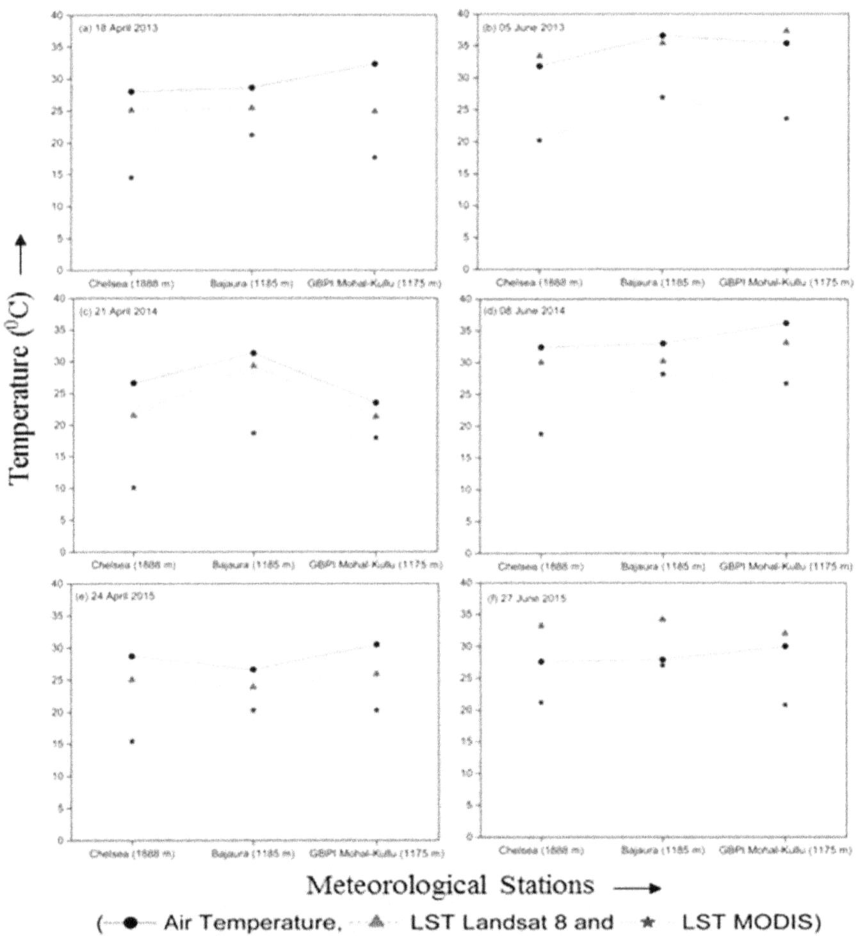

6. CONCLUSION

Researchers in the Himalayas have found that air temperature has a crucial role in the amount of runoff water from snowfall. Observational or point data are not indicative of the whole landscape since meteorological stations that gather air temperature data are seldom established in high terrains like the Himalayas. The air temperature is also measured manually, which might lead to inaccuracies. When this occurs, it's preferable to use LST maps derived from satellite imagery. As a continuous data source, LST maps are invaluable. When a proper radiometric calibration of the sensor channels is accomplished, they are free of human error since they are derived from satellite data, which automatically monitors radiance values.

We use the SW method to estimate LST for the Beas River basin, and we use elevation maps to determine the lapse rate of LST for 18 April, 5 June 2013, 21 April, 8 June 2014, and 24 April, 27 June 2015. For LST, the TLR ranged from 0.71°C/100m to 0.87°C/100m. Researchers studying snowmelt runoff may feel comfortable using the lapse rate range established in the current work. Seasonal shifts in LST TLR may also be estimated using this approach. Landsat-8 pictures and MODIS LST maps will be made freely available for research purposes, reducing the financial burden of accessing the data. Also, the aforementioned datasets may be accessed in a timeframe very close to real-time, which is a huge plus. It is possible to use LST data to calculate the amount of water that melts from snow, making this method particularly useful for scientific inquiry.

ACKNOWLEDGMENT

We gratefully thank the SAC MOSDAC for providing the air temperature data, the USGS earth explorer for giving the Landsat-8 data, and the NSIDC for supplying MODIS satellite data which were used in this study.

REFERENCES

Aggarwal, K. C., Kumar, V., & Dass, T. (1983). Snowmelt run-off for a catchment of Beas Basin. In *Proceedings of the First National Symposium on Seasonal Snowcover* (*Vol. 2*, pp. 43-63). Academic Press.

Bagchi, A. K. (1981). *Snowmelt runoff in Beas basin using satellite images* [PhD Thesis]. Department of Civil Engineering, University of Roorkee, India.

Becker, F. (1987). The impact of spectral emissivity on the measurement of land surface temperature from a satellite. *International Journal of Remote Sensing*, *8*(10), 1509–1522. doi:10.1080/01431168708954793

Becker, F., & Li, Z. L. (1990). Towards a local split window method over land surfaces. *International Journal of Remote Sensing*, *11*(3), 369–393. doi:10.1080/01431169008955028

Bolstad, P. V., Swift, L., Collins, F., & Régnière, J. (1998). Measured and predicted air temperatures at basin to regional scales in the southern Appalachian mountains. *Agricultural and Forest Meteorology*, *91*(3-4), 161–176. doi:10.1016/S0168-1923(98)00076-8

Coll, C., & Caselles, V. (1997). A split-window algorithm for land surface temperature from advanced very high resolution radiometer data: Validation and algorithm comparison. *Journal of Geophysical Research*, *102*(D14), 16697–16713. doi:10.1029/97JD00929

Haritashya, U. K. (2005). *Glacio-Hydrological Studies of the Gangotri Glacier Basin, Himalayas* [Doctoral dissertation]. Indian Institute of Technology, Roorkee, India.

Jain, S. K. (2001). *Snowmelt runoff modeling and sedimentation studies in Satluj basin using remote sensing and GIS* [Unpublished PhD thesis]. University of Roorkee.

Jain, S. K., Goswami, A., & Saraf, K. (2008). Determination of Land Surface Temperature and Its Lapse Rate in the Satluj River Basin Using NOAA Data. *International Journal of Remote Sensing*, *29*(11), 3091–3103. doi:10.1080/01431160701468992

Jeyram, A., Tiwari, R. S., & Sharma, K. P. (1983). Snowmelt Runoff Using Landsat Imageries. *Proceedings of the First National Symposium on Snowcover, 2*, 11–21.

Jiang, G. M., Zhou, W., & Liu, R. (2013). Development of split-window algorithm for land surface temperature estimation from the VIRR/FY-3A measurements. *IEEE Geoscience and Remote Sensing Letters*, *10*(4), 952–956. doi:10.1109/LGRS.2013.2255859

Jiménez-Muñoz, J. C., Sobrino, J. A., Skoković, D., Mattar, C., & Cristóbal, J. (2014). Land surface temperature retrieval methods from Landsat-8 thermal infrared sensor data. *IEEE Geoscience and Remote Sensing Letters*, *11*(10), 1840–1843. doi:10.1109/LGRS.2014.2312032

Key, J., & Haefliger, M. (1992). Arctic ice surface temperature retrieval from AVHRR thermal channels. *Journal of Geophysical Research*, *97*(D5), 5885–5893. doi:10.1029/92JD00348

Key, J. R., Collins, J. B., Fowler, C., & Stone, R. S. (1997). High-latitude surface temperature estimates from thermal satellite data. *Remote Sensing of Environment*, *61*(2), 302–309. doi:10.1016/S0034-4257(97)89497-7

Lakshmi, V., Czajkowski, K., Dubayah, R., & Susskind, J. (2001). Land surface air temperature mapping using TOVS and AVHRR. *International Journal of Remote Sensing*, *22*(4), 643–662. doi:10.1080/01431160050505900

Rajeshwari, A., & Mani, N. D. (2014). Estimation of land surface temperature of Dindigul district using Landsat 8 data. *International Journal of Research in Engineering and Technology*, *3*(5), 122–126. doi:10.15623/ijret.2014.0305025

Ruescas, A. B., Danne, O., Fomferra, N., & Brockmann, C. (2016). The Land Surface Temperature Synergistic Processor in BEAM: A Prototype towards Sentinel-3. *Data*, *1*(3), 18. doi:10.3390/data1030018

Saraf, A. K., Mishra, B. P., Choudhury, S., & Ghosh, P. (2005). Digital elevation model (DEM) generation from NOAA–AVHRR night-time data and its comparison with USGS-DEM. *International Journal of Remote Sensing*, *26*(18), 3879–3887.

Sattari, F., & Hashim, M. (2014). A brief review of land surface temperature retrieval methods from thermal satellite sensors. *Middle East Journal of Scientific Research*, *22*(5), 757–768.

Seth, S. M. (1983). Modelling of daily snowmelt runoff during pre-monsoon month for Beas basin up to Manali. In *Proceedings of the First National Symposium on Seasonal Snowcover* (*Vol. 2*, pp. 104-115). Academic Press.

Singh, P. (1991). A temperature lapse rate study in Western Himalayas. *Hydrology Journal Indian Association of Hydrologists*, *14*, 156–163.

Skoković, D., Sobrino, J. A., Jimenez-Munoz, J. C., Soria, G., Julien, Y., Mattar, C., & Cristóbal, J. (2014). *Calibration and Validation of land surface temperature for Landsat8-TIRS sensor*. Land Product Validation and Evolution.

Sobrino, J. A., El Kharraz, J., & Li, Z. L. (2003). Surface temperature and water vapour retrieval from MODIS data. *International Journal of Remote Sensing*, *24*(24), 5161–5182. doi:10.1080/0143116031000102502

Sobrino, J. A., Li, Z. L., Stoll, M. P., & Becker, F. (1996). Multi-channel and multi-angle algorithms for estimating sea and land surface temperature with ATSR data. *International Journal of Remote Sensing*, *17*(11), 2089–2114. doi:10.1080/01431169608948760

Sobrino, J. A., & Raissouni, N. (2000). Toward remote sensing methods for land cover dynamic monitoring: Application to Morocco. *International Journal of Remote Sensing*, *21*(2), 353–366. doi:10.1080/014311600210876

Stroeve, J., & Steffen, K. (1998). Variability of AVHRR-derived clear-sky surface temperature over the Greenland ice sheet. *Journal of Applied Meteorology*, *37*(1), 23–31. doi:10.1175/1520-0450(1998)037<0023:VOADCS>2.0.CO;2

Tak, S., & Keshari, A. K. (2020). Investigating mass balance of Parvati glacier in Himalaya using satellite imagery based model. *Scientific Reports*, *10*(1), 1–16. doi:10.1038/s41598-020-69203-8 PMID:32699284

Tang, H., & Li, Z. L. (2014). Land surface temperature retrieval from thermal infrared data. In *Quantitative Remote Sensing in Thermal Infrared* (pp. 93–143). Springer. doi:10.1007/978-3-642-42027-6_5

Thapa, K. B. (1980). *Analysis for snow-melt runoff during premonsoon months in Beas basin using satellite imageries* [ME Dissertation].

Ulivieri, C. M. M. A., Castronuovo, M. M., Francioni, R., & Cardillo, A. (1994). A split window algorithm for estimating land surface temperature from satellites. *Advances in Space Research*, *14*(3), 59–65. doi:10.1016/0273-1177(94)90193-7

Wan, Z. (2006). *MODIS land surface temperature products users' guide*. Institute for Computational Earth System Science, University of California.

Zhao, S., Qin, Q., Yang, Y., Xiong, Y., & Qiu, G. (2009). Comparison of two split-window methods for retrieving land surface temperature from MODIS data. *Journal of Earth System Science*, *118*(4), 345–353. doi:10.1007/s12040-009-0027-4

Chapter 8
Future of Energy Transition Relies on Prosumer-Based Smart Grid-Integrated Renewable Distributed Generation System

Manish Kumar
ICFAI Business School, Hyderabad, India

Bharat Singh Patel
https://orcid.org/0000-0001-5155-997X
Prestige University, Indore, India

Cherian Samuel
IIT (BHU), Varanasi, India

ABSTRACT

The availability of reliable power without interruption is critical for a fast and sustained economic growth. No country can develop and sustain itself beyond subsistence levels unless it has access to modern energy (electrical) services at a minimum. According to studies, despite the country's vast renewable energy potential, access to alternate unconventional types of energy is extremely limited. More than half of the world lack access to electricity and continue to rely primarily on traditional energy sources. This may be significantly advanced by incorporating prosumers in present smart-grid infrastructure to generate power with renewable distributed generation technologies to meet expanding demand while minimising environmental risk. The authors assessed the current state of awareness of the potential positive implications of the energy transition for various communities and socioeconomic groups on the frontlines of the change using a fast growing corpus of academic and policy research.

DOI: 10.4018/979-8-3693-1396-1.ch008

1. INTRODUCTION

New emerging perspectives are required to face the challenges that global warming, limited reservoirs, and urbanisation will provide in the next decades. The energy system is undergoing significant transformation. Traditional outdated models will be disintegrated. Electricity demand will grow at an unprecedented rate, particularly in emerging countries where the role of metropolitan and smart-cities will be vital to their success. Furthermore, countries and their energy systems are under pressure to cut greenhouse gas emissions in order to meet the Paris Climate Agreement (Bodansky, 2016). The energy transition will be important in achieving this, with an increasing number of scattered renewable energy sources. Indeed, a new unconventional, flexible distributed generation model based on renewable distributed generation is already displacing energy systems that have relied for decades on a traditional, conventional energy-based generating and distribution paradigm (Akorede et al., 2010). The new model of unconventional renewable distributed generation is gaining traction in scientific discussions, thanks in large part to an increase in power distribution system digitisation. The modern smart-grid allows for more data accessibility and unprecedented network stability control, particularly in the face of periodic disruption and customer interaction (Ahmad et al., 2022). The electrical generation and distribution system cannot function properly without the active participation of customers as an important stakeholder, who are also emerging into "prosumers" (both producers and consumers) (Gautier et al., 2018). Despite this progress, there are challenges to managing intermittent power supply, sociopolitical acceptance of prosumers, and the development of the smart-grid integrated renewable distributed generation system, all of which can be addressed through an appropriate regulatory framework that takes this transition into account.

One critical component of this change is that not everyone will benefit equally unless clear steps to encourage an equitable transition are implemented. International climate treaties have long recognised the significance of incorporating equitable distribution of energy access facilities into energy and climate policy (Markard, 2018).

While the concept's technological feasibility and cost-effectiveness are debatable. However, there are other gaps in energy transition-related distribution inequalities that we can see, such as demand side management, activities needed to assist consumers and communities in transitioning, engagement of prosumers as decision makers, and development of an effective energy model. As a result, there is enormous opportunity to raise collective awareness of the negative consequences of the energy transition, as well as opportunities for scholars to perform basic scientific study to influence future policymakers.

The interrelated theories on energy transition in smart-grid integrated renewable distributed generation systems, as well as how the energy transition will effect prosumers, were investigated in this study. Our goal is to raise a series of topics that, when expanded, continuous, and properly assessed, can provide critical insights into why, how, and where to establish energy programmes, as well as how to integrate these programmes into wider energy and climate policy efforts. We feel that a lack of infrastructure makes both the transition to decentralised power generation and the possibilities for prosumer engagement extremely challenging. While investing in research & development will result in the establishment of policies to solve these concerns, theories on such policy efforts and the extent to which policymakers ease challenges are still missing.

2. ENERGY TRANSITION

An energy transition is the transfer from one main energy resource or collection of resources to another. Examples include the late-nineteenth-century substitution of whale oil for paraffin and the move from wood to coal during the industrial revolution. The most current energy transition act is distinguished by a shift away from traditional energy sources and towards renewable energy sources such as wind, solar, biomass, and other renewable energy sources. Although energy transitions in which one resource completely replaces another are rare, it is more normal for one resource to begin with a little percentage and progressively develop to a major share of the energy mix. One interpretation is a shift in energy consumption from secondary to primary for a specific energy supply or technology. The literature on energy transition is vast and growing (Markard, 2018; Solomon & Krishna, 2011; Blazquez, 2020). Previously, research focused on the rate of the energy transition; however, the literature has expanded to investigate variations in transition trajectories.

Increasing the capacity of smart-grid integrated renewable distributed generation has the potential to lower the levelized cost of energy, which represents the net present value of the electricity unit cost throughout the lifetime of the power generation unit (Phuangpornpitak & Tia, 2013). This is due to a significant decrease in operational costs when compared to a baseline that assumes no renewable distributed generation units are installed. Furthermore, it has been demonstrated that implementing a demand-side management plan has synergistic effects, resulting in further savings (Puttamadappa & Parameshachari, 2019). Reduced operating expenses translate into significant overall cost savings. Following the discovery of the scientific result, we conclude that it is critical to develop modern smart and flexible power distribution networks, energy policies, and sociopolitical acceptance in order to balance the presence of prosumers with increasing renewable distributed generation technology penetration in the current conventional grid network. Such strategic policy implications would enable a formally organised prosumer group to pick the best mix of customers to participate in the new energy models developed by a formally organised prosumer group (Rosen & Madlener, 2016). Figure 1 demonstrates the impacts of an energy transition in which the proportionate contribution of renewable energy sources to conventional energy sources is increased in order to limit the potential for carbon emissions. This indicates that increasing conventional energy consumption results in higher carbon emissions, but growing renewable energy consumption results in fewer carbon emissions.

Figure 1. Energy transition to reduce carbon emissions through increased use of renewable energy options

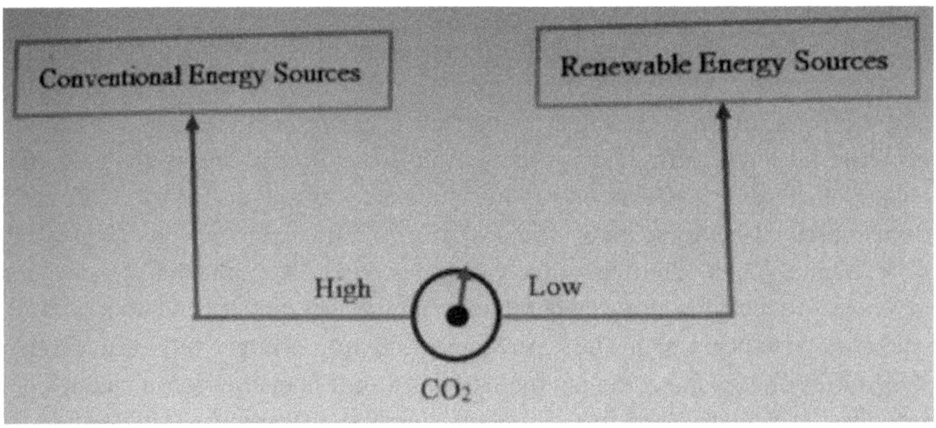

A number of scholars have already addressed broad sustainability concerns about the energy transition, such as climate policy, which takes into account popularly held normative assumptions from well-established systems in order to change the more sustainable energy system in the long run (Solomon & Krishna, 2011). Working on an energy transition may appeal to a diverse group of scholars. Scientific researchers and engineers value the inclusion of socio-political analysis; energy planners value valuable energy transition studies and their uncertainty; energy economists are influenced by emerging phenomena and institutional change; and socialists value the emphasis on material aspects, such as technologies and infrastructures.

Certain features of energy transitions provide severe theoretical, strategic, and policymaking challenges. Five of these qualities are crucial in emphasising some of the phenomenon's general aspects (Markard, 2018).

First and foremost, public policies are crucial in energy transitions. Energy concerns, as well as long-term sector growth goals and specific strategies to attain them, are frequently stated in policy programmes, white papers, or other comparable materials.

Another defining aspect is the high amount of complication and unpredictability. Sustainability concerns are wicked problems because they are poorly defined, potential remedies and effects are unknown, societal choices are vague and/or diversified, and technological-economic growth and political processes are frequently unpredictable. As a result, politicians and other decision-makers face unanticipated consequences, opposition and disputes, rapid shifts, challenges, and so on.

In a third point, we stressed the value-laden benefits of sustainability transitions. Objectively defining transition goals is impossible. Instead, they rely on public preferences for environmental, social, and economic factors. Concerns with valuation become increasingly apparent when trade-offs are involved, such as carbon emissions vs nuclear power hazards.

The fourth point demonstrates that energy transitions are contentious. There are winners and losers in every transformation. Furthermore, as a result of the second and third characteristics, there are varying views about what the problem is, how important it is, and how to address it. As a result, energy transitions are characterised by fierce battles over the intended level of energy, the transformation's potential, as well as its future implications and guidelines.

A fifth factor is condition dependency; as a result of varying social benefits, political systems and choosing the best one, today's industries and existing infrastructures, available natural resources, well-defined practises, and so on, energy transitions can differ significantly across areas and sectors. As a result, different transition paths in different places in a given sector can be predicted.

An important transition is the continual change in the electrical sector. This demonstrates their long-term growth as improved renewable energy capacity, increased energy efficiency, reduced carbon emissions, and reduced nuclear hazards, among other things. 'Clean' electricity, because of its versatility, can also play a major part in energy transitions in nearby industries such as heating and transportation.

3. SMART-GRID

The implementation of smart-grids will result in a shared platform for modern technology, customer responses, and decision-makers. According to the IEEE organisation, smart-grid is a revolutionary endeavour that necessitates new communications and control capabilities, energy sources, generating models, and adherence to cross-jurisdictional regulatory systems (Yan et al., 2012). A smart-grid may control instantaneous demand-supply balance by utilising modern communication technologies such as the Internet of Things and providing real-time data from the electricity system. One significant distinction between conventional grid and smart-grid is their information sharing; smart-grid allows for two-way information communication, whilst conventional grid only allows for unidirectional exchange. Smart-grid technology allows users to link numerous energy sources to grids and enhance the percentage contribution of renewable energy sources, ushering in a new paradigm of demand-side energy supply management. Energy supply is no longer conventional, and the role of decision science in managing the complexity of multi-energy sources is growing. We may see in Figure 2 the integration of various energy options such as fossil fuel-based thermal power plants, renewable energy sources (RES), and nuclear power plants to meet the diverse needs of consumers such as industrial demand, residential demand, and charging stations for electric vehicles (EV). If we examine this critically, we can see that following the ground level implementation of smart-grid technology, it will generate a number of future scientific research questions, particularly for management to discover the best solution among technological, economic, and social scales.

Figure 2. Smart-grid enables the connection of many energy choices to benefit users at various levels

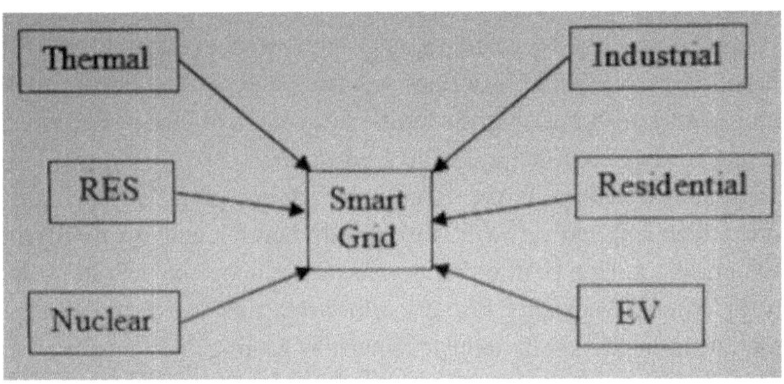

The contemporary electricity distribution network is the product of recent notable urbanisation and infrastructural growth in various parts of the world. Despite operating in different geographical regions, utility companies use comparable technology. The foundation and evolution of a modern energy system are influenced by political, economic, and geographical factors. Despite these changes, the fundamental structure of the traditional energy system has not changed. The energy sector has operated with clear boundaries between its primary subsystems, namely generation, transmission, and distribution, since the introduction of smart-grid. This results in expansion, with many elements such as automation, modification, and upgradation emerging at each level (Fang et al., 2011). The existing energy utility grid is a unidirectional system in which power plants at the top of the chain ensure power delivery to users at the bottom of the chain. Because systems operate in a one-way fashion, the source does not have real-time access to information on the service characteristics of the terminating point. As a result, the power distribution system has been the subject of intense research and technological analysis in order to endure the largest prediction of the peak demand time over the actual load of that duration. Peak demand occurs infrequently, implying that a system designed with peak demand in mind is inefficient. As a result of the high electricity use and low distribution network investment, system stability may suffer. A case of irregularity in the power distribution network or any unexpected fast surge in demand would result in component failures and may trigger catastrophic blackouts due to the lack of a safe margin (Andersson et al., 2005). By investing in infrastructure upgrades, power distribution companies have established a range of facilities to protect distribution networks through troubleshooting and prevention measures (Hassani et al., 2022).

By embedding smart-grid technology into the infrastructure of electrical power utility grids (Dileep, 2020),

(1) Improves utility grid reliability by minimising power quality problems and the repercussions and probability of extensive blackouts.
(2) Allows for future advancements and efficiency to be realised.
(3) Lowers consumer electricity bills by applying downward pressure.
(4) Energy users' affordability is maintained.
(5) Consumers have more options for supply and information.
(6) Renewable distributed generation that is renewable and nonconventional is integrated.
(7) Enhances security by lowering the consequences and likelihood of natural disasters and man-made assaults.
(8) Increase the penetration of alternate energy producing sources.
(9) Reduces the number of deaths and injuries caused by utility grid-related events, hence reducing safety concerns.
(10) Integrates electric vehicles as power generation and storage devices, revolutionising the transportation sector.
(11) Reduces energy losses and waste to increase overall efficiency.
(12) Smart-grids reduce pollution by lowering greenhouse gas and carbon particle emissions and deliver cleaner power by encouraging the deployment of more renewable distributed generations.

According to the findings of the preceding study, smart-grids enable the employment of modern communication technologies. Modernization of technologies and information systems involves the redesign of traditional power distribution networks. Because of its importance in economic growth, making large modifications to the conventional power distribution system will require careful study and alternative selection to reduce any additional long-run investment costs. Environmental scientists, on the other hand, are unanimous in their belief that human-caused greenhouse gas emissions are dangerous and will harm our ecosystem. In terms of climate change, smart-grid can help to reduce it by implementing behavioural and institutional changes alongside technological shifts (Lamnatou et al., 2022). A smart-grid integrated renewable distributed generation system reduces carbon footprints and greenhouse gas emissions, so contributing to climate change mitigation. Through effective demand side management schemes, smart-grid aids in energy conservation and reduces energy consumption. By considering the seriousness of climate change in smart-grid planning and implementation, electrical stakeholders may argue that smart-grid will not contribute to the reduction of greenhouse gas emissions and will have no impact on climate change-related damage due to grid vulnerability. Accurate data should be gathered to determine energy losses, waste, and effective load management. After examining all power system factors, the grid utility's position must be assessed, and a variety of control possibilities emerge. The future of sustainable green energy will be largely determined by the energy mix of renewable distributed generation, nuclear power plants, and traditional fossil-fuel based supply with the lowest possible carbon emissions and storage. Combining several sources of energy through different generation modules will make the system more complex, making it more difficult to operate the power system and run it at constant output. It is extremely difficult to manage and monitor the cost-effectiveness of a power system in the absence of a smart-grid. So, we can say future of the power system is with smart-grid technology.

4. RENEWABLE DISTRIBUTED GENERATION

The rapid expansion of renewable distributed generation systems has been witnessed all over the world as a result of the ongoing depletion of fossil fuel energy sources, the scarcity of adequate transmission channels, and the gradually rising global temperature. The concept of renewable distributed generation developed from the use of renewable energy sources in distributed power generation (Abdmouleh et al., 2017). Because of the environmental benefits and the flexibility of onsite power generation to increase power quality and dependability, this rule became effective. Because of their benefits, this is given significant consideration when formulating global energy strategy. The utilisation of renewable distributed generation units benefits deferring gearbox investment, lowering primary energy use, cutting greenhouse gas emissions, and thereby lowering global warming. Because renewable distributed generation systems are becoming more prevalent in distribution systems, proper renewable distributed generation unit placement in the distribution system is becoming increasingly crucial.

The demand for more adaptable electrical systems, changing regulatory and economic scenarios, energy savings, and environmental effect are pushing the development of renewable distributed generation systems, which are likely to play a significant part in the future power system (Manditereza & Bansal, 2016). The use of small-modular residential or commercial onsite generation units is one of the principal uses of the renewable distributed generation system. Renewable distributed generation technology can be chosen to meet the load requirements of the client at the lowest possible cost. Adequate management is essential for power grid integration systems to last longer and perform better (Dkhili et al., 2020).

A distributed generating system is a small-scale, renewable energy-based energy conversion system with power ratings ranging from 500KW to 1MW (Cardell & Tabors, 1997). This can be used to replace or improve the present power producing system. The location of energy generating units in this system is motivated by and close to client needs. Renewable distributed generation units are now being directly integrated into power distribution systems to meet randomly distributed demand more reliably. Renewable distributed generation units would be connected to the same transmission system as central stations to ensure reliability (Mahmud & Zahedi, 2016). For industrial, commercial, and residential purposes, it becomes a viable alternative to standard power generation systems (Adefarati & Bansal, 2019). When structuring the distribution of economic power, this makes selecting the best option from a plethora of options more difficult.

The following factors contribute to the significant adoption of renewable distributed generation in today's power market:

- By installing several units of varied capacities at ideal sites or adjacent to demand areas, the renewable distributed generation system can construct a new, enhanced power system with minimal transmission and distribution losses.
- We can simply assemble renewable distributed generation units anywhere we need to generate electricity.
- Renewable distributed generation has a substantial advantage in situating its unit where there is a high demand due to the decentralisation of the power system. As a result, it enables flexible installation to fulfil the needs of customers.

- Renewable distributed generation is in favour of demand control and "peak load shaving."
- In the event of an emergency or system failure, renewable distributed generation can be used as a backup power source on-site.
- Renewable distributed generation allows for the release of gearbox capacity, system stability, and the necessary spinning reserve.
- In comparison to large units, a location on the ground level is easier to find for the installation of a small unit.
- It aids in the utilisation of the potential of local renewable energy sources.

It is best to examine the system impact when employing renewable distributed generation units on a big scale during the planning stage. Engineers in distribution systems must investigate how the future of renewable distributed generation technologies will affect the modern integrated smart-grid (Shaukat et al., 2018). Figure 3 depicts the integration of a variety of renewable distributed generation sources based on the interests of the consumer or prosumer. It is dependent on the availability of a sufficient number of alternatives and their competitively lower per unit generation costs in order to entice consumers or prosumers to invest in the best option. This provides customers with the option and freedom to move to greater renewable energy sources, which can aid in the reduction of carbon emissions. Strategic socio-political activities will assist the country in raising public understanding and acceptance of renewable distributed energy. This will improve user accountability by transforming their behaviours from consumers to prosumers.

Renewable distributed generation planning entails determining an appropriate installation location and size that fulfils technical, financial, and environmental restrictions. It also requires choosing the most dependable and efficient equipment that is digitally networked through the smart-grid system. Planners place a larger emphasis on the financial measure because all variables are measured in terms of money (Ehsan & Yang, 2018). Power consumption has increased as a result of increased population and better living conditions. Human existence now would be impossible to imagine without electricity; indeed, we cannot envision surviving without electricity in the modern day. A new investment approach will be necessary to fulfil the increasing stochastic demand for electricity (Montoya-Bueno et al., 2016). A planner will have the best strategy to utilise their resources or avoid losses in order to fulfil the load demand in an environmentally friendly manner for long-term planning (more than a year). A planner can reduce power losses while simultaneously controlling the irregular nature of load demand and variable renewable energy sources by adopting renewable distributed generation technologies as an investment strategy.

Figure 3. Renewable distributed generation enables access to variable energy sources in small-scale decentralised generation capacity

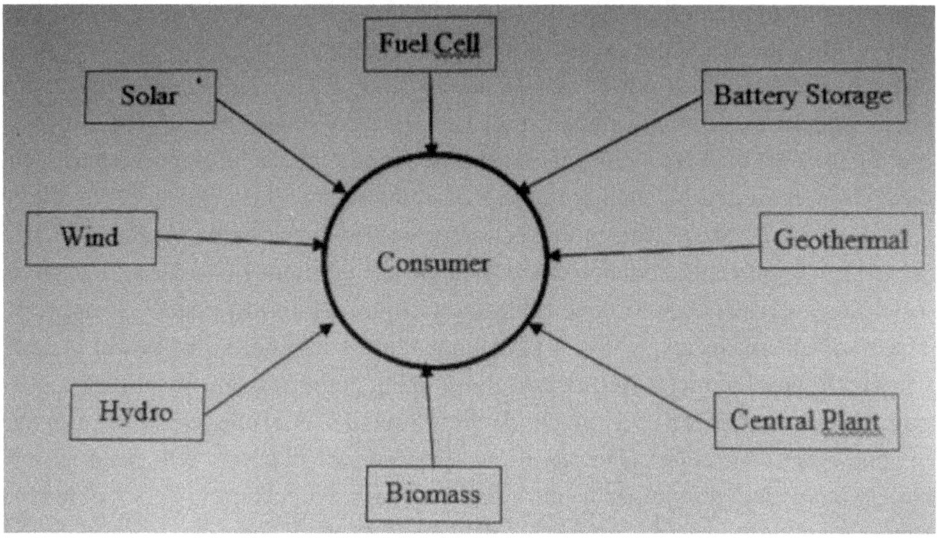

When considering the renewable distributed generation system and its effects on the distribution system, we should prioritise the best renewable distributed generation system planning, because an unplanned renewable distributed generation system by the decision maker may result in losses and a drop in reliability level (Roggatz et al., 2020). It is vital to select the appropriate instrument and technique in order to optimise the renewable distributed generation system and save project costs while meeting all technical criteria (Fathabad et al., 2020). A planner or decision-maker should consider the long term rather than only the short term or aiming for an instant result. As a result, using existing design and technology, they will be able to examine the long-term effects of renewable distributed generating adoption in the electrical market. Long-term and short-term planning will increase the stability of the power generation system and reduce the probability of future failure. Given the fact that the future is absolutely unknown, we can only make a few educated guesses or assumptions based on our prior experiences. As a result, renewable distributed generation power generation should be able to meet variable demand and provide the best possible power flow under stochastic future demand.

5. PROSUMERS' INVOLVEMENT

The paper emphasises the author's unique vision on the future of energy transition, highlighting the relevance of prosumers' participation. By taking the first steps towards a decentralised power market, the existing power system is progressively transitioning to a consumer-oriented model, in which consumers will have peer-to-peer trading freedom with renewable distributed generation. Because of the advancement of modern technologies, such as Internet of Things enabled Blockchain technology (Hahnel et al., 2020), power trading is now possible with a decentralised, open, and secure market. Simultaneously, a new type of virtual money based on blockchain technology was launched for renewable energy transactions in a smart-grid network. These advancements are swiftly establishing a future internet-based power system, which will pose an increasing risk to the existing conventional model that energy policymakers have been operating under for a long time. To fully realise the promise of renewable distributed generation capacity, energy planners will need to rely largely on modern regulatory frameworks and market procedures. This will lead to a significantly different future market than what we were experiencing previously.

Figure 4. Involvement of prosumers in energy sharing with dual role

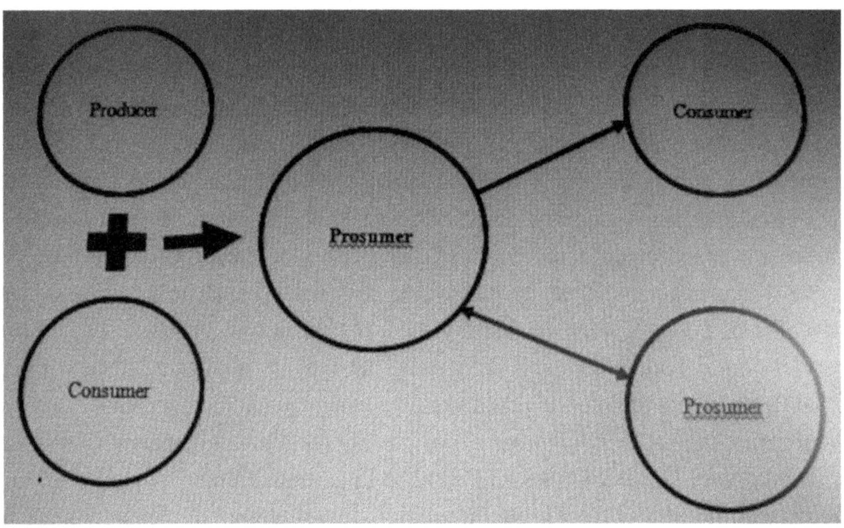

The function of the prosumer is not only to consume power, but also to share excess electricity generated by renewable distributed generation with the smart-grid or the consumer community (Espe et al., 2018). Figure 4 demonstrates the presence of prosumers in a two-way energy and information flow in a smart-grid integrated renewable distributed generation system, as opposed to the typical one-way flow to consumers. We can see the function of the prosumer as a producer as well as a consumer, and any excess energy capacity that prosumers have can be shared with consumers or other prosumers via energy sharing communities enabled by smart-grid networks. This also allows prosumers to generate money by sharing their excess capacity through peer-to-peer trading and to participate in the energy market as a decision-makers. This enables prosumers to appreciate their social significance and to minimise losses on the technology front through economic investments. This phenomenon helps to address the environmental, social, and economic difficulties brought on by increased energy demand. In order to effectively manage prosumers in energy sharing groups, smart-grid allows customers to form communities based on several characteristics such as power consumption behaviour (Zafar et al., 2018).

Energy could be shared by consumers through a local energy sharing group with a control centre linked to the smart-grid. Local power control system functions include keeping statistics on energy generation and consumption in order to respond to user demands effectively. If a member of a community sharing group need more energy than usual but does not wish to connect to the grid, they can send a request to their local control centre, which will meet their extra demand based on availability.

Modern information and communication technology, as well as optimisation approaches, promote energy sharing among prosumers, grid, and consumers. There are options for sharing information between prosumers and communities, the bulk of which rely on wireless connection. Various optimisation algorithms have been developed in the literature for optimising energy trade (Parag & Sovacool, 2016).

6. POLICY IMPLICATIONS

Scholars from several disciplines are becoming increasingly interested in modern smart-grid integrated renewable distributed generation related to continuous energy transfer. We examined current research on prosumers' participation in smart-grid integrated renewable distributed generation systems in this study, which helps us understand the nature of challenges, accessibility gaps in modern technologies, and effective efforts to improve adaptability in assisting communities to manage large economic, sociopolitical, and cultural changes.

One major implication is that the benefits of shifting to renewable distributed power have far-reaching repercussions beyond the fate of energy planners. This is not to argue that attention to energy planners and policymakers is unimportant; rather, it is to emphasise that other segments of society will play a role in the energy transition, which is primarily public and political in nature.

The study reported in this paper verifies significant earlier findings, but further research is needed to gain a comprehensive understanding of the nature and extent of benefits from the energy transition, as well as how they can be interpreted differently in different society groups. In general, extensive assessments of climate change mitigation and adaptation strategies, as well as additional measures to accelerate the global energy transition, are required. Essentially, this form of assessment focuses on efficiency and cost effectiveness, with multidimensional issues tackled only as a last resort, if at all.

Among the main scientific difficulties that must be tackled further is making renewable distributed generation systems and energy efficiency technologies more realistic, particularly for people with limited financial resources. A distinct requirement is the selection and examination of alternative models, such as public and non-profit efforts, as well as public-private partnerships, whose explicit goal is to make these technologies available to all. Furthermore, there have been few formal, thorough studies of energy aid efforts to reduce energy insecurity. Important questions remain, such as how and to what extent developed-country energy assistance, policy revision, and technology support will help to alleviate energy poverty in developing countries; what types of initiatives will help to increase the widespread availability and accessibility for a greater number of needy residents; and how energy policy implications can be effectively combined with housing development and health awareness programmes to reach out to the most vulnerable residents. More comprehensive and diversified research can help tackle these problems while also informing decisions on how to allocate efforts and optimise trade-offs.

Any large-scale change during the energy transition is inevitably riddled with obstacles. Every country and global organisation has and is creating strategies to restructure their energy economy in order to reduce greenhouse gas emissions; nevertheless, these strategies must have policy implications. The scientific research group can assist in ensuring that activities are evidence-based and based on the experiences of those who will be most impacted: individuals, households, and communities. This is a quantifiable difficulty that will demand an inter-disciplinary study utilising a range of research methodologies carried out at multiple levels. The challenge is large, but it will provide opportunities for improvement in scientific research and policy implications.

7. CONCLUSION

This research looked at emerging countries' energy policy and infrastructure upgrades in conjunction with energy transition. According to previous studies, energy is a key factor for a country's economic growth; however, a large number of developing-country residents do not have adequate access to clean energy, and many of them still rely on old-fashioned diesel generators as backups due to frequent power outages. As a result, there is an urgent need for policy adjustments in order to further expand developing economies' energy infrastructure and prioritise their activity towards improving the living conditions of the poorer sections through rural electrification schemes, in order to ensure rapid economic growth. This can be significantly improved by making significant use of the country's plentiful natural renewable energy resources. To reduce losses, efficient utilisation of renewable energy resources necessitates participation by prosumers in a smart-grid integrated renewable distributed generation system.

The current capacity of world to generate electricity from renewable distributed generation shows that a significantly very high capacity can be added with an effective use of smart-grid integrated renewable distributed generation in a bigger participation of prosumers. However, in order to advance smart-grid integrated renewable distributed generation capacities and accelerate the energy transition, developing countries' existing energy infrastructure requires immediate technical and financial assistance from proactive developed nations, particularly increasing industrialisation in developing countries.

Finally, for the system to be sustainable, decision-makers must be proactive and committed to the country's economic progress through the use of environmentally friendly clean energy, and actions must be continuously evaluated to determine their outcomes. The system must be adequately monitored in order for policy implications to be successful.

REFERENCES

Abdmouleh, Z., Gastli, A., Ben-Brahim, L., Haouari, M., & Al-Emadi, N. A. (2017). Review of optimization techniques applied for the integration of distributed generation from renewable energy sources. *Renewable Energy*, *113*, 266–280. doi:10.1016/j.renene.2017.05.087

Adefarati, T., & Bansal, R. C. (2019). Reliability, economic and environmental analysis of a microgrid system in the presence of renewable energy resources. *Applied Energy*, *236*, 1089–1114. doi:10.1016/j.apenergy.2018.12.050

Ahmad, T., Madonski, R., Zhang, D., Huang, C., & Mujeeb, A. (2022). Data-driven probabilistic machine learning in sustainable smart energy/smart energy systems: Key developments, challenges, and future research opportunities in the context of smart grid paradigm. *Renewable & Sustainable Energy Reviews*, *160*, 112128. doi:10.1016/j.rser.2022.112128

Akorede, M. F., Hizam, H., & Pouresmaeil, E. (2010). Distributed energy resources and benefits to the environment. *Renewable & Sustainable Energy Reviews*, *14*(2), 724–734. doi:10.1016/j.rser.2009.10.025

Andersson, G., Donalek, P., Farmer, R., Hatziargyriou, N., Kamwa, I., Kundur, P., Martins, N., Paserba, J., Pourbeik, P., Sanchez-Gasca, J., Schulz, R., Stankovic, A., Taylor, C., & Vittal, V. (2005). Causes of the 2003 major grid blackouts in North America and Europe, and recommended means to improve system dynamic performance. *IEEE Transactions on Power Systems*, *20*(4), 1922–1928. doi:10.1109/TPWRS.2005.857942

Blazquez, J., Fuentes, R., & Manzano, B. (2020). On some economic principles of the energy transition. *Energy Policy*, *147*, 111807. doi:10.1016/j.enpol.2020.111807

Bodansky, D. (2016). The Paris climate change agreement: A new hope? *The American Journal of International Law*, *110*(2), 288–319. doi:10.5305/amerjintelaw.110.2.0288

Cardell, J., & Tabors, R. (1997). Operation and control in a competitive market: Distributed generation in a restructured industry. *The Energy Journal (Cambridge, Mass.)*, *18*(1_suppl), 18. doi:10.5547/ISSN0195-6574-EJ-Vol18-NoSI-6

Dileep, G. J. R. E. (2020). A survey on smart grid technologies and applications. *Renewable Energy*, *146*, 2589–2625. doi:10.1016/j.renene.2019.08.092

Dkhili, N., Eynard, J., Thil, S., & Grieu, S. (2020). A survey of modelling and smart management tools for power grids with prolific distributed generation. *Sustainable Energy. Grids and Networks*, *21*, 100284.

Ehsan, A., & Yang, Q. (2018). Optimal integration and planning of renewable distributed generation in the power distribution networks: A review of analytical techniques. *Applied Energy*, *210*, 44–59. doi:10.1016/j.apenergy.2017.10.106

Espe, E., Potdar, V., & Chang, E. (2018). Prosumer communities and relationships in smart grids: A literature review, evolution and future directions. *Energies*, *11*(10), 2528. doi:10.3390/en11102528

Fang, X., Misra, S., Xue, G., & Yang, D. (2011). Smart grid—The new and improved power grid: A survey. *IEEE Communications Surveys and Tutorials*, *14*(4), 944–980. doi:10.1109/SURV.2011.101911.00087

Fathabad, A. M., Cheng, J., Pan, K., & Qiu, F. (2020). Data-driven planning for renewable distributed generation integration. *IEEE Transactions on Power Systems*, *35*(6), 4357–4368. doi:10.1109/TPWRS.2020.3001235

Gautier, A., Jacqmin, J., & Poudou, J. C. (2018). The prosumers and the grid. *Journal of Regulatory Economics*, *53*(1), 100–126. doi:10.1007/s11149-018-9350-5

Hahnel, U. J., Herberz, M., Pena-Bello, A., Parra, D., & Brosch, T. (2020). Becoming prosumer: Revealing trading preferences and decision-making strategies in peer-to-peer energy communities. *Energy Policy*, *137*, 111098. doi:10.1016/j.enpol.2019.111098

Hassani, H., Razavi-Far, R., & Saif, M. (2022). Real-time out-of-step prediction control to prevent emerging blackouts in power systems: A reinforcement learning approach. *Applied Energy*, *314*, 118861. doi:10.1016/j.apenergy.2022.118861

Lamnatou, C., Chemisana, D., & Cristofari, C. (2022). Smart grids and smart technologies in relation to photovoltaics, storage systems, buildings and the environment. *Renewable Energy*, *185*, 1376–1391. doi:10.1016/j.renene.2021.11.019

Mahmud, N., & Zahedi, A. (2016). Review of control strategies for voltage regulation of the smart distribution network with high penetration of renewable distributed generation. *Renewable & Sustainable Energy Reviews*, *64*, 582–595. doi:10.1016/j.rser.2016.06.030

Manditereza, P. T., & Bansal, R. (2016). Renewable distributed generation: The hidden challenges–A review from the protection perspective. *Renewable & Sustainable Energy Reviews*, *58*, 1457–1465. doi:10.1016/j.rser.2015.12.276

Markard, J. (2018). The next phase of the energy transition and its implications for research and policy. *Nature Energy*, *3*(8), 628–633. doi:10.1038/s41560-018-0171-7

Montoya-Bueno, S., Muñoz-Hernández, J. I., & Contreras, J. (2016). Uncertainty management of renewable distributed generation. *Journal of Cleaner Production*, *138*, 103–118. doi:10.1016/j.jclepro.2016.02.135

Parag, Y., & Sovacool, B. K. (2016). Electricity market design for the prosumer era. *Nature Energy*, *1*(4), 1–6. doi:10.1038/nenergy.2016.32

Phuangpornpitak, N., & Tia, S. (2013). Opportunities and challenges of integrating renewable energy in smart grid system. *Energy Procedia*, *34*, 282–290. doi:10.1016/j.egypro.2013.06.756

Puttamadappa, C., & Parameshachari, B. D. (2019). Demand side management of small scale loads in a smart grid using glow-worm swarm optimization technique. *Microprocessors and Microsystems*, *71*, 102886. doi:10.1016/j.micpro.2019.102886

Roggatz, C., Power, M., & Singh, N. (2020). Power system restoration: Meeting the challenge to resiliency from distributed generation. *IEEE Power & Energy Magazine*, *18*(4), 31–40. doi:10.1109/MPE.2020.2985438

Rosen, C., & Madlener, R. (2016). Regulatory options for local reserve energy markets: Implications for prosumers, utilities, and other stakeholders. *The Energy Journal (Cambridge, Mass.)*, *37*(2_suppl), 37. doi:10.5547/01956574.37.SI2.cros

Shaukat, N., Ali, S. M., Mehmood, C. A., Khan, B., Jawad, M., Farid, U., Ullah, Z., Anwar, S. M., & Majid, M. (2018). A survey on consumers empowerment, communication technologies, and renewable generation penetration within Smart Grid. *Renewable & Sustainable Energy Reviews*, *81*, 1453–1475. doi:10.1016/j.rser.2017.05.208

Solomon, B. D., & Krishna, K. (2011). The coming sustainable energy transition: History, strategies, and outlook. *Energy Policy*, *39*(11), 7422–7431. doi:10.1016/j.enpol.2011.09.009

Yan, Y., Qian, Y., Sharif, H., & Tipper, D. (2012). A survey on smart grid communication infrastructures: Motivations, requirements and challenges. *IEEE Communications Surveys and Tutorials*, *15*(1), 5–20. doi:10.1109/SURV.2012.021312.00034

Zafar, R., Mahmood, A., Razzaq, S., Ali, W., Naeem, U., & Shehzad, K. (2018). Prosumer based energy management and sharing in smart grid. *Renewable & Sustainable Energy Reviews*, *82*, 1675–1684. doi:10.1016/j.rser.2017.07.018

Chapter 9
Extraction of Urban Targets Using Fusion of Spectral and Shape Features in AVIRIS-NG Hyperspectral Data:
Use of Hyperspectral Data for Detecting Roads and Roofs

Shalini Gakhar
International Rice Research Institute, India

K. C. Tiwari
Delhi Technological University, India

ABSTRACT

Hyperspectral imagery holds the essence of spectral and spatial attributes, making it a rich source of information for applications like target detection in the field of agriculture, urban areas, etc. The chapter provides an insight of extraction of shape-based features fused with the spectral characteristics for detection of urban targets particularly roads and roofs. Roads are considered as linear structures whereas roofs occur in clusters. Therefore, to detect them, spectral information is not sufficient. The work proposes integration of both spectral and shape aspects of a target, which has given better results using SVM. The experimentation involves results in the form of spectral analysis, shape analysis, fusion of spectral and shape-based features along with the detection rate and supporting results. Delineating the roads and roofs of urban topography is an essential fragment for city planning, urban sprawl estimation, population studies, sustainable development, etc. The work done here may help numerous governmental and non-governmental organizations for conducting related studies.

DOI: 10.4018/979-8-3693-1396-1.ch009

1. INTRODUCTION

Hyperspectral data has an advantage of high spectral resolution along with spatial information forming a three-dimensional structure often referred as hypercube, where first two dimensions corresponds to arrangement of pixels and third dimension is a continuous function of reflectance values with respect to different wavelengths. The spectral signature is unique and can be treated as fingerprint to detect, classify, identify, or map a particular target. Hyperspectral imagery is expansively used for urban forestry monitoring (Zhang, 2011), agriculture (Teng, 2009), marine research (Lou, 2008), mineralogy (Pei, 2007) and many more. Apart from the aforesaid applications, hyperspectral remote sensing extends its scope over target detection domain, where target may be confined to a single pixel (full-pixel targets) or within a pixel (sub-pixel targets). This process is based on binary hypothesis where target pixels are kept active with respect to the non-target or background pixels. The targets to be detected vary according to the application areas, such as for defense purposes (troops, bunkers), wildlife conservation (animals), similarly for urban areas, the major portion comprises of man-made infrastructure (roads and roofs).

Target detection algorithms are broadly categorized into spectral matching and anomaly detection. The former is a supervised technique working on labelled data available from standard spectral libraries, ground spectral data collection or from the image itself, whereas latter is an unsupervised approach, which extracts the abnormalities assuming them as targets. Algorithms like Spectral Angle Mapper (SAM), Spectral Correlation Angle (SCA) (De Carvalho, 2000), Matched filter (MF), Adaptive Cosine Estimator (ACE) (Truslow, 2013) belong to spectral matching and Orthogonal subspace projection (OSP), Reed-Xiaoli (RX) and uniform target detector (UTD) (Tiwari, 2011) are types of anomaly detection.

Urban advancement is happening at an exponential pace, having significant effect on national and international level. Urban expansion has far-reaching consequences on vegetative cover, water and air quality, and surface temperature, ultimately impacting the microclimate of human habitats and contributing to environmental degradation (Shafri, 2012). Fortunately, advancements in space-borne and air-borne technologies have simplified the process of mapping urban areas. This modern approach is not only more efficient but also significantly reduces the time and effort required, providing a comprehensive bird's-eye view of larger areas simultaneously. During past decade, researchers have proposed various techniques for determination of robust spectral and spatial features to detect the urban targets (Bo, 2009) (Heiden, 2007).

In recent times, there has been a notable exploration of machine learning and deep learning methods for urban target detection. This surge in interest is attributed to the remarkable ability of these methods to autonomously learn intricate relationships within data. By leveraging these advanced techniques, researchers and practitioners can unravel complex patterns and information relevant to urban environments, enhancing the accuracy and efficiency of target detection processes. Support Vector Machine (SVM) by Vapnik (Cortes, 1995) is an extensively used method for hyperspectral data analysis. The method operates based on the principle of maximizing margins, creating a decision boundary that maximally separates data points belonging to different classes. The nature of the decision boundary, whether linear or non-linear, is determined by the separability analysis of the utilized data. Support Vector Machines (SVMs) excel in their performance, consistently achieving higher classification accuracies when compared to alternative pattern recognition techniques like maximum likelihood and multilayer neural networks. This superiority is attributed to SVMs' robust ability to discern complex relationships in the data, making them particularly effective in scenarios where the boundaries between classes are intricate or nonlinear. SVMs may prove helpful when multiple heterogenous classes are available with few training samples in a scene (Melgani F. a., 2004).

Till now, many researchers have explored detection and analysis related to natural targets such as vegetation (Farooq, 2019), forests (Kumar, 2013), waterbodies (Igamberdiev, 2011), and soil (Peng, 2013), but applications within the urban domain have to explore newer challenges and complexities. Standard target detection methods conduct analysis with spectral data, considering pixel wise evaluation approach without taking spatial details into account (Shafri, 2012). In urban areas, the spectral signatures of built-up materials often exhibit similarities in material composition. For instance, materials such as asphalt and concrete are commonly used in the construction of roads and roofs. This similarity in spectral characteristics poses a challenge in spectrally discriminating these materials. The difficulty is compounded by the within-class variability, which is influenced by both spectral and spatial resolutions determined by sensor characteristics.

The observed within-class variability is contingent upon the sensor's ability to capture fine details in both spectral and spatial dimensions. The challenges are further exacerbated by the heterogeneous and complex attributes inherent in urban built-up areas. The presence of mixed pixel patterns adds a layer of practical complexity, making it challenging to precisely discern individual materials or features solely based on spectral information. In essence, the signal received by the sensor is a composite of multiple features, contributing to the background complexity. This complexity, coupled with the similarities in spectral signatures of certain materials, underscores the importance of employing advanced analytical techniques, such as those rooted in machine learning and deep learning, to enhance the accuracy of material discrimination in urban environments.

There is currently no universally accepted method for mapping road surfaces and assessing their conditions. While some researchers have emphasized the extraction of road centerlines, challenges persist, especially in differentiating between newly paved dark roads and dark tile roofs (Segl K. U., 2003), (Mohammadi, 2012). Attempts have been made to address this issue by classifying various roof materials in urban areas using hyperspectral data. This involves creating a spectral library of diverse roofing materials and employing multiple supervised approaches for their detection (Chisense, 2012). In an effort to enhance the detection of buildings and improve the differentiation between buildings and open spaces, Segl et al. extended the analysis of reflective hyperspectral data to include thermal bands. This extension aims to bolster the reliability of distinguishing between buildings and open areas, ultimately leading to more accurate results. However, it's noteworthy that the shape features employed in these approaches, such as rectangular or quadratic features for smaller buildings, may not be universally applicable to the entire area under consideration (Segl K. S., 2003).

Therefore, the present work focusses on exploitation of shape-based features, based on standard deviation and range of the neighboring pixels for detection of urban targets (roads and roofs). These methods can accommodate spatial information in terms of shape of the target and enhance the desired response by overpowering the background. The analysis is three-fold, at first only spectral features are trained and investigated by variants (Linear, Quadratic, Cubic and Gaussian) of SVM, followed by shape features and lastly, both of them are fused together to increase the detection accuracy. The rest of the chapter is organized in multiple segments, the immediate next section gives a brief insight of the theoretical background. Section 3 explains the dataset used for the entire study; Section 4 expounds on the methodology opted for detection of urban targets; Section 5 discusses the details of results generated and Section 6 concludes the work done along with future scope.

2. THEORETICAL BACKGROUND

2.1 Spectral Features

A spectral pixel vector x_i constitutes all the pixel values in third domain, $x_{1i}, x_{1i}, \dots\dots\dots x_{1N}$ at every corresponding spatial location of hyperspectral data cube, where N is the total number of pixels. For a hyperspectral image having r rows and c columns, then the number of image vectors present would be $N=r\times c$. Equation 1 represents the hyperspectral data in higher dimension,

$$\left\{x_i\right\}_{i=1}^{N}, x_i \in \mathfrak{R}^d \tag{1}$$

Where, d is the number of spectral bands. This implies that the dimension of every spectral feature is $1 \times d$, as shown in Fig. 1.

Figure 1. Three-dimensional hyperspectral data cube depicting pixel vector

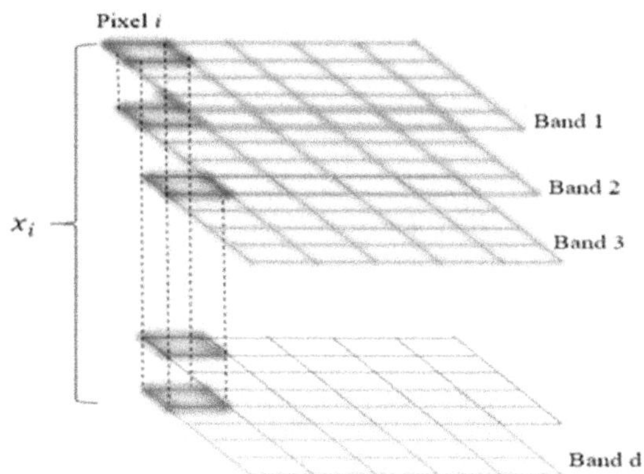

Here, x_i denotes a spectral signature of a particular pixel i, which holds the reflectance value at every wavelength yielding a spectral plot. Methods utilizing spectral information obtained from individual pixels are referred to as spectral or pixel-based techniques. SAM, for instance, is a method within this category that calculates the spectral angle similarity between a reference and a target spectrum. These approaches are characterized by their computational simplicity and offer a straightforward means of extracting information from the spectral components of hyperspectral data.

2.2 Shape Features

Shape features constitute a component of the spatial characteristics within hyperspectral imagery, encompassing attributes such as texture, size, and morphological operators computed relative to neighboring pixels. Han et al. introduced the shape-size index (SSI), a metric that integrates homogeneous regions between a central pixel and its adjacent pixels by leveraging spectral similarity, specifically designed for multispectral images (Han, 2012). Grey level co-occurrence matrix (GLCM) (Haralick, 1973) is one of the common spatial texture extraction approaches which transforms the grey scale value of an image by various statistical measures like homogeneity, energy, entropy and contrast. Another technique mentioned in literature for performing shape-based analysis is Histogram of Oriented Gradient (HoG) (Dalal, 2005). Farooq et al used

HOG features for automatic grass weed detection and classification in hyperspectral data (Farooq, 2019). The methods proposed for the study to extract shape are based on standard deviation and range of the central pixel with the neighboring pixels instead of other geometric shapes which are unable to fit the asymmetric urban targets (roads and roofs). Table 1 gives a brief description about these features, each of them considers a window of a size *s*, where *s*=3,5,7 and so on.

Table 1. Shape extraction features used for proposed work

S No.	Shape Extraction Feature	Description	Formula
1.	Standard deviation	• The value of output pixel is the standard deviation of neighborhood window of size *s*x*s* around the corresponding input pixel. • Brighter pixels in the output image corresponds to the higher standard deviation thus extracting the shape of the target. • The central element of the neighborhood is computed as *floor*((*size*(*s*)+1)/2	$\sqrt{\frac{\Sigma(x_i - x)^2}{s*s-1}}$ Where, x_i is the individual pixel value of input; x is the mean value of pixels; s is the neighborhood size.
2.	Range	• The value of output pixel is the range (maximum – minimum) of neighborhood window of size *s*x*s* around the corresponding input pixel. • Pixels having greater values of difference belongs to other feature class, discriminating the target from the background. • Range is calculated by dilation and erosion morphological attributes.	$\delta_E(X) = \{x \in R^D \mid E_x \cap X \neq \varnothing\}$ $\varepsilon_E(X) = \{x \in R^D \mid E_x \subseteq X\}$ Range=$\delta_E - \varepsilon_E$ Where, δ_E – Dilation with respect to Structuring Element E ε_E – Erosion with respect to Structuring Element E X – Image on which dilation is performed x – Origin where the locus of points is taken

2.3 Fusion of Spectral and Shape Features

Using either spectral or spatial component of hyperspectral data does not completely exploit the rich information it stores. Particularly for urban areas, integrating spectral and spatial information is necessary, as spectral component helps in classification of heterogeneous materials and spatial component delineates the shape of roads and roofs. Fauvel et al proposed a method based on fusion of morphological information and original hyperspectral data by concatenating the two vectors into single feature vector followed by classification of urban data using SVM (Fauvel M. J., 2008). Fusing spectral and spatial aspect of hyperspectral data is not only limited to classification but mixed pixel characterization is also addressed in the literature (Khodadadzadeh, 2014). Recently deep learning approaches have also been presented for better classification accuracies as they automatically learn high level features from the image itself (Zhao, 2016). However, there are few standard fusion models available from previous studies, such as component replacement (Vivone, Rocco, Giorgio, Mauro Dalla, & Jocelyn, 2014), multiresolution analysis (Yokoya, Claas, & Jocelyn, 2017) (Wilson, Steven K., & Lemuel R., 1995), Bayesian matrix decomposition (Lin, Xiaoming, Mai, Linhao, & Jianhua, 2017) which have been used in different studies to detect minute target details. Lately, machine learning and deep learning methods (Ghamisi, Bernhard, & Xiao, 2016)have been widely used for target detection, due to availability of large sample datasets, ease of automatically extracting the features and advancement in supporting infrastructures.

2.4 SVM and Its Variants

In case of SVM, if the data is linearly separable then a hyperplane can be used to separate the considered classes, otherwise the algorithm maximizes the margin adjusting few misclassifications (Kuo, 2013). SVM deals with two classes at a time, therefore real-time problems are transformed to series of binary ones. It is a machine learning algorithm that utilizes labelled data, trains the associated model, and predicts for unlabeled data. Fig 2 illustrates the basics of SVM.

The key advantage that SVM has over other machine learning algorithms is that the features can be transfigured using multiple kernel functions. The kernel enables data to be mapped (dot product of two vectors) in preferably higher dimension where it is assumed that the features may be separated by a linear boundary. This technique enhances the capability of SVM and makes it feasible for multi-class problems also. The kernel functions used for the current work are tabulated in Table 2 with their description.

Figure 2. Support vector machine showing the margin between two classes

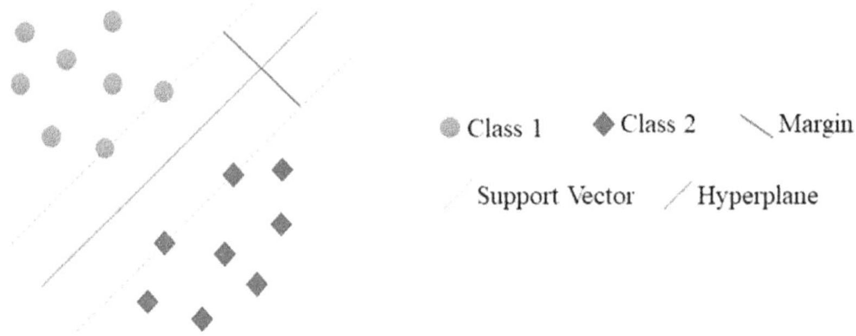

Table 2. SVM and its variants with kernel functions

S No.	Type of SVM	Variants	Details	Kernel Functions
1.	Linear (Melgani & Bruzzone, 2004)	-	A single hyperplane holds the efficiency to classify the two classes, but the optimal decision boundary is required to maximize the margin between them.	$K(x_i, x_j) = x_i^T \cdot x_j$ x_i = feature vector 1 x_i = feature vector 2 x^T= transpose
2.	Polynomial (Fauvel M. J., 2009)	Quadratic	Kernel trick refers to calculating the dot product of feature vectors. Here the degree is 2.	$K(x_i, x_j) = (x_i^T \cdot x_j + 1)^d$
		Cubic	Here the power is increased as 3.	Value of d is 3
3.	Gaussian (Fauvel M. J., 2008)		The value depends upon the distance of any data point to the margin. Highly effective when no knowledge of data is there	$K(x_i, x_j) = \exp\left(-\frac{\lVert x_i - x_j \rVert^2}{2\sigma^2}\right)$ σ = width of kernel

3. DATASET DESCRIPTION

3.1 AVIRIS-NG Data Collection Campaign

The hyperspectral data used in this analysis was collected as part of the Airborne Visible and Infrared Imaging Spectrometer – Next Generation (AVIRIS-NG) data collection campaign in February 2016. This mission is a collaboration between the Jet

Propulsion Laboratory (JBL) of NASA and the Indian Space Research Organization (ISRO). The data was collected across 27 regions of India covering various terrains such as forests, agricultural land, urban areas, water bodies, snow, and glaciers.

For the specific analysis of urban targets, the study focuses on the Udaipur scene in Rajasthan. The Udaipur scene consists of imagery with dimensions of 1295 pixels by 388 pixels and comprises 425 spectral bands ranging from 376 nm to 2500 nm. The spectral data has a resolution of 5 nm, while the spatial resolution is 8.1 meters.

A Level 2 spatial subset has been manually selected for analysis, consisting of 394 lines and 385 samples. This subset is chosen to encompass the maximum urban area within the scene and provides derived geospatial variables at the same resolution and location as the Level 1 source data. Figure 3 displays the RGB subset of the Udaipur image, with the red (R), green (G), and blue (B) channels set to specific values (R=51, G=33, B=20). This subset is sized 394 pixels by 385 pixels and is a portion of the complete Udaipur image. In the top middle of the subset, there is a rectangle representing the target roof, which stands out distinctly from the surrounding background. The second rectangle located in the bottom left corner is an enlarged view of a highway road, while the third rectangle positioned at the bottom right corner depicts another type of roof. These rectangles highlight specific features within the subset for further analysis and identification.

Figure 3. Image dataset of AVIRIS-NG hyperspectral data

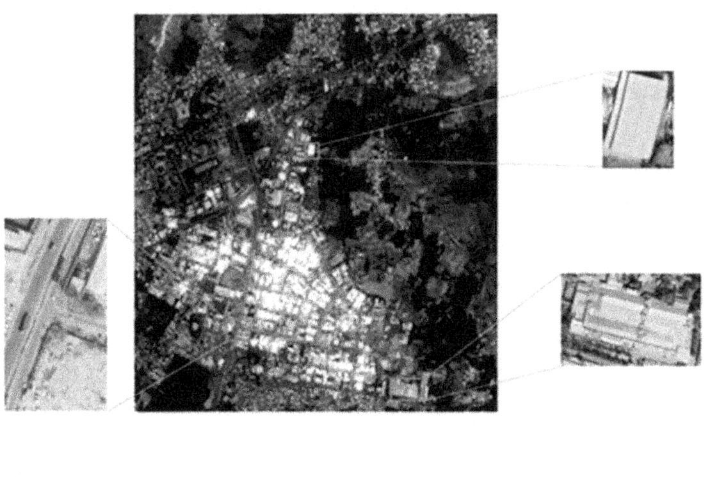

3.2 Ground Data Collection

The spectroradiometer utilized for ground spectra collection within the same region operates across 2151 channels with 1nm spectral sampling. Table 3 outlines the key parameters for image and ground data collection pertinent to this study. Roughly 1000 spectral signatures characterizing urban built-up features, such as parking lots, roads, roofs, pavements, and sports infrastructure, have been incorporated for analysis. The ground data has been used to correlate with the image spectral data, due to the limited scope of the chapter, the same is not presented here. Also, spatial coordinates have been considered to identify the roads and roofs a potential target for the study.

Table 3. Description of image and ground data collection

Parameters	Image Data Description	Ground Data Description
Location	Udaipur, Rajasthan, India	
Date of Data acquisition	02/02/2016	16/07/2017 - 21/07/2017
Spatial Resolution	8.1 meter	-
Spectral Resolution	5 nm	3nm(700nm),8nm(1500nm), 6nm(2100nm)
Channels	425	2151
Level of data	Level 1 and Level 2	
Main features in the images	Urban features such as roads, bridges, different types of roofs (Concrete, bitumen, sandstone, CGI sheet), vehicles, railway tracks etc.	

4. METHODOLOGY

4.1 Preprocessing

During this phase, the atmospherically corrected Level 2 AVIRIS-NG data, consisting of 425 spectral bands, undergoes processing to obtain a transformed hypercube with 387 bands. In this process, bands influenced by water vapors, atmospheric gases, and noise are excluded before proceeding with the implementation.

4.2 Spectral-Based Features

To maintain the integrity of the hyperspectral data, no dimensionality reduction techniques are employed for the analysis. Spectral signatures corresponding to each class are extracted at the pixel level, resulting in a feature vector of size $1 \times d$, where d represents the number of bands.

4.3 Shape-Based Features

The detection of urban targets, including roads and roofs, involves the utilization of standard deviation and range as spatial features. These features operate on small window sizes, capitalizing on the spatial attributes present in hyperspectral data. Initially computed band-wise to discern the shapes of significant targets, they are subsequently applied at the pixel level for precise analysis. Figure 4 illustrates the process of extracting shape features in further detail.

Figure 4. Fusion of spectral and spatial features

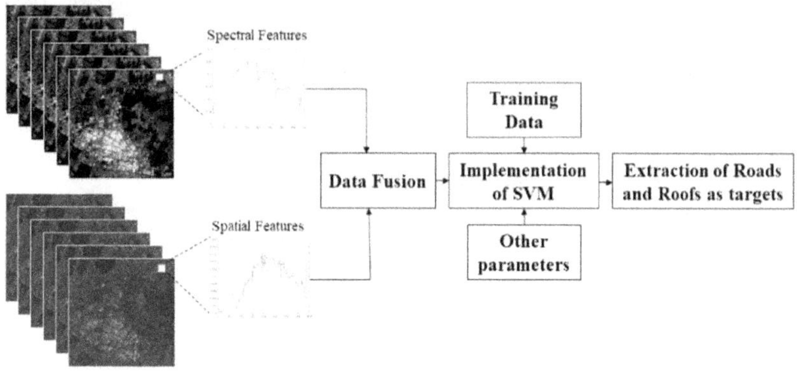

4.4 Training File Generation

There are three different training files used for the study comprising of spectral, shape and fused feature vectors belonging to each class considering Vegetation, Impervious surfaces, and Soil (VIS Model) (Ridd, 1995). Each training file contains 2000 labelled samples of vegetation, imperious surfaces (roads and roofs) and soil surface, where every individual category has 500 observations. The first training file contains only the spectral information, the second training file comprises of the shape attributes and the third one has the average of spectral and shape-based feature considered for a particular pixel. Other pixels encompassing water, bare lands are already masked out by visually interpreting the image data, field examination and validating the geo-locations from Google Earth software.

4.5 SVM Implementation

The previously generated training files are set as an input for SVM model for its learning step followed by applying the result to the complete image for detecting the presence of urban targets. Cross-validation is set to avoid overfitting by partitioning the training dataset into k-folds, and value of k is kept as 5.

4.6 Performance Evaluation

The experiments are designed to evaluate the results using the spectral, shape and fused features to evaluate the accuracy by considering the variants of SVM (Linear, polynomial and gaussian). An exhaustive comparative assessment has been performed to extract the urban targets and explore the suitable approach for the same. The flow chart of the approach and the detailed steps are shown in Fig. 5.

Figure 5. Flow chart of methodology of proposed approach

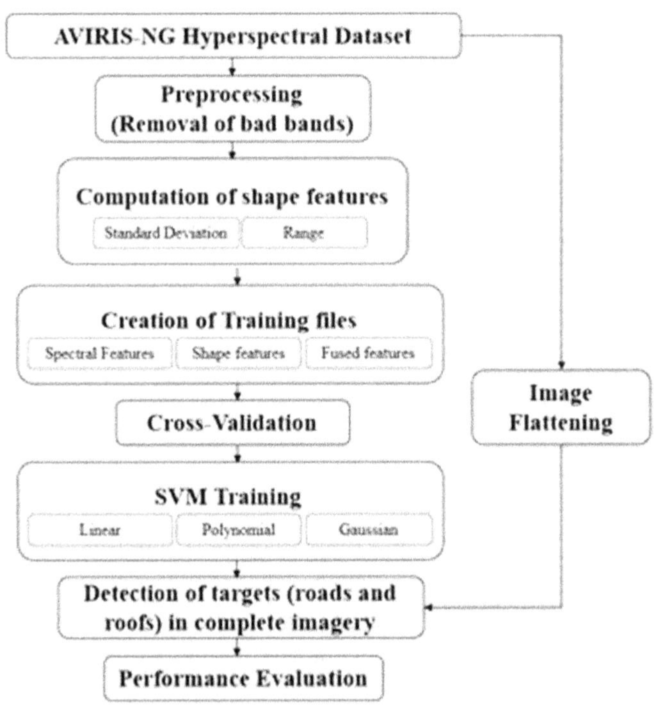

5. RESULTS AND DISCUSSION

The work is an improved method to detect urban targets using spectral and spatial components of hyperspectral data. The study is an attempt to analyze AVIRIS-NG data collected during this joint campaign and unfurl the treasure the information stored in it. As mentioned, with advent of new materials being used for built-up paves way for other challenges to be addressed therefore this work is done to exploit AVIRIS-NG hyperspectral dataset for urban analysis. This part deals with three set of experiments based on the training files created, that is, Spectral Features, Shape Features and Fused features (spectral and spatial both). Shape features are again computed using two local windowing methods, standard deviation and range. All three training files are then fed into kernel variants of SVM (Linear, polynomial with degree 2 and 3, gaussian) to generate road and roof extraction results. The white pixels correspond to the presence of targets whereas the black pixels represent background pixels. The software used for the proposed work are ENVI 4.8 and MATLAB 2018b along with a system configuration of 16GB RAM, with Intel i7 processor.

5.1 Spectral Features

The spectral signatures at multiple confirmed locations are extracted from the image itself to create the training file. Ground data spectral collection was synchronized with the hyperspectral airborne campaign; therefore, the labelled data is created, taking the geo-locations of urban targets as reference. The resultant images are illustrated in Fig. 6 and the corresponding accuracies are shown in Table 4. Due to the limited scope of the chapter and to avoid complexities only the results with maximum accuracy are shown with respective SVM kernel method.

Table 4. Accuracy assessment using spectral features

Linear SVM	Polynomial SVM		Gaussian SVM		
	Quadratic	Cubic	Fine	Medium	Coarse
89.1%	89.6%	***89.8%***	88.2%	88.4%	87.9%

When the spectral signatures are given as input, the variation in accuracies by using different kernel methods in SVM is very less. All the models perform equally consistent and later the trained model with maximum accuracy (here, cubic) is used for extraction of roads and roofs from the hyperspectral subset of Udaipur region.

Figure 6. Spectral approach: (a) detection of roads, (b) detection of roofs, (c) scatter plot for both urban targets

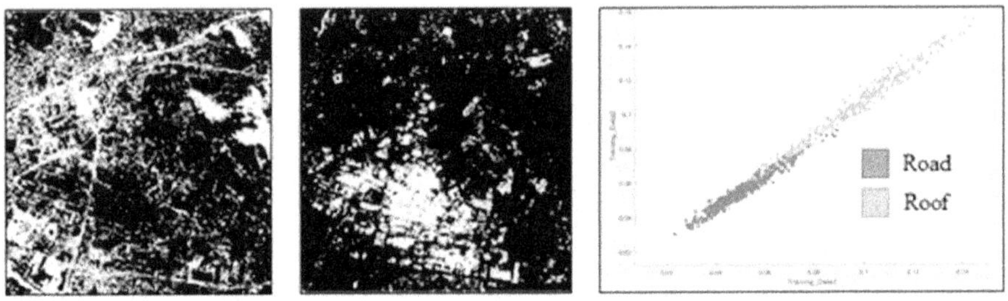

(a) (b) (c)

Upon visual examination of Fig. 6(a), it is evident that road surfaces closely resemble bare soil, particularly in the top-right region of the image. Consequently, extracting roads using solely spectral features proves challenging. In the subsequent result depicted in Fig. 6(b), significant roof areas are identified; however, the edges and boundaries lack distinctiveness due to the exclusive consideration of spectral features. Furthermore, Fig. 6(c) illustrates a scatter plot of roads and roofs, revealing a considerable overlap between the two categories. This overlap indicates a spectral similarity at the material composition level.

5.2 Shape Features

Various spatial features are employed based on the specific attributes of the targets under detection. Utilizing simple mathematical operators like standard deviation and range can effectively reduce computational load during the training process. In this context, a neighborhood value of 3 is selected, resulting in a window size of 3×3 and providing 8 neighbors for the central pixel being detected, with provisions made for cell padding around boundary pixels. Concurrently, standard deviation and range calculations are performed for each band, and feature vectors are derived from the exact same geo-locations considered for spectral attributes. The ensuing results are presented below:

The accuracy of shape features is expected to be less, which is also evident from Table 5. Polynomial kernel SVM gives a marginal detection percentage for both standard deviation and range-based analysis. In Fig. 7 (a) (b) and Fig. 8 (a) (b) neither the road nor roof appears to be detected, a very faint boundary is visible for both. Also, the data points for most of the samples are having very high overlap, unable to distinguish them on spatial grounds.

Table 5. Accuracy assessment using shape features

Shape Features	Linear SVM	Polynomial SVM		Gaussian SVM		
		Quadratic	Cubic	Fine	Medium	Coarse
Standard Deviation	39.1%	***41.3%***	37.4%	38.2%	37.5%	34.3%
Range	38.8%	***40.4%***	36.9%	39.1%	38.4%	34.6%

Figure 7. Standard deviation: (a) detection of roads, (b) detection of roofs, (c) scatter plot for both urban targets

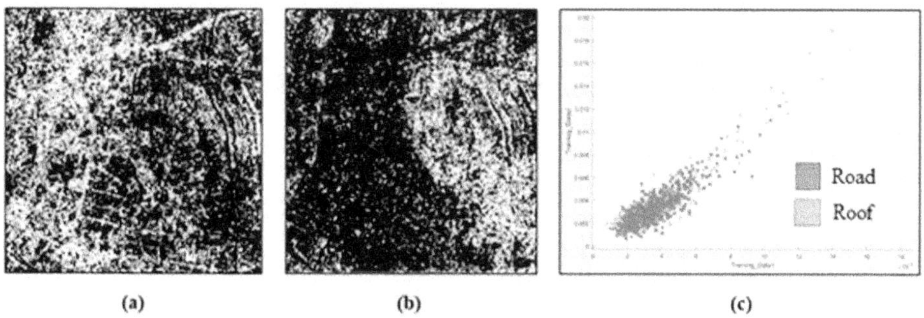

Figure 8. Range: (a) detection of roads, (b) detection of roofs, (c) scatter plot for both urban targets

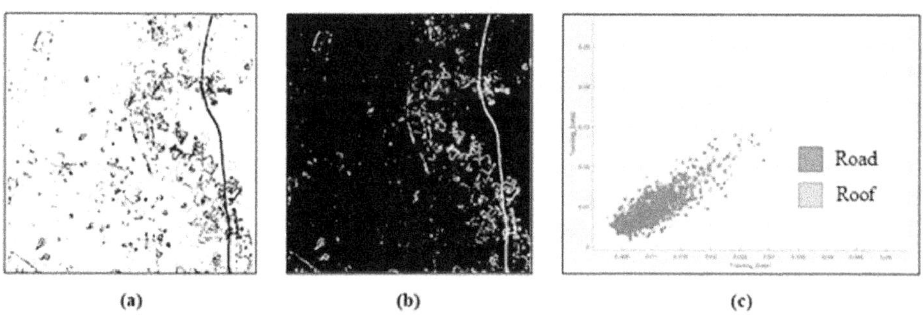

5.3 Spectral and Spatial Features

Combining the spectral and spatial components to extract the urban targets appears to give good results as compared to spectral or spatial attributes considered individually. The very same inference can be drawn from the literature also, but varying the process and methods to extract spectral or spatial features differs. Spectral features may be transformed using Principal Component Analysis (PCA) (Bakken, 2019), Minimum Noise Fraction (MNF) (Luo, 2016) etc. whereas, spatial features include multiple morphological operations (Benediktsson, 2005), texture-based analysis (Seifi Majdar, 2017), shape extraction etc. Here in this study, in order to preserve the spectral shape of spectral signature, no dimensionality reduction method is used. Therefore, the spectral signatures are combined with shape-based features to enhance the detection percentage. Appending spectral features with shape features is also tried as part of study, but due to average performance the results are not shown in the chapter. Simply, averaging both the attributes has yielded better results which are shown in Fig. 9 and 10 and the corresponding results are mapped in Table 6.

Table 6. Accuracy assessment using spectral and shape features

Spectral and Shape Features	Linear SVM	Polynomial SVM		Gaussian SVM		
		Quadratic	Cubic	Fine	Medium	Coarse
Standard Deviation	96.5%	***97.3%***	96.9%	96%	96.7%	95.6%
Range	91.3%	***91.5%***	89.8%	88.9%	90.6%	88%

Figure 9. Spectral features and standard deviation: (a) detection of roads, (b) detection of roofs, (c) scatter plot for both urban targets

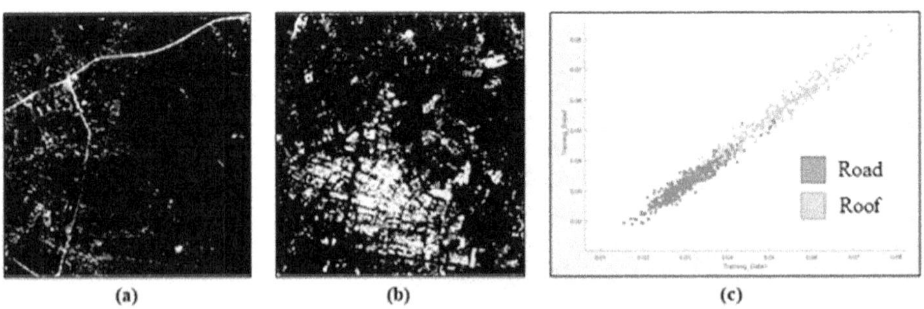

Upon analyzing the outcomes presented in Table 6, it becomes evident that spectral features combined with standard deviation yield higher accuracy compared to those combined with range. This discrepancy can be attributed to the fact that while range, as a straightforward statistical measure, operates on the minimum and maximum values within the neighborhood, standard deviation computes the mean value and then utilizes deviation to determine the presence or absence of the target. Moreover, employing a polynomial kernel method with a degree of 2 produces superior results in comparison to other types of Support Vector Machines (SVM). As illustrated in Fig. 9(a), road boundaries and cross-sections are clearly delineated, while roof clusters are identified in Fig. 9(b). Conversely, Fig. 10(a) and (b) depict instances of misdetections when spectral signatures are fused with range, as observed through visual examination of the images.

Figure 10. Spectral features and range: (a) detection of roads, (b) detection of roofs, (c) scatter plot for both urban targets

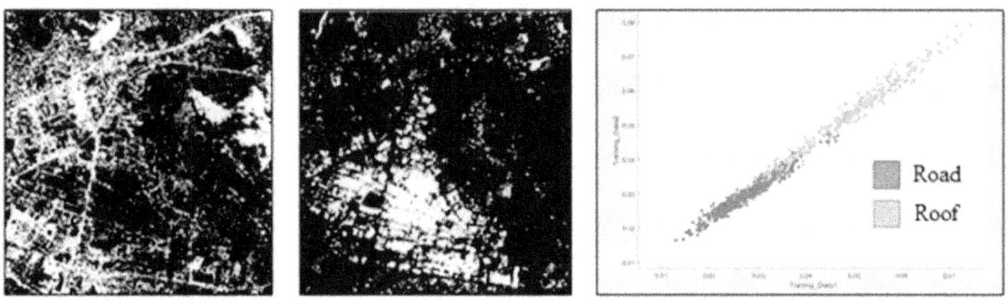

(a) (b) (c)

6. CONCLUSION

A fusion approach leveraging both spectral signatures and shape properties of urban features has been successfully applied using hyperspectral data. This method focuses on averaging spectral attributes with shape-based features to detect roads and roofs, which constitute a significant portion of urban land covers. Training files containing (i) spectral, (ii) shape, and (iii) combined spectral and shape feature vectors serve as inputs for various variants of Support Vector Machines (SVM). The best-performing model is then utilized to predict the availability of targets. Pixels containing the target exhibit brightness in contrast to background pixels, resulting in a binary result image. This approach not only utilizes spectral information but also aids in detecting the shapes of urban targets. Overall, the polynomial kernel trick, particularly with quadratic and cubic degrees, has shown superior performance compared to other methods examined in this study. Furthermore, standard deviation, functioning as a two-level statistical approach operating locally on small windows, yields better results than range. Given the challenge of spectral variability in hyperspectral remote sensing across different locations, this approach proves beneficial in detecting linear features such as roads and clustered features like roofs by analyzing neighboring pixels locally.

Funding

The work is in collaboration with Defence Research and Development Organization (DRDO) and is sponsored by the Department of Science and Technology (DST). The authors extend their gratitude to ISRO (Indian Space and Research Organization) for supporting the project. This research was funded by Ministry of Science & Technology, Department of Science and Technology (DST), grant number BDID/01/23/2014-HSRS.

Conflict of Interest

The authors declare no conflict of interest. The funders had no role in the design of the study; in the collection, analyses, or interpretation of data; in the writing of the manuscript, or in the decision to publish the results.

Code, Data, and Materials Availability

The AVIRIS-NG Phase 1 hyperspectral dataset campaign held in 2016 have a Non-Disclosure Agreement signed with ISRO and the Principal Investigator (PI) of the project, therefore the data cannot be shared on public platforms, but the research associated with it is highly encouraged to successfully explore the high spectral and spatial resolution of the imagery. Further information is available at https://avirisng.jpl.nasa.gov/dataportal/

REFERENCES

Bakken, S. M. (2019). *The effect of dimensionality reduction on signature-based target detection for hyperspectral remote sensing. In CubeSats and SmallSats for remote sensing III*. International Society for Optics and Photonics.

Benediktsson, J. A., Palmason, J. A., & Sveinsson, J. R. (2005). Classification of hyperspectral data from urban areas based on extended morphological profiles. *IEEE Transactions on Geoscience and Remote Sensing*, *43*(3), 480–491. doi:10.1109/TGRS.2004.842478

Bo, D. Z. (2009). A target detection algorithm for urban areas using HSRH imagery. *Joint Urban Remote Sensing Event*, 1–5.

Chisense, C. M. (2012). Classification of roof materials using hyperspectral data. *The International Archives of the Photogrammetry, Remote Sensing and Spatial Information Sciences, 39*(B7), 103–107. doi:10.5194/isprsarchives-XXXIX-B7-103-2012

Cortes, C., & Vapnik, V. (1995). Support-vector networks. *Machine Learning, 20*(3), 273–297. doi:10.1007/BF00994018

Dalal, N. a. (2005). Histograms of oriented gradients for human detection. *IEEE Computer Society Conference on Computer Vision and Pattern Recognition, 1*, 886-893.

De Carvalho, O. A. (2000). Spectral correlation mapper (SCM): an improvement on the spectral angle mapper (SAM). In *Summaries of the 9th JPL Airborne Earth Science Workshop.* JPL Publication.

Farooq, A. X. (2019). Texture and Shape Features for Grass Weed Classification Using Hyperspectral Remote Sensing Images. In *IEEE International Geoscience and Remote Sensing Symposium* (pp. 7208-7211). IEEE. 10.1109/IGARSS.2019.8900132

Fauvel, M. J., Benediktsson, J. Ó. A., Chanussot, J., & Sveinsson, J. R. (2008). Spectral and spatial classification of hyperspectral data using SVMs and morphological profiles. *IEEE Transactions on Geoscience and Remote Sensing, 46*(11), 3804–3814. doi:10.1109/TGRS.2008.922034

Fauvel, M. J., Chanussot, J., & Benediktsson, J. A. (2009). Kernel principal component analysis for the classification of hyperspectral remote sensing data over urban areas. *EURASIP Journal on Advances in Signal Processing, 2009*(1), 1–14. doi:10.1155/2009/783194

Ghamisi, P., Bernhard, H., & Xiao, X. Z. (2016). Hyperspectral and LiDAR data fusion using extinction profiles and deep convolutional neural network. *IEEE Journal of Selected Topics in Applied Earth Observations and Remote Sensing, 10*(6), 3011–3024. doi:10.1109/JSTARS.2016.2634863

Han, Y. H., Kim, H., Choi, J., & Kim, Y. (2012). A shape–size index extraction for classification of high resolution multispectral satellite images. *International Journal of Remote Sensing, 33*(6), 1682–1700. doi:10.1080/01431161.2011.599348

Haralick, R. M., Shanmugam, K., & Dinstein, I. H. (1973). Textural features for image classification. *IEEE Transactions on Systems, Man, and Cybernetics, 6*(6), 610–621. doi:10.1109/TSMC.1973.4309314

Heiden, U. K., Segl, K., Roessner, S., & Kaufmann, H. (2007). Determination of robust spectral features for identification of urban surface materials in hyperspectral remote sensing data. *Remote Sensing of Environment*, *111*(4), 537–552. doi:10.1016/j.rse.2007.04.008

Igamberdiev, R. M., Grenzdoerffer, G., Bill, R., Schubert, H., Bachmann, M., & Lennartz, B. (2011). Determination of chlorophyll content of small water bodies (kettle holes) using hyperspectral airborne data. *International Journal of Applied Earth Observation and Geoinformation*, *13*(6), 912–921. doi:10.1016/j.jag.2011.04.001

Khodadadzadeh, M. J.-D., Jun Li, Plaza, A., Ghassemian, H., Bioucas-Dias, J. M., & Xia Li. (2014). Spectral–spatial classification of hyperspectral data using local and global probabilities for mixed pixel characterization. *IEEE Transactions on Geoscience and Remote Sensing*, *52*(10), 6298–6314. doi:10.1109/TGRS.2013.2296031

Kumar, T. S., Panigrahy, S., Kumar, P., & Parihar, J. S. (2013). Classification of floristic composition of mangrove forests using hyperspectral data: Case study of Bhitarkanika National Park. *Journal of Coastal Conservation*, *17*(1), 121–132. doi:10.1007/s11852-012-0223-2

Kuo, B.-C. H.-H.-H.-C.-S., Ho, H.-H., Li, C.-H., Hung, C.-C., & Taur, J.-S. (2013). A kernel-based feature selection method for SVM with RBF kernel for hyperspectral image classification. *IEEE Journal of Selected Topics in Applied Earth Observations and Remote Sensing*, *7*(1), 317–326. doi:10.1109/JSTARS.2013.2262926

Lin, B., Xiaoming, T., Mai, X., Linhao, D., & Jianhua, L. (2017). Bayesian hyperspectral and multispectral image fusions via double matrix factorization. *IEEE Transactions on Geoscience and Remote Sensing*, *10*(10), 5666–5678. doi:10.1109/TGRS.2017.2711640

Lou, Q. S. (2008). The application and prospecting of the hyperspectral remote sensing technology in marine research. *Trans Oceanol Limno*, 168-173.

Luo, G. G.-E., Chen, G., Tian, L., Qin, K., & Qian, S.-E. (2016). Minimum noise fraction versus principal component analysis as a preprocessing step for hyperspectral imagery denoising. *Canadian Journal of Remote Sensing*, *42*(2), 106–116. doi:10.1080/07038992.2016.1160772

Melgani, F., & Bruzzone, L. (2004). Classification of hyperspectral remote sensing images with support vector machines. *IEEE Transactions on Geoscience and Remote Sensing*, *42*(8), 1778–1790. doi:10.1109/TGRS.2004.831865

Mohammadi, M. (2012). Road classification and condition determination using hyperspectral imagery. *The International Archives of the Photogrammetry, Remote Sensing and Spatial Information Sciences*, *39*(B7). doi:10.5194/isprsarchives-XXXIX-B7-141-2012

Pei, C. a. (2007). Status and prospect of hyperspectral remote sensing technique in rock and mineral identification. *World Nuclear Geoscience*, *24*(1), 32–38.

Peng, J. H., Shen, H., He, S. W., & Wu, J. S. (2013). Soil moisture retrieving using hyperspectral data with the application of wavelet analysis. *Environmental Earth Sciences*, *69*(1), 279–288. doi:10.1007/s12665-012-1955-x

Ridd, M. K. (1995). Exploring a VIS (vegetation-impervious surface-soil) model for urban ecosystem analysis through remote sensing: Comparative anatomy for cities. *International Journal of Remote Sensing*, *16*(12), 2165–2185. doi:10.1080/01431169508954549

Segl, K. S., Roessner, S., Heiden, U., & Kaufmann, H. (2003). Fusion of spectral and shape features for identification of urban surface cover types using reflective and thermal hyperspectral data. *ISPRS Journal of Photogrammetry and Remote Sensing*, *58*(1-2), 99–112. doi:10.1016/S0924-2716(03)00020-0

Segl, K. U. (2003). Endmember detection in urban environments using hyperspectral HyMap data. *Third EARSeL Workshop on Imaging Spectroscopy.*

Seifi Majdar, R., & Ghassemian, H. (2017). A probabilistic SVM approach for hyperspectral image classification using spectral and texture features. *International Journal of Remote Sensing*, *38*(15), 4265–4284. doi:10.1080/01431161.2017.1317941

Shafri, H. Z. (2012). Hyperspectral remote sensing of urban areas: An overview of techniques and applications. *Research Journal of Applied Sciences, Engineering and Technology*, *4*(11), 1557–1565.

Teng, A. G. (2009). Progress of hyperspectral technology application research in agriculture. *Jiangsu Nongye Kexue*, 8–11.

Tiwari, K. C., Arora, M. K., & Singh, D. (2011). An assessment of independent component analysis for detection of military targets from hyperspectral images. *International Journal of Applied Earth Observation and Geoinformation*, *13*(5), 730–740. doi:10.1016/j.jag.2011.03.007

Truslow, E. D., Manolakis, D., Pieper, M., Cooley, T., & Brueggeman, M. (2013). Performance prediction of matched filter and adaptive cosine estimator hyperspectral target detectors. *IEEE Journal of Selected Topics in Applied Earth Observations and Remote Sensing*, *7*(6), 2337–2350. doi:10.1109/JSTARS.2013.2272697

Vivone, G., Rocco, R., Giorgio, L., Mauro Dalla, M., & Jocelyn, C. (2014). Multiresolution analysis and component substitution techniques for hyperspectral pansharpening. *IEEE Geoscience and Remote Sensing Symposium*, 2649-26. 10.1109/IGARSS.2014.6947018

Wilson, T. A., Steven, K. R., & Lemuel, R. M. J. (1995). Perceptual-based hyperspectral image fusion using multiresolution analysis. *Optical Engineering (Redondo Beach, Calif.)*, *34*(11), 3154–3164. doi:10.1117/12.213617

Yokoya, N., Claas, G., & Jocelyn, C. (2017). Hyperspectral and multispectral data fusion: A comparative review of the recent literature. *IEEE Geoscience and Remote Sensing Magazine*, *5*(2), 29–56. doi:10.1109/MGRS.2016.2637824

Zhang, H. S. (2011). Application of hyper spectral remote sensing for urban forestry monitoring in natural disaster zones. In *International Conference on Computer and Management (CAMAN)* (pp. 1-4). IEEE. 10.1109/CAMAN.2011.5778867

Zhao, W., & Du, S. (2016). Spectral–spatial feature extraction for hyperspectral image classification: A dimension reduction and deep learning approach. *IEEE Transactions on Geoscience and Remote Sensing*, *54*(8), 4544–4554. doi:10.1109/TGRS.2016.2543748

Chapter 10
Understanding Climate Smart Agriculture (CSA) Practices by Smallholder Farmers to Cope With Risks in Tanzania

Joel Johnson Mmasa
https://orcid.org/0000-0001-5711-7937
The University of Dodoma, Tanzania

ABSTRACT

This chapter used a systematic review to organize the data by a systematic review of existing literature. The chapter documented climate smart agriculture (CSA) practices that exist in the country to cope with various production risks. These included the adaption of better seed varieties, application of weather information, efficient fertilizer use, intercropping, and crop residue and canal irrigation. Food system issues, affecting food production, processing, distribution, and consumption, are among the main causes of food insecurity and hunger in Tanzania. Tanzanian rural households are particularly susceptible to environmental pressures.

1. INTRODUCTION

This chapter focuses on documenting the Climate Smart Agriculture (CSA) practices adopted by smallholder farmers in Tanzania as a means to mitigate risks. Despite climate change being a prominent concern on Tanzania's political and economic agenda, there is limited available data regarding the specific CSA strategies employed as coping mechanisms. The agricultural sector in Tanzania contributes 28.2% to the

DOI: 10.4018/979-8-3693-1396-1.ch010

Gross Domestic Product, supplies 70% of industrial raw materials, and employs 66% of the country's population in agricultural activities (URT, 2018). The anticipated impact of climate change on the livelihoods of rural communities in Tanzania is significant, with frequent occurrences of floods and droughts posing threats to food security.

1.1 Background Information

As climate change and whether variability impose risks on economies, food systems and livelihoods, understanding on how smallholders' farmers use climate smart agricultural technologies occupy considerable attention in policy discussion. Beside climate change threats on food supplies to meet food supplies from growing population, agriculture face pressure from increase in demand for Industrial Raw Materials, and Biofuels. For instance, it is estimated that more than 2.8 billion individuals worldwide reside in regions prone to multiple physical manifestations of climate change, such as floods, storms, droughts, and rising sea levels (FAO, 2020). The most vulnerable populations are found in countries situated within the semi-arid dryland belt, Sub-Saharan Africa, South and Southeast Asia, Latin America, Small Island Developing States, and the Arctic. The anticipated consequences of climate change will particularly impact the livelihoods of impoverished rural communities in developing nations. According to the Fourth Assessment Report of the Intergovernmental Panel on Climate Change (IPCC), agricultural production in numerous African countries is likely to be significantly affected by climate change.

In some parts of Africa, the rise in agricultural production is not keeping up with this demand (Ojo, et al., 2022). Due to its proximity to the equator, climate change has been predicted to be dangerous for Africa (Thinda, et al., 2021); (Sarr, et al., 2021); (Onyeneke, 2022). In Africa, climate change manifests through various physical impacts such as increasing sea levels, shifts in temperature patterns, and alterations in rainfall behavior (WB, 2010), all of which will have a negative influence on agricultural production, farm revenue, food security, and economic development. The impact of climate change on Sub-Saharan Africa's underprivileged population, whose livelihoods rely on agriculture, is concerning. The alarming statistic of 811 million people globally experiencing chronic hunger highlights the inadequacy of food systems and supply chains in meeting the needs of significant portions of society. To achieve the goal of Zero Hunger, it is crucial that food systems improve and have the capacity to provide nourishment to even the most vulnerable individuals (WFP, 2021).

In Tanzania, the effects of climate change have become evident. For instance, there has been a noticeable rise in mean annual temperatures, with an increase of 1°C recorded since 1960. Additionally, there has been a decline in rainfall, with an average decrease of 2.8 mm per month and 3.3% per decade. The southern region of Tanzania has experienced a more pronounced reduction in rainfall compared to other areas (Mashingo, 2010).

Tanzania possesses a tropical climate and is characterized by four primary climatic zones: the hot and humid coastal plain, the semi-arid central plateau, the high rainfall lake regions, and the temperate highlands. In the highlands, temperatures range between 10ºC and 20ºC during the cool and warm seasons, respectively. The impacts of climate change in Tanzania are evident in both its natural environment and the well-being of its residents. Rising temperatures in the country contribute to an increased likelihood of extreme rainfall events, leading to floods, as well as extended periods of dry spells, resulting in droughts. To assess and address the risks associated with climate change, organizations can utilize the Climate Change Vulnerability Index (CCVI) developed by Maplecroft, a global risks advisory firm. This index evaluates 42 factors encompassing social, economic, and environmental aspects to determine national vulnerabilities across three core areas. These areas include exposure to climate-related natural disasters and sea-level rise, human sensitivity indicated by population patterns, development levels, natural resource availability, agricultural dependency, and conflicts, and future vulnerability by assessing a country's adaptability through government policies and infrastructure in combating climate change. (Climate-ADAPT, 2022). The agriculture sector in Tanzania encounters several challenges that impact land productivity and overall production. Some key factors contributing to these challenges include the utilization of outdated agricultural practices, where farmers predominantly rely on manual labor using hand hoes. Additionally, limited access to modern technologies and essential production inputs further hinders productivity in the sector (IFAD, 2018).

Food insecurity and hunger in Tanzania stem from challenges within the food system, which encompass various stages of food production, processing, distribution, and consumption. The term "food systems" refers to the interconnected activities and actors involved in the production, aggregation, processing, distribution, consumption, and waste management of food derived from agriculture, forestry, or fisheries. These systems are intricately linked to the broader economic, societal, and natural contexts in which they operate. By addressing issues within the food system, such as inefficiencies or inequalities, it becomes possible to tackle the root causes of food insecurity and hunger in Tanzania (WFP, 2021). The food system in Tanzania and the broader sub-region faces numerous challenges that impact food and nutrition security, as well as opportunities for increased production and demand. This research also considers scenarios that account for processing and aggregation within the food system. Inefficient functioning of the food system leads to several main consequences, including reduced profitability for producers due to the sale of poor-quality or insufficient quantities of products. Additionally, there is limited availability of fresh and nutritious foods, which affects their affordability and accessibility. Moreover, an inefficient food system puts additional strain on the environment through wasteful utilization of scarce production inputs, such as water, and the ongoing conversion of land, often at the expense of vulnerable ecosystems (WFP, 2021). The current barriers make it challenging to establish a sustainable food system that simultaneously supports viable livelihoods, ensures adequate and affordable nutrition, protects natural resources, and minimizes negative impacts on the climate and environment. This study focuses on analyzing the Tanzanian food system, with particular emphasis on climate change as a significant threat. The study explores the potential of investing in Climate Smart Agriculture (CSA) technologies for smallholder farmers as a means to address these challenges.

Addressing the adverse effects of climate change on crop production has become a global concern, prompting discussions on potential solutions. Many experts believe that implementing sustainable agriculture practices could offer a promising approach to mitigating the impacts of climate change (Kassie, et al., 2013): (Ndiritu, et al., 2014); (Ogemah, 2017); (Zhao, et al., 2018); (Adenle, et al., 2015); (Zeweld, et al., 2017); (Ehiakpor, et al., 2021). This strategy is anticipated to increase agricultural output performance while reversing the agroecosystem's detrimental degradation processes, especially in smallholder farming systems. It is an improvement on the green revolution, which raised agricultural output significantly across the board and is credited with launching economies in Asia out of poverty, but has left behind certain problems. The main support in many facets of human life is the availability of sufficient and sustainable food (Taylor, et al., 2016); (Rajaram & Van Ginkel, 2019). The agriculture industry in Tanzania plays a crucial role in the nation's provision of the main staple food, employment, raw materials for industries, and foreign exchange revenues ((Lyatuu & Urasa, 2016); (Mkonda & He, 2016): (Kuzilwa, et al., 2017).

It is crucial for Tanzania to invest in terms of financial resources and research on adaptation measures farmers could implement to cope with climatic changes especially in the agriculture sector (Guthiga & Newsham, 2011) Recent years have seen a growing recognition of indigenous people as valuable repositories of knowledge regarding climate change and as crucial contributors to the development of policies aimed at coping with and adapting to its effects. Utilizing local knowledge can be an effective and sustainable approach to mitigation and adaptation strategies that are cost-effective and participatory. The success or failure of adaptation efforts is influenced by several factors, such as seasonal climatic variations, the specific agricultural production system in place, socio-economic factors, the involvement of governmental bodies, non-governmental organizations (NGOs), and private companies, as well as long-term changes in climate and market conditions. (Nhemachena & Hassan, 2007). Therefore, this study through systematic review focusing on the topics on the CSA practices farmers can adapt due to effect of climate change in Tanzania. This chapter will cover introduction, methodology, findings which includes: food system concept, Tanzania's food systems, factors affecting food system, CSA practices, the food system approach, food security and rainfall forecast.

2. METHODOLOGY

Traditional research methods, such as value-chain analysis, supply-chain assessments, and field-based interviews, may not fully capture the intricacies of a complex system characterized by feedback loops, delays, nonlinearity, and non-rational elements, as well as the policy-path dependency that shapes sustainable development. In this study, a systematic review methodology was employed to gather data by examining existing documents. This approach was utilized to document and conceptualize the complexity of the food system in Tanzania, specifically focusing on investigating the effects of climate change on small-scale farmers in the country. The study uses a review of the literature (Use of reports, journal articles, thesis etc). The literature was retrieved from Google scholar regarding to the topic of interest and limiting at least ten years back (2020-2022) was used. The study also analyses report from national (Tanzania Meteorological Authority (TMA), Ministry of Agriculture Reports etc) and international organizations (World Food Programme, Food and Agriculture Organization of the United Nations (FAO), United Nations etc). This study excluded unrelated papers, duplicated, unavailable full texts, or abstract-only papers.

3. RESULTS AND DISCUSSION

3.1 Food System Concept

A "food system" refers to the complete range of components and operations involved in producing, processing, distributing, preparing, and consuming food (Willett, 2022). This pertains to everything that goes into getting food from farms to tables, including the environment, people, inputs, procedures, infrastructure, and institutions (IFPRI, 2016). The food system approach integrates various disciplines, such as Agronomy, Anthropology, Economics, Sociology, Political Sciences, and Nutrition, and describes the various components of our food system as well as the connections between them (Van Berkum, 2018). A food systems approach is thus gaining increasing traction as a way to achieve a balance between healthy families, healthy communities, and sustainable livelihoods, particularly following the 2021 UN Food Systems Summit (IIED, 2022).

Food systems have the potential to produce food in a way that promotes healthy diets, but when they don't, it costs both people and the environment dearly. First, according to (Swinburn, et al., 2019) malnutrition is the primary global cause of poor health. For the fourth year in a row, malnutrition is increasing (FAO, et al., 2019) but no country has accomplished to stop the global overweightness epidemic (Zwicker, et al., 2015). Second, the food systems help to exceed a number of "planetary boundaries" that try to elect a region where mankind can exist carefully (Springmann, et al., 2018). Food systems, and in particular food production, are significant contributors to ecosystem degradation, land use change, biodiversity loss, climate change, freshwater resource depletion, and land use change. For instance, up to 37% of anthropogenic greenhouse gas emissions are produced through food systems (IPCC, 2019). Food is one of the major problems for human health and the environment in the twenty-first century. Due to these difficulties, there is an prerequisite for a global change to supportable food systems and healthful diets (Willett, et al., 2019). Food system change is very context-specific and differs significantly across industrialized and developing nations. Even within regions, local factors that affect the sustainability of food systems, such as consumption habits, input usage, and poverty, can differ greatly. The political determination and policy interplanetary to endorse, modify, and enforce required policies vary widely amongst countries, in addition to the resources and capacities to do so. In order to achieve the necessary transformation, interdisciplinary research is needed to produce context-specific programs and policies that are owned or controlled by resident actors.

3.2 Tanzania's Food Systems

Tanzania's food systems deal with a complicated web of interconnected problems and opportunities that either make it difficult or advantageous to attain economic, environmental, and social sustainability. A lack of nutrient-rich foods and a focus on staples characterize Tanzanian diets. Malnutrition, particularly in children, is so widespread. Since agriculture is the main foundation of income, Tanzanian households are particularly susceptible to environmental pressures. Already, flooding, droughts, and more severe environmental degradation. Climate change and environmental deterioration reinforce each other, harming vulnerable communities (Campbell, et al., 2016). Tanzania is predictable to import 75% of its cereal needs in 2050 due to forecast population growth, historical production increases, and changes in food preferences (Makoi, et al., 2017).

3.3 Factors Affecting Food System

In Africa, various factors such as shifts in dietary patterns, urbanization, population growth, insufficient livelihoods, and climate change pose ongoing challenges to food systems. It is imperative for African food systems to undergo transformation in order to address these issues effectively and align with the objectives of the Sustainable Development Goals (FAO, 2018). The upcoming chapter in the book will center on examining how climate change affects the food system, with a specific emphasis on documenting the implementation of Climate-Smart Agriculture (CSA) technologies by small-scale farmers.

According to (UNDP, 2022), The Climate Change Vulnerability Index (CCVI) offers a quick and cost-efficient method for assessing the comparative vulnerability of plant or animal species to climate change. The CCVI is determined by considering three key factors: the degree of exposure to climate change (Climate Exposure), the level of sensitivity to climate change (sensitivity), and the capacity of a system to adapt to climate change (adaptive capacity). In Tanzania, the distribution of the population is highly disparate. The majority of people reside along the northern border or the eastern coast, while large portions of the country have sparse populations. (Lake, 2013). Density varies from 12 per square kilometre (31/sq mi) in the Katavi Region to 3,133 per square kilometre (8,110/sq mi) in Dar es Salaam (NBS, 2022). Around 70 percent of the Tanzanian population resides in rural areas, although this proportion has been decreasing since at least 1967. Dar es Salaam serves as the de facto capital and is the largest city in the country. Dodoma, situated in the central region of Tanzania, is the de jure capital, although efforts to transfer government institutions to Dodoma are progressing steadily. As of now, the estimated population of the United Republic of Tanzania stands at 63,904,925 (NBS, 2022). Tanzania experiences an annual population growth rate of 2.9%, which is associated with a relatively high-risk index ranging from 2.5 to 5. According to the 2019 Climate Risk Index data, Tanzania scored 66.5 on the Climate Risk Index, indicating a significant level of climate-related risks. In terms of fatalities in 2019, Tanzania ranked 27th, with a reported number of deaths attributed to the effects of climate change. Specifically, there were 47 deaths per 100,000 inhabitants in that year. (Eckstein, and Schäfer, 2021). The index evaluates various factors such as climate-related natural disasters, population distribution, development levels, natural resource availability, reliance on agriculture, and conflicts to determine the vulnerability of a country. It also considers the adaptive capacity of a country's government and infrastructure to address climate change when assessing future vulnerability. Additionally, the study highlights that Africa and Oceania consistently experience a relatively high percentage of population changes, which contributes to the challenges posed by climate change (Figure 1).

Figure 1. Climate change vulnerability index
Source: UNDP (2022)

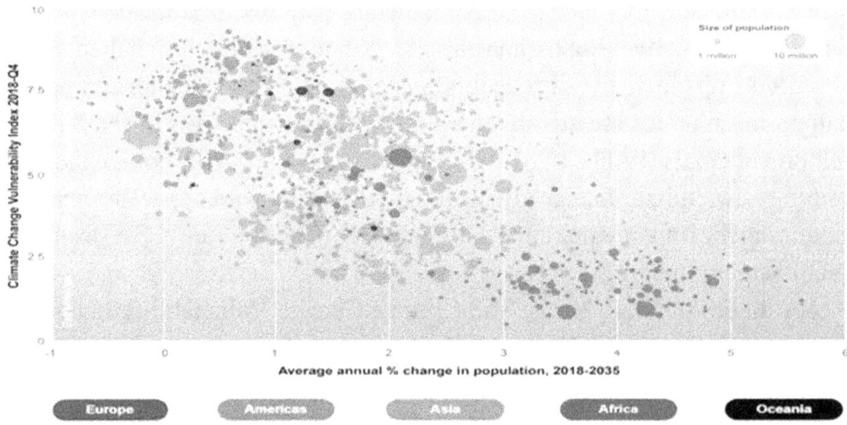

3.3.1 Climate Impact on the Business of Value Chain Actors

Smallholders, who typically cultivate less than 5 ha per year and consume survival farming as their primary objective, grow between 85 - 90 percent of the entire amount of land that is used for agriculture (13 million ha) (WFP, 2021). Some of Tanzania's most significant crops include cereals (Maize, Rice, Sorghum) and pulses (Beans), which account for about 40% of the nation's total arable land. Concerning the effect of environment change on the business of value chain actors it was noted that all actors in the chain are affected (Table 1). There is an effect on agro-dealers like poor seed multiplications, cooperatives both owned by males and female and processors like having insufficient raw materials (SNV, 2019). Climate Change Vulnerability Index, Europe and America are at low and medium risk, this is also, attributed to the average annual change in population. Asia is at high and medium risk. Africa countries is at extreme risk, which is attributed by the average annual change in population at (3%-6%) (UNDP, 2022).

3.4 Climate Smart Agriculture (CSA) Practices

Climate change can be promoted by climate-resilient agricultural practices. The vulnerability of smallholder farmers has increased as a consequence of climate and weather changes finished the past few decades, particularly because less than 2% of their field is irrigated (WFP, 2021). The agricultural sector is significantly affected by the detrimental consequences of climate change, such as diminished crop yields resulting from droughts, floods, and limited water availability. In practical terms,

Table 1. Climate impact to business of value chain actors

Agro-Dealers/Service Providers	Cooperatives/Female Farmers	Cooperatives/Males Farmers	Processors
• Low sales of Inputs • Defaults of Credited inputs • Low seeds Multiplication	• Low Production/ Food Insecurity • Unemployment • Collapse of Cooperatives • Weak Cooperatives • Low Soil Fertility • Soil Erosion	• Crop drying • Water Sources Getting Dry • Reduce Raw Materials to transport	• Insufficient Raw Materials • Seasonal Operation • Increased Cost of Production • Poor Quality Products • High Default Rate • Low Motive to agriculture lending • Increased Demand

Note: Every actor in the Chain is Affected
Source: SNV (2019)

smallholder farmers are particularly vulnerable to weather fluctuations, as both excessive rainfall and droughts contribute to an increased prevalence of pests and diseases. These factors have a negative impact on productivity, leading to reduced income and a disruption in their way of life. As an example, it has been projected that droughts alone result in an annual loss of $200 million for Tanzania's agricultural industry (WFP, 2021). The primary objective of CSA is to foster a food system that ensures both food security and overall development objectives in the face of a changing climate and growing food demand. CSA initiatives are designed to enhance productivity, build resilience, and minimize greenhouse gas emissions in a sustainable manner. It is crucial to engage in comprehensive planning to effectively manage the trade-offs and synergies that exist among the three core pillars of CSA: productivity, adaptation, and mitigation. By carefully considering these aspects, CSA can contribute to the long-term sustainability and success of agricultural systems (FAO, 2020). CSA encompasses agricultural practices that promote sustainable increases in productivity and income, enhance adaptive capacity, foster community resilience to climate change, and improve food and nutrition security. It also aims to achieve co-benefits related to mitigation in line with national economic development priorities. Although the concept of CSA is relatively new and still evolving, many of the practices and technologies that constitute CSA are already in use by farmers worldwide to mitigate various production risks.

To mainstream CSA, it is important to conduct a comprehensive assessment of successful ongoing practices, completed initiatives, and future requirements in terms of institutional and financial support. A recent study conducted in Tanzania identified key CSA practices adopted by smallholder farmers (SHFs) in the country. These practices include the adoption of improved seed varieties, utilization of weather information for decision-making, efficient use of fertilizers, intercropping, the incorporation of crop residues, and the implementation of canal irrigation.

The impact of climate change on crop production in Tanzania and the resulting severe consequences are evident through the analysis of seasonal rainfall patterns in various cities across the country. By examining trends in rainfall and temperature, the data substantiates the assertion that climate change is responsible for the alteration in the timing of seasonal rainfall. This shift often leads to rainfall occurring at inappropriate times, such as when crops have already been damaged or when they are not in need of additional water (Levira, 2009). One of the proposed approaches to mitigate the effects of climate change is to focus on investing in irrigation agriculture due to insufficient and unpredictable seasonal rainfall patterns, as well as the need for mechanization (Levira, 2009).

3.4.1 Adaption of Better Seed Varieties

According to Figure 2 results show that, 3,042,749 households planted their crops in Tanzania during short rainy season, out of which 1,040,995 households equivalent to 34.2 percent reported improved seeds use (994,299 households were in Mainland Tanzania and 46,696 households in Tanzania Zanzibar) (URT, 2021). Class seed is at the core of farm inputs desirable to surge crop production, nutrition, and rural wellbeing. In Tanzania, smallholder farmers face significant challenges in accessing affordable, high-quality seeds. The majority of seeds used for planting, accounting for over 90 percent, are saved by farmers from previous harvests. However, the quality of these seeds is often subpar. When it comes to maize cultivation, the adoption of Improved Maize Varieties (IMV), such as hybrids and Open Pollinated Varieties (OPV), can have a substantial positive impact on maize yield per unit of land (Monela, 2014). The selection of improved varieties by farmers is a critical factor that significantly influences crop productivity. Numerous factors come into play and influence the decision-making process regarding variety adoption by farmers (Rogers, 2003). Similarly, in Tanzania, the cultivation of rice has traditionally focused on local varieties that trace their origins back to seeds brought in by Arab traders prior to 1960. These varieties, such as *Supa, Behenge, Kula na Bwana, Kalamata,* among others, have demonstrated excellent adaptation to the local climate and cater to the taste preferences of Tanzanians. However, they tend to have relatively low yields, averaging around 1 to 1.5 tons per acre (RLDC, 2009).

Figure 2. Number of agricultural households reported to use improved seed and area planted by season during 2019/20 agricultural year, Tanzania

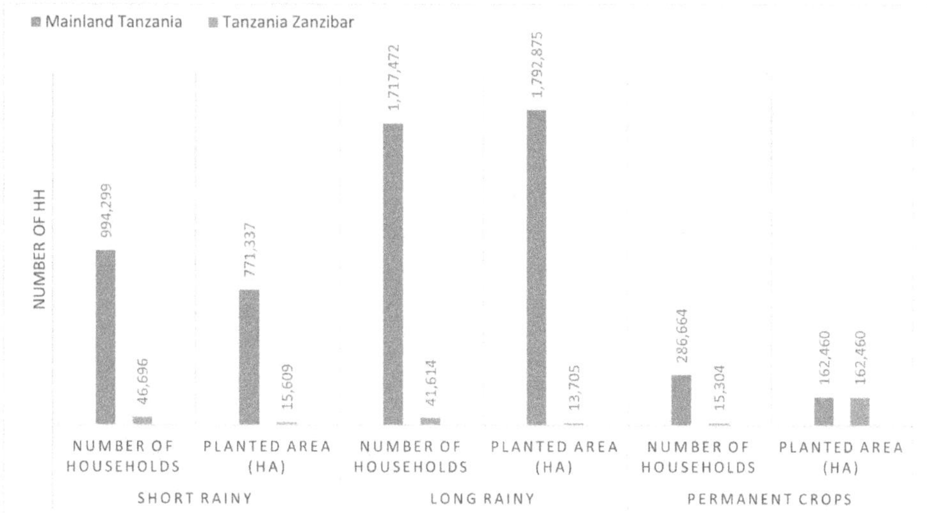

3.4.2 Weather Information Services

Information has always played a crucial role in the advancement of agricultural practices. Throughout history, as people began cultivating crops and raising livestock, they sought out information to improve their farming techniques (Meyer, 2015). Tanzania, like the rest of the world, is affected by global warming. It is projected that the country will experience a temperature increase of 2-4 °C by 2100, with dry seasons warming more significantly than wet seasons, and inland regions experiencing greater warming compared to coastal areas (Paavola, 2008). A study has shown that farmers in Tanzania primarily receive weather information and forecasts through television or radio broadcasts (CropMon, 2017). These services are typically provided by national weather services and local radio stations, as radio and television are the preferred means of communication (Phahlane, et al., 2016). However, a limitation of radio and television is the challenge of tailoring forecasts to specific locations or individual user needs.

Mobile services present a potential solution to this issue. Given the relatively low adoption of smartphones in rural Africa, services accessible via feature phones are more likely to be successful and have a greater impact. Examples of such services include text messaging (SMS), interactive voice response (IVR), and interactive messaging services (such as Unstructured Supplementary Service Data, or USSD). IVR services allow farmers to communicate with a computer system through phone calls, while USSD enables interactive text messaging between a computer and a farmer. Several mobile services providing farm advisories or weather forecasts have already been effectively implemented (AGRA, 2016), and mobile service providers can play a significant role in supporting weather services for agriculture (GSMA, 2016).

Mobile technology has emerged as a significant platform for digital solutions in Africa. In this region, mobile phones are utilized for activities that individuals in other parts of the world often rely on computers for. The cost of mobile devices and data is decreasing, making them more accessible to a broader population, including those in rural areas (GSMA, 2016). Basic mobile phones, commonly known as "feature phones," with text messaging (SMS) and telephone functionality, are prevalent in rural Africa, while the adoption of smartphones remains lower in comparison to the global average. This can be attributed to the relatively high costs of smartphones and subscriptions for individuals in rural communities, which is expected to continue hindering widespread adoption of such services in the foreseeable future (GSMA, 2017). Nevertheless, mobile services accessible through basic mobile phones are increasingly prevalent in developing countries. While the most common applications are related to mobile money and banking, there has been a growing presence of agricultural information services that combine finance, market insights, weather updates, and cultivation tips in recent years. The significance of such services is underscored by a study conducted by Anderson et al. (2016), which reveals that only one in ten smallholder farmers possess an official bank account, while half of them are financially included through mobile services. The relatively high penetration of mobile banking services in rural Tanzania suggests an opportunity for leveraging mobile phones for agricultural and climate information services as well.

3.4.3 Effective use of Fertilizer

The utilization of fertilizers in Tanzania has historically been low but has shown an upward trend in recent years. Fertilizer consumption in the United Republic of Tanzania has experienced fluctuations, but overall it has increased from 1971 to 2020, reaching 16.4 kilograms per hectare in 2020. Farmers who utilize improved maize and rice varieties with fertilizers achieve significantly higher yields compared to those using unimproved varieties (URT, 2021). Approximately 2.8 million hectares, which accounts for 20.2 percent of the total planted area, were treated with fertilizers. Out of this, 2.7 million hectares were in Mainland Tanzania, and 40,024 hectares were in Tanzania Zanzibar. Among the total area treated with fertilizers, 39.6 percent received organic fertilizers, while 60.4 percent received inorganic fertilizers (URT, 2021). Farmers in regions with higher rainfall tend to have higher adoption rates of improved maize varieties and chemical fertilizers (IFDC, 2021).

In 2020, the fertilizer consumption rate for the United Republic of Tanzania was 16.4 kilograms per hectare. However, it is important to note that Tanzania's fertilizer consumption has historically remained below recommended rates, and the use of fertilizers in the country is limited, especially in low-input and rain-fed subsistence farming, which is prevalent in Tanzania's agriculture. The high cost of fertilizers hinders access for small-scale farmers, as more than 90 percent of fertilizers used in the country are imported and expensive. To address this issue, the Ministry of Agriculture of Tanzania introduced the fertilizer bulk procurement system (FBPS) in 2017 to lower retail prices and improve access to fertilizers. Due to subsidies, fertilizer utilization increased to over 140,000 tons compared to the 360,000 tons utilized in the 2021/22 agricultural season.

Figure 3 illustrates the fertilizer consumption trend in Tanzania, ranging from a minimum of 0.5 kilograms per hectare in 1961 to a maximum of 16.4 kilograms per hectare in 2020, with an average value of 5.1 kilograms per hectare of arable land during that period. The most recent figure available is 16.4 kilograms per hectare in 2020. In contrast, the global average for fertilizer consumption on arable land in 2020 is 180.1 kilograms per hectare, based on data from 161 nations.

Figure 3. Tanzania fertilizer consumption

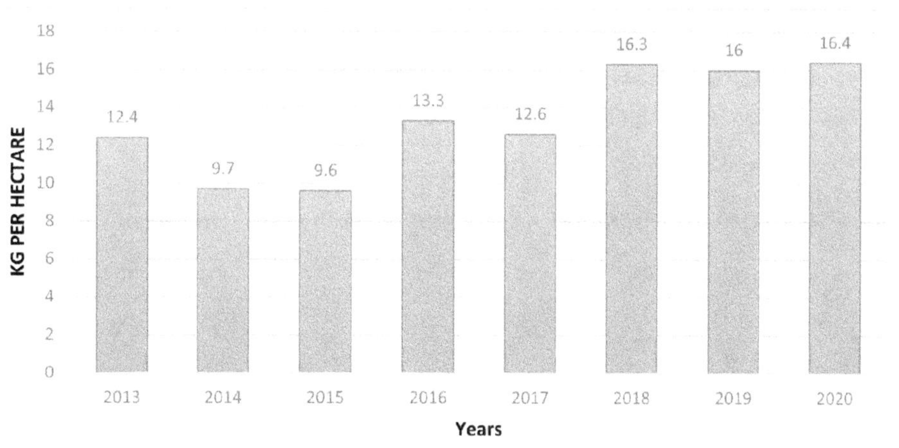

3.4.4 Intercropping

According to research literature, intercropping can be categorized into five distinct planting patterns (Jeruto & Ouma, 2010). Row intercropping involves growing two or more crops together, with at least one crop planted in rows. Strip intercropping entails planting multiple crops in wide strips, allowing for crop interaction. Mixed intercropping refers to planting two or more crops without a specific row arrangement. Relay intercropping involves planting a second crop into an established crop before harvesting the initial crop. Hedgerow intercropping (or alley cropping) entails planting rows of annual crops between rows of trees, which are periodically pruned to minimize shading effects (Kang, et al., 1985). Intercropping is a widely practiced farming strategy in Tanzania. According to (EPAR, 2013). 81 percent of households in Tanzania engaged in intercropping on at least one plot during the long or short rainy season. This includes intercropping seasonal crops with fruit or permanent crops. During either the long or short rainy season, or with fruit or permanent crops, 63 percent of plots were intercropped.

Among the intercropped plots, maize was the most commonly grown priority crop in Tanzania. It was also the most frequently cultivated crop in intercropped plots, with 73 percent of such plots being planted with maize. Figure 4 provides a visual representation of the crops most commonly found in intercropped plots alongside maize.

Following maize, beans were the next most commonly cultivated crop in intercropped plots, present in 32 percent of the plots. Groundnuts and sweet potatoes were also commonly grown on intercropped plots, with 16 percent and 11 percent of the plots respectively. Sorghum, cowpeas, rice, millet, cassava, and yams were planted on 10 percent or fewer of the intercropped plots, indicating that these crops are more likely to be grown as monocrops. Some crops, such as cowpeas, beans, and sweet potatoes, were frequently intercropped. For instance, 93 percent of plots where cowpeas were planted were intercropped with other crops.

Figure 4. Frequent intercropping combinations with maize, all seasons

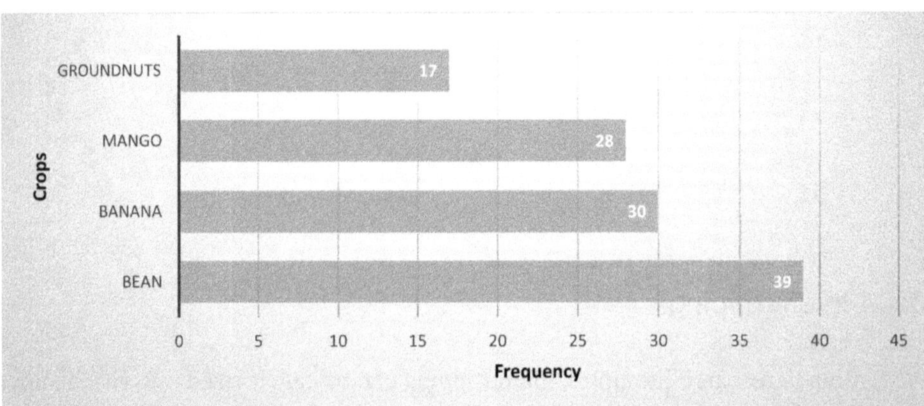

Every year, farmers plant both long and short rainy season crops, whereas fruit and permanent crops are not replanted annually. Among plots planted with annual crops, an average of 57 percent of plots were intercropped. Intercropping was more common during the short rainy season, with 67 percent of plots being intercropped, compared to 54 percent during the long rainy season. On the other hand, permanent crops and fruit crops were even more likely to be intercropped, with 78 percent and 82 percent of plots, respectively, practicing intercropping (Figure 5).

Figure 5. Percentage of plots intercropped by season and crop type

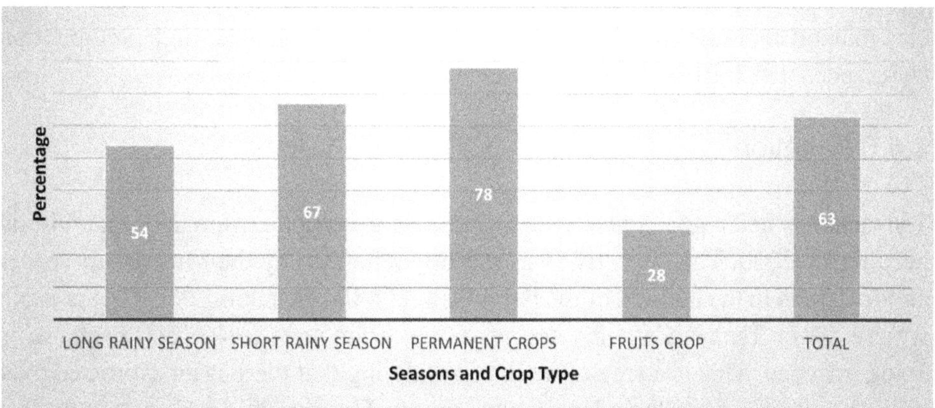

3.4.5 Crop Residue

The fertility of acid soils heavily relies on organic matter, particularly in areas with intense weathering and high precipitation, resulting in the predominance of the clay mineral kaolinite. However, traditional agricultural practices have led to the depletion of inherent soil nutrients over the years, primarily due to leaching and continuous nutrient removal by crops. Burning is a common practice in Tanzanian farms, often considered necessary for facilitating planting and controlling weeds and pests (IAEA, 2013). This practice, along with the replacement of perennial vegetation with short-season crops, results in reduced organic inputs to the soil. Conversely, the application of animal and crop residues, the use of cover crops, and reduced tillage can increase organic inputs or reduce losses, thereby aiding in the maintenance or increase of soil organic matter content. In tropical conditions, organic matter is rapidly lost through accelerated oxidation caused by the hot and moist environment (IAEA, 2012). The management of crop residues significantly influences nutrient cycling and the dynamics of Soil Organic Carbon (SOC) in cereal-based systems, which are predominant in smallholder farming across much of sub-Saharan Africa. In many of these systems, nutrient inputs, both organic and inorganic, are often insufficient compared to nutrient outputs, leading to negative nutrient and carbon balances for each cropping season and consequently severe soil degradation (Giller, et al., 1997). Although the importance of organic matter is widely recognized, the success of crop residue management in ensuring agricultural sustainability depends on the local biophysical and socio-economic conditions (Erenstein, 2002). Therefore, the significance of crop residue management for improved soil fertility varies according to farming systems, production orientation, crop types, and production intensity. Organic matter's role in maintaining the fertility of acid soils is well-established, particularly due to the prevalence of the clay mineral kaolinite resulting from intensive weathering under high precipitation (Alexander, 1977; Elliot, 2012).

3.4.6 Irrigation

Tanzania has had a goal since 2006 to increase irrigation coverage to one million hectares by 2020. This objective was recently reiterated by the Minister for Water and Irrigation in the autumn of the previous year (Allafrica.com, 2016). At present, a mere 5% of Tanzania's cultivated land, which amounts to 461,000 hectares, is being irrigated. This is a small fraction considering that there is an estimated 29.4 million hectares of available land in the country. One possible explanation for this situation is the lack of emphasis on irrigation in Sub-Saharan Africa, despite the renewed focus on agricultural development following the Maputo declaration of 2003 (AU, 2003).

Irrigation is primarily practiced in paddy farming areas such as Kilombero, Kilosa, and Mvomero Districts in Tanzania. These regions have fertile soil suitable for rice cultivation, and the potential for irrigation to enhance productivity is recognized. While many farmers currently rely on traditional methods like buckets and watering cans for irrigation, there has been a growing trend in recent years towards the use of motorized pumps with a power range of 1.5-6.7 horsepower, coupled with flood or basin irrigation techniques. These motorized pumps, which are relatively easy to install and the most cost-effective option among available technologies (in comparison to solar, electricity, and wind), have gained popularity. However, they come with significant maintenance costs and have a relatively short lifespan.

In terms of irrigation systems, open canal systems, which are commonly used, have the lowest efficiency rates, ranging from 15-25 percent. In contrast, lined canals have efficiency rates of 60-80 percent, while piped systems exhibit efficiency rates as high as 90 percent. On the other hand, flooding/basin application systems, although cheaper, also have relatively lower efficiency rates of 60-70 percent compared to drip irrigation with a 90 percent efficiency rate and sprinkler systems with an 80 percent efficiency rate (WB, 2018).

3.4.7 In Situ Rainwater Harvesting

Soil-water conservation, also referred to as in-situ Rain Water Harvesting (RWH), encompasses a set of methods aimed at preventing runoff and promoting water infiltration. The objective is to retain rainwater that would otherwise be lost as runoff from cultivated areas. With in-situ RWH, the focus is on conserving rainwater at the location where it falls, without introducing additional runoff from external sources. The primary techniques employed include conservation tillage, contour barrier strip catchment, tillage, and hillside systems. These approaches are designed to effectively manage water by minimizing runoff and maximizing water infiltration into the soil. (Gowing, et al., 1999).

3.5 The Food System Approach

Social fairness, environmental sustainability, and food security concerns are interconnected and are rarely best addressed separately (Springmann, et al., 2018). Policy circles and research are turning more and more to the idea of food systems to comprehend the connections between these challenges and food and to address the complexity that results. An approach to food that links food production, processing, distribution, and consumption while acknowledging that these activities are supported by intricate political, economic, social, and ecological interactions is known as a "food system approach" (Dekeyser, 2019). Feedback loops, in which specific interactions reinforce or balance each other out, are a crucial component of food systems. For instance, greater agri-inputs may result in greater income, which may result in greater agri-inputs. These feedback loops indicate that there are unclear cause-and-effect interactions in food systems and that the result is unpredictable.

Each food system exhibits a unique set of characteristics, comprising various elements. For instance, a food system may be capable of offering nutritious and affordable food while also providing decent livelihoods. However, it may simultaneously deplete local water resources, rendering it environmentally unsustainable. The sustainability of food systems can be assessed by evaluating their performance across social, economic, and environmental dimensions. Consequently, a sustainable food system can be defined as one that explicitly addresses the requirements of society, the economy, and the environment over time, ensuring sustainable outcomes across all three dimensions and effectively managing the inherent trade-offs between them (TFRA, 2019). Every food system possesses distinct qualities and components. For example, a food system might be able to provide healthy and affordable food while also supporting livelihoods. However, it could also have a detrimental impact on local water resources, making it environmentally unsustainable. The sustainability of food systems can be evaluated by considering their performance in social, economic, and environmental aspects. In this context, a sustainable food system can be defined as one that actively considers the needs of society, the economy, and the environment over time, striving for sustainable outcomes in all three dimensions and effectively managing the trade-offs that may arise between them (Pinstrup & Watson, 2011). In this report, we build on the conceptualization.

3.6 Food Security

The adoption of the Integrated Phase Classification (IPC) marks a significant milestone in addressing food insecurity. Universally recognized by the global community, the IPC serves as a framework for assessing the severity of food emergencies. By employing standardized criteria and terminology, this five-phase scale aims to facilitate a rapid comprehension of a crisis or potential crisis by governments and other humanitarian organizations, enabling them to promptly respond and intervene (NETFEW, 2022). During the current analysis period (November 2021 - April 2022), approximately 437,000 individuals, which accounts for 13% of the analyzed population of 3.4 million, are estimated to be facing significant levels of acute food insecurity, classified as IPC Phase 3 or higher. Within this group, it is estimated that around 22,000 people (1% of the analyzed population) fall under IPC Phase 4 (Emergency), while approximately 415,000 people (12% of the analyzed population) are classified under IPC Phase 3 (Crisis). Additionally, about 785,000 people (23% of the analyzed population) are categorized under IPC Phase 2 (Stressed) (Figure 6).

Figure 6. People with insufficient food consumption and those using crisis level food-based coping strategies (millions)
Source; World Food Programme (WFP) (2022)

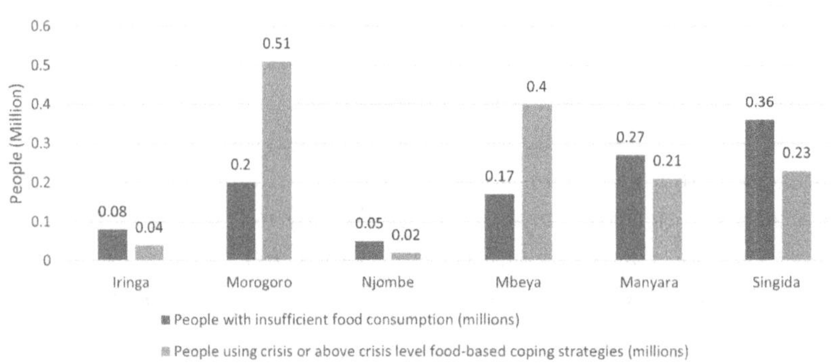

Among the 14 analyzed councils in Tanzania, consisting of 12 district councils and the rural areas of two town councils (TCs), four councils (Handeni DC, Longido, Mkinga, and Monduli) are classified under IPC Phase 3 (Crisis). These councils have 20% to 30% of their respective populations experiencing significant levels of acute food insecurity (Phase 3 or above). The remaining councils are categorized under IPC Phase 2 (Stressed) (NETFEW, 2022).

3.7 Raifall Forecast

According to the statistics, rainfall in the Iringa, Njombe, Mbeya, and Morogoro areas will be average to below average from November 2022 to April 2023, while it will be average to above average in the Singida region (TMA, 2022). However, it is anticipated that rainfall will increase starting in February 2023 and that it will begin to rain out of season in May 2023. Low rainfall in November and December 2022 will have an influence on plant development and the imal folder as well. A decrease in the depth of water in lakes, rivers, and marshes will have an impact on irrigation activities and the generation of hydroelectric electricity in Mtera and Nyumba ya Mungu. The TMA recommends infrastructure for water collecting, the sustainable use of water sources, and strategy for sustainable agriculture like engaging in agroforestry. The figure 20,30 and 50 potryas the rainfall below average and 20,45 and 35 average and below avaerage level (Fig 7).

4. CONCLUSION

This chapter highlighted the key CSA practices by SHFs within Tanzania were the adaption of better seed varieties, application of weather information, and efficient fertilizer use, intercropping, crop residue and canal irrigation. The impact of climate also affects food system. In conclusion, the food system approach offers an analytical framework that offers fresh perspectives on intervention strategies that broaden the “menu of chances” for tying together important food policy tools and incorporating many stakeholders. This study has concluded that, Tanzania food systems are frequently threatened by factors such as dietary changes, urbanization, population expansion, inadequate livelihoods, and climate change. The review also, found that climate change affects the business of value chain actors including small-scale farmers, effect to agro-dealers like poor seed multiplications, cooperatives and processors like having insufficient raw materials. In addition, Tanzania already has existing agricultural practices and technologies that fall under the umbrella of Climate Smart Agriculture (CSA). These practices are actively employed by farmers to mitigate different production risks. In order to fully integrate CSA into mainstream agricultural practices, it is crucial to identify and map successful ongoing practices, as well as identify the necessary institutional and financial mechanisms for future implementation.

Figure 7. Rainfall forecast in (November, 2022 - April, 2023)
Source: TMA (2022)

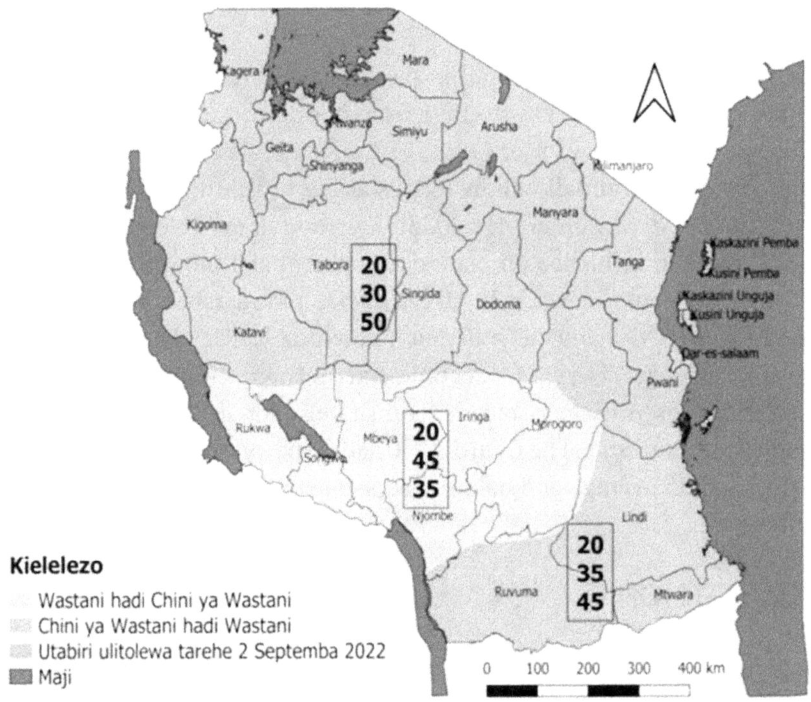

One recommendation outlined in the chapter is for the national agriculture policy to place greater emphasis on subsidized agriculture, particularly considering that many farmers in the country are economically disadvantaged. However, it is also suggested that the removal of national credit and an increase in subsidies could potentially enhance farmers' adaptation strategies, making them less vulnerable to climate change and its effects.

REFERENCES

Adenle, A., Azad, H., & Joseph. (2015). *Global Assessment of Technological Innovation for Climate Change Adaptation and Mitigation in Developing World.* https://www.researchgate.net/publication/280135149_

AGRA. (2016). *Annual Report Towards Africa Agriculture Transformation.* https://agra.org/wp-content/uploads/2017/06/AGRA-2016-AR-final.pdf

Alexander. (1977). *Introduction to Soil Microbiology.* https://scirp.org/reference/referencespapers.aspx?referenceid=1453835

Allafrica.com. (2016). *Tanzania: Irrigation Plots Increase By 233,715 Hectares in Five Years.* https://allafrica.com/stories/202008040846.html

Anderson, L., Ostrom, A., & Mende, G. M. (2016). *Transformative Service Research: An Agenda for The Future.* https://www.sciencedirect.com/science/article/abs/pii/S0148296312002287

AU. (2003). *Maputo Declaration Agriculture and Food Security.* https://www.nepad.org/caadp/publication/au-2003-

Campbell, B. M., Vermeulen, S., Aggarwal, P., & Dolloff, C. (2016, June). *Reducing Risks to Food Security from Climate Change.* https://www.researchgate.net/publication/304619592_

Climate-ADAPT. (2022). *IPCC Sixth Assessment Report.* https://www.ipcc.ch/report/ar6/wg2/

CropMon. (2017). *Mobile Weather Services for Small Scale Farmers.* https://www.weatherimpact.com/wpcontent/uploads/2019/10

Dekeyser. (2019). *The Food Systems Approach in Practice Our Guide for Sustainable Transformation.* https://www.researchgate.net/publication/343255882

Eckstein & Schäfer. (2021). *Global Climate Risk Index 2021.* GermanWatch. https://www.germanwatch.org/en/19777

Ehiakpor, Yussif, & Gideon. (2021). *Adoption of Interrelated Sustainable Agricultural Practices Among Smallholder Farmers in Ghana.* https://www.researchgate.net/publication/346633908

Erenstein. (2002). *Crop Residue Mulching in Tropical and Semi-Tropical Countries: An Evaluation of Residue Availability and Other Technological Implications.* https://www.sciencedirect.com/science/article/pii/S0167198702000624

FAO. (2018). *Transforming Food and Agriculture to Achieve the SDGs*. https://www.fao.org/3/I9900EN/i9900en.pdf

FAO, IFAD, WHO, UNICEF, & WFP. (2019). *The State of Food Security and Nutrition in the World: Safeguarding Against Economic Slowdowns and Downturns*. https://www.fao.org/3/ca5162en/ca5162en.pdf

FAO. (2020). *Climate-Smart Agriculture: capturing the synergies among mitigation, adaptation and food security*. https://www.fao.org/3/aq548e/aq548e.pdf

Giller, Smaling, & Buresh. (1997). *The Influence of Organic and Inorganic Nutrient Inputs on Soil Organic Carbon Functional Groups Content and Maize Yields*. https://www.ncbi.nlm.nih.gov/pmc/articles/PMC8405894/

Gowing, Mzirai, & Hatibu. (1999). *Review of Rainwater Harvesting Techniques and Evidence for their Use in Semi-Arid Tanzania*. https://www.ajol.info/index.php/tjags/article/view/115921/105468

GSMA. (2016). *Weather Forecasting and Monitoring Mobile Solutions for Climate Resilience*. https://www.gsma.com/mobilefordevelopment/wp-content/uploads/2016/02/

IFDC. (2021). *IFDC Annual Report 2021*. https://ifdc.org/wpcontent/uploads/2022/06/IFDC-Annual-Report-2021.pdf

IFPRI. (2016). *Food Systems*. International Food Policy Research Institute. https://www.ifpri.org/topic/food-systems

IPCC. (2019). *Food Security*. https://www.ipcc.ch/site/assets/uploads/2019/08/2f.-Chapter-5_FINAL.pdf

Jeruto & Ouma. (2010). *Sustainable Horticultural Crop Production Through Intercropping: The case of fruits and vegetable crops: A review*. https://www.researchgate.net/publication/273009988_Sustainable_horticultural_crop_production_through_intercropping_The_case_of_fruits_and_vegetable_crops_A_review

Kang, G. H., & Lawson, T. (1985). *Alley Cropping Sequentially Cropped Maize and Cowpea with Leucaena on A Sandy Soil in Southern Nigeria*. https://www.jstor.org/stable/42934819

Kassie, M., Jaleta, M., Shiferaw, B., & Frank, M. (2013). *Adoption of Interrelated Sustainable Agricultural Practices in Smallholder Systems Evidence from Rural Tanzania*. https://www.researchgate.net/publication/256859403_

Kuzilwa, J., & Mpeta, D. M., & Niel, F. (2017). *Strategies and Policies for Agricultural Development in Tanzania*. https://www.taylorfrancis.com/chapters/edit/10.4324/9781315651422-3/strategies-policies-agricultural-development-tanzania-joseph-andrew-kuzilwa-mpeta-daniel-marianne-nylandsted-larsen-niels-fold

Levira. (2009). *Climate Change Impact in Agriculture Sector in Tanzania and Its Mitigation Measure*. https://iopscience.iop.org/article/10.1088/1755-1307/6/37/372049/pdf

Lyatuu, P. M., & Urasa. (2016). *Land Access, Livelihood Strategies and Rural Household Well-being in Mvomero District, Tanzania*. https://journals.ub.uni-koeln.de/index.php/JNRD/article/view/739/762

Makoi, J., Martin, V., & Wiebe, K. (2017). *Can Tanzania Feed Itself 2050 Estimating Cereal, Self Sufficiency 2050*. International Food Policy Research Institute. https://www.ifpri.org/publication/

Mashingo. (2010). *Vulnerability Assessment on Pastoral Livestock and Pasture Development to the Extreme Climatic Events in Tanzania*. Dar es Salaam.

Mkonda, M., & He, X. (2016). Production Trends of Food Crops: Opportunities, Challenges and Prospects to Improve Tanzanian Rural Livelihoods. *Horizon Research Publishing Coperation*, *4*(4), 51–59. doi:10.13189/nrc.2016.040402

Monela. (2014). *Access to and Adoption of Improved Seeds by Smallholder Farmers in Tanzania: Cases of Maize and Rice Seeds in Mbeya and Morogoro Regions Sokoine University of Agriculture*. https://www.suaire.sua.ac.tz/handle/123456789/601?show=full

Ndiritu, S. W., Bekele, S., & Kassie, M. (2014). *Are There Systematic Gender Differences in the Adoption of Sustainable Agricultural Intensification Practices Evidence from Kenya*. https://www.researchgate.net/publication/264242026

Nhemachena & Hassan. (2007). *Micro-level Analysis of Farmers' Adaptation to Climate Change in Southern Africa*. International Food Policy Research Institute.

Ogemah, V. (2017). *Effect Of Modified Spacing Arrangements, Fertilizer Use and Legume Intercrop on Prevalence of Cassava Brown Streak Disease in Western Kenya*. https://academicjournals.org/journal/AJAR/article-stat/6D2B91066572

Ojo, Ogundeji, & Belle. (2022, May). *Climate Change Perception and Impact of On-Farm Demonstration on Intensity of Adoption of Adaptation Strategies Among Smallholder Farmers in South Africa*. https://www.sciencedirect.com/science/article/pii/S0040162521004637

Onyeneke. (2022, November 9). *Impact of Climate Information Services on Crop Yield in Ebonyi State, Nigeria*. https://www.mdpi.com/2225-1154/11/1/7/pdf

Paavola. (2008). *Livelihood Vulnerability and Adaptation to Climate Change in Morogoro Tanzania*. https://www.researchgate.net/publication/254354050

Phahlane, Setshedi, & Phuthi. (2016). *Communication of Seasonal Weather Forecast Information in two Selected Provinces of South Africa*. Agricultural Research Council and Department of Africulture, Fisheries and Forestry.

Pinstrup & Watson. (2011). *Food Policy for Developing Countries the Role of Government in Global National and Local Food Systems*. https://www.researchgate.net/publication/239522608

Rajaram & Van Ginkel. (2019). *Prioritizing Achievable Goals for Food Security in the Developing World*. https://www.researchgate.net/publication/351250964

RLDC. (2009). *Farmers Adoption of Recommended Rice Varieties: A Case of Kilombero District of Morogoro Region, Tanzania*. https://www.ajol.info/index.php/sajae/article/view/121487/110948

Rogers. (2003). *Detailed Review of Rogers' Diffusion of Innovations Theory and Educational Technology-Related Studies Based on Rogers' Theory*. https://files.eric.ed.gov/fulltext/ED501453.pdf

Sarr, M., Ayele, B., Kimani, M., & Ruhindoka, R. (2021). *Who Benefits from Climate Friendly Agriculture, The Marginal Returns to a Rainfed System of Rice Intensification in Tanzania*. https://www.researchgate.net/publication/347879128

SNV. (2019). *Horticulture Cooperatives Final Report 2019*. https://snv.org/assets/explore/download/profiling_horticulture_cooperatives_final_report.pdf

Springmann, M., Clarck, M., Wiebe, K., & Calson, K. (2018, October 10). *Options For Keeping the Food System Within Environmental Limits*. https://www.nature.com/articles/s41586-018-0594-0

Taylor, L., Rezai, A., & Foley, D. (2016). *An Integrated Approach to Climate Change, Income Distribution, Employment and Economic Growth*. https://www.sciencedirect.com/science/article/abs/pii/S0921800915002323

Thinda, Ogundeji, & Belle. (2021). *Determinants of Relevant Constraints Inhibiting Farmers' Adoption of Climate Change Adaptation Strategies in South Africa*. https://www.researchgate.net/publication/342569250

V. I., Allender, S., & Aktins, V. J. (2019). *The Global Syndemic of Obesity, Undernutrition and Climate Change the Lancet Commission Report*. https://www.researchgate.net/publication/330676764_

Van Berkum, D. (2018). *Food Systems Approach Sustainable Solutions for a Sufficient.* Wageningen University and Research. https://research.wur.nl/en/publications/

WB. (2018). *Developing An Irrigation Fund in Tanzania.* World Bank Group. https://2030wrg.org/

WFP. (2021). *Global Hunger Crisis World Food Program.* https://www.wfp.org/global-hunger-crisis

Willett. (2022). *Food in the Anthropocene the EAT Lancet Commission on Healthy Diets from Sustainable Food Systems*. https://www.researchgate.net/publication/330443133_

Zeweld, Guido, V., Girmay, & Speedman. (2017). *Smallholder Farmers' Behavioral Intentions Towards Sustainable Agricultural Practices*. Academic Press.

Zhao, D., Yu, C., Jian, Y., & Wang, Y. (2018). *An Overview of Climate Change Impacts on The Society in China*. https://www.sciencedirect.com/science/article/pii/S1674927821000411

Zwicker, L., Swinburn, B., Hawkes, C., & Brownel, K. D. (2015, June). *Patchy Progress on Obesity Prevention: Emerging Examples, Entrenched Barriers and New Thinking*. National Library of Medicine. https://pubmed.ncbi.nlm.nih.gov/25703111/

Chapter 11
Traffic Sign Detection for Real-World Application Using Hybrid Deep Belief Network Classification

K. Aravinda
https://orcid.org/0000-0001-9497-8074
New Horizon College of Engineering, India

B. Santosh Kumar
New Horizon College of Engineering, India

Balasubramanian Prabhu Kavin
https://orcid.org/0000-0001-6939-4683
SRM Institute of Science and Technology, India

Arunadevi Thirumalraj
https://orcid.org/0009-0003-5396-6810
K. Ramakrishnan College of Technology, India

ABSTRACT

By integrating automated driving systems (ADS) and AI-driven advanced driver assistance systems (ADAS) like the traffic sign detection (TSD) technology, the automotive sector can develop smart and self-driving cars. Traffic signs (TS) play a crucial role in avoiding accidents and traffic congestion. Motorists need to understand the visual representations of various data elements incorporated in traffic symbols. There are often instances where drivers neglect TS located ahead of their vehicles, resulting in severe outcomes. This research offers an automatic TSD forecast utilising the hybrid deep belief network (HDBN) model for classification to address this issue.

DOI: 10.4018/979-8-3693-1396-1.ch011

When it comes to forecasting the future world of smart urban cities, the given HDBN model primarily focuses on high-precision traffic prediction. The rider sunflower optimization (RSFO) technique is utilised to improve the hyper parameter tuning, which improves the overall effectiveness of the traffic flow prediction process. Overall, the suggested TSD system is found to be a highly efficient method of detecting TS, performing exceptionally well in relation to precision, recall, accuracy, and F1. The suggested solution under evaluation appears to perform better in terms of accuracy than other current methods stated in this chapter.

1. INTRODUCTION

Road safety, regulating and managing traffic flow, and assisting drivers are all made possible by TS. According to a World Health Organization (WHO) study on the Worldwide Global Status report, over 1.35 million individuals died from starvation as a result of road traffic accidents in 2018 (Khan et al., 2023). As a result, the scientific and research community has been very interested in TSD and categorization because of its many applications in fields like automated vehicles, automated driving, and security monitoring (Tao et al., 2020). A traffic accident may be caused directly or indirectly by neglecting or failing to recognize these TS. However, in poor road conditions, the driver could purposefully or unintentionally fail to see TS (Saadna et al., 2019). Many automated driving employ high-definition (HD) mapping supply comprehensive information about the surrounding of the road (Zhang et al., 2021). However, using HD mapping costs money because of the time consuming and hard production process (Kim et al., 2021). Further significantly, HD mapping could be impacted by discrepancies among saved TS and recent updates (Rajendran et al., 2019). Along with guiding drivers, smart object recognition systems can make it simpler to manage the surrounds of the road, such as speed limit signs, road markings, and guard rails (Fazekas et al., 2022). Traffic surveillance systems, for instance, can rapidly analyse harm or defects using automated vehicles for monitoring purposes since it is challenging to manually examine every part of a road scene (Kortmann et al., 2022). As a consequence, traffic surveillance equipment is essential for keeping an eye across both road administration and vehicle choice systems. There are two substantial challenges to solve even if actual driving vehicles should have steady backing from a traffic monitoring system. The first problem is the poor picture clarity brought on by a variety of external elements, including the climate, light, and diffraction (Liu et al., 2022).

In real-world situations, it is extremely difficult to quickly and accurately identify TS when the car is moving at a random pace. Weather, ambient illumination, similar objects, panel quality decline, least partial blockage, etc are only a few of the factors that affect the detection and classification algorithms. In order to meet all of these obstacles, the device must be trustworthy and offer real time (RT) response before approaching the sign. Various kinds of signs, particularly the regulatory one, are generally addressed by the solutions that are available. However, there are other sorts of TS that need to be acknowledged (Babi'c et al., 2021). For instance, TESLA's Autopilot system analyses speed limits and can recognise them by modelling them on the instrument panel (Lambert, 2020). Today, to obtain the positively and negatively samples, Deep Learning (DL)-based traffic sign identification systems frequently use random sampling. TSD is currently dominated by DL. The algorithms are based on the enhanced network structure improve detection precision for small-sized TS (Zhang et al., 2020). The identification and recognition of common objects have shown encouraging results thanks to recent developments in DL. For TSD, DL techniques have already been used in several prior works, but their evaluation was only applicable to a very small subset of traffic-sign categories (Tabernik & Skočaj, 2019).

The following are the paper's major contributions:

- The new hybrid-DBN model for classification that is suggested in this research allows us to attain a test accuracy rating of 99.01 percent.
- The RSFO algorithm is used as a hyperparameter tuning optimizer to improve the prediction accuracy of the DBN framework, thereby improving overall effectiveness of the traffic flow forecasting process.
- The GTSRB dataset is used in this paper to train the proposed TSD model, which is subsequently evaluated on various TS. The suggested technique is then examined and verified on an embedded system (Raspberry Pi 4).
- Overall, the suggested TSD system is likely to be a high efficiency method of detecting TS, performing exceptionally well in terms of precision, recall, accuracy, and F1. The suggested solution under evaluation appears to perform best in terms of accuracy than other current methods stated in this paper.

The remaining sections of the study are organised in the form of shadows: The relevant works are summarised in Section 2, the suggested model is briefly explained in Section 3, the results and validation analysis are shown in Section 4, and conclusion is provided in Section 5.

2. RELATED WORKS

Table 1.

Authors	Methodology	Performance
Zhang, X., and Zhang, Z. (2023)	To provide a few shots object learning (FSOD) approach for spotting traffic signs. This technique modifies the original model's backbone network and incorporates dropout, resulting in improved detection precision and reduced overfitting risk. It is also proposed that a region proposal network with an improved attention framework be utilised to provide more particular focus candidate boxes.	The reworked algorithm optimizes the 5-way 3-shot and 5-way 5-shot activities by 4.27 and 1.64 percent, respectively, when compared to the baseline model. In this study, the PASCAL VOC dataset and model framework are used. The results demonstrate that the proposed outperforms several recent few-shot image detection methods.
N. Triki et al. (2023)	To provide a system for identifying the most efficient TSD and classification (TSC) methods in terms of accuracy and speed of processing by evaluating several TSD and classification (TSC) methods. As a result, the approach they offer mixes a deep CNN structure classifier and the Haar cascade algorithm. Following testing on the GTSRB dataset, the resulting TSC model is evaluated on various categories of traffic signs. Overall accuracy is 98.56 percent. We suggest the use of an unique attention-based deep CNN to improve classification performance.	We suggest the use of an unique attention-based deep CNN to improve classification performance. The acquired results outperform those published in earlier studies on traffic sign categorization, with 99.91 percent testing accuracy and 99 percent F1-measure thresholds, respectively, exceeding those reported in previous studies.
Sharma, V., et al. (2023)	A new method for identifying traffic signs based on Yolov4-tiny has been developed Aiming to resolve the problems with existing sign recognition models in complicated situations is the goal of this project. The performance of the system is poor because of a significant number of factors, low detection accuracy, and poor RT functionality. The suggested model has undergone several important changes. In order to reduce low-frequency feature redundancy and minimize parameter count, frequency convolution is incorporated into the spine network.	our method improves precision by 3.5 percent while increasing frame rate by 12.5 frames per second, all while keeping the model size relatively small at 16MB. With regards to recognition rate and detection speed, it outperforms yolov4-tiny in terms of RT detection and tracking.
Dalborgo, V., et al. (2023)	The current study goes into the creation of databases for AI learning as well as the design of Traffic Sign Recognition (TSR) devices. A new class is included in these datasets known as "foliage opacity," which will be used in TSR. Considering that even this software converts motorists into sensing devices in the landscape, the results illustrate the viability of this strategy in accelerating traffic sign servicing. Adhering to the proposed methodology allows anomalies in traffic laws to be recognised and submitted to the proper authorities for future repair, hence improving vehicle safety.	Furthermore, in light of several case studies, this research investigates the use and efficiency of several YOLO models.

continues on following page

Table 1. Continued

Authors	Methodology	Performance
Dang, T.P., et al. (2023)	use DL techniques in their study to create a model capable of determining and categorising traffic signs underneath a variety of severe weather circumstances. Climate data is acquired in a variety of ways and analysed using a variety of approaches. By using collected data, the YOLOv5s and YOLOv7 designs are trained. To improve accuracy, these YOLOv5s models are further enhanced by inserting the Worldwide Context (GC) block or the Squeeze-and-Excitation (SE) attention module.	YOLOv5s achieves 76.8 percent accuracy, YOLOv7 achieves 78 percent accuracy, YOLOv5s+SE media exposure module achieves 78.4 percent accuracy, and YOLOv5s+C3GC achieves 79.2 percent accuracy. The results show that the YOLOv5s+C3GC modelling improves precision in recognising blurred-distant objects significantly.
Kheder, M.Q., and Mohammed, A.A. (2023)	The primary goal of Kheder, M.Q., and Mohammed, A.A.'s research is to develop and improve a CNN-based framework that can be used effectively and accurately in embedded devices. This goal is accomplished through the use of various preprocessing methods. An improved network model known as LeNet-5 is being developed to evaluate traffic indicators. Furthermore, the proposed architecture network is trained using the enlarged GTSRB (EGTSRB) and German Traffic Surveillance Benchmark datasets.	The test results show that the upgraded LeNet-5 architecture achieves 99.12% and 99.78% accuracy on the GTSRB and EGTSRB datasets, respectively. This is a significant improvement over other cutting-edge publications in terms of accuracy.
Li Y, Li J, Meng P et al. (2023)	To achieve real-time and accuracy detection for traffic signs, we treat traffic sign detection as an end-to-end problem in this paper and proposed an improved one-stage traffic sign detection algorithm: Attention-YOLO V4. In order to achieve real-time and high-accurate detection of traffic signs so we analyzed the principle of YOLO V4 for small objects detection:	In order to improve the ability of YOLO Head detect traffic signs we combine channel attention mechanism with YOLO head. We used TT100K dataset to evaluate Attention-YOLO V4 algorithm, compared with the existing methods, our method achieve real-time and accurately performance in complex backgrounds.
William P et al., (2023)	This framework is limited to speed signs and achieves outstanding displays due to rapid but hearty computations. The division is based on shading data, and the recognition is based on a model coordinating computation. The human–computer interface is a voice that announces which sign has been discovered. Driver carelessness is a key element that commonly results in distortion of surroundings, for example, erroneous recognition or disregard of street and traffic signs.	Although significant progress has been made in developing a vehicle capable of autonomous guided driving, development has been slow because to concerns of speed, safety, and the ever-changing complexity of the on-road situation. Street plays a key role for the motorist in gathering data from the sign and then acting in a similar fashion.

3. PROPOSED WORKS

Only TSD systems that manage to process data quickly and with a high accuracy rate are considered performant. In fact, especially when driving at high speeds, the TSD system must recognise and identify a TS on the side of the road in order of appearance. Thus, the core of this situation is our methodology. The input frame is pre-processed to furthering and assist for the recognition system to identify the indicators. In our case, the RGB image is converted to greyscale using a colour space conversion algorithm, and the greyscale image is then resized to 32×32 pixels.

Figure 1. Flowchart of proposed work

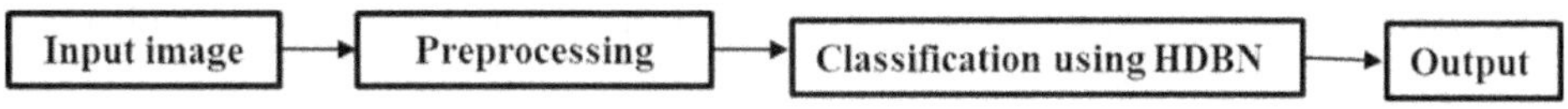

3.1. Dataset Description

The GTSRB dataset were collected by the Institute of Neural Information Processing (INI) at the University of Ulm in Germany (Schematic structure of DBN, n.d.). This dataset's innovative parameters can be used to define a variety of its components, including:

- Size: This dataset contains 51,839 road sign classes available in real traffic conditions in Germany.
- image pixelation: The image characteristics with in GTSRB dataset range from 15×15 to 250 ×250 pixels;
- Annotation: Experienced professionals who analyzed the dataset carefully labelled all image in the dataset with the relevant TS class. The annotation were made via an internet annotations software designed specifically for identifying TS, and they include both the TS adjacency matrix and the accompanying class value.
- Diversity: There is a diverse range of TS, including caution and prohibition signs, as well as speed limitation signs. Some TS may well be heavily represented or undervalued in the dataset due to differences in how frequently they arise in actual traffic scenarios. In reality, how many samples have been used to train a model for each class varies slightly. The number of photographs of TS is ten times more in the largest classes than in the lowest. Since some indications manifest more frequently than others in actual life, this is expected. To discuss this, the dataset is divided into three sections: training, recognition, and testing, ensuring that the model trained solely on the dataset is tested on a relevant data set.
- Technique of collection: The GTSRB data - set was gathered using a special camera mounted on a vehicle and recorded under such a variety of conditions to ensure the consistency of simulation results.
- Availability: This database is free to use in educational and technical settings.
- Usage: It is frequently used to assess the contribution of different TSD systems, including those based on deep learning. The dataset has now been utilised in a variety of identification-related tasks, competitions, and research. As a

consequence of the use of this dataset, numerous reducing TSD algorithms with potential applications in fields such as automated vehicles and traffic security have been developed.

- A number of classes in the GTSRB dataset make the dataset difficult to analyze due to the variance in traffic sign appearance caused by lighting, weather, and occlusion. These traffic indicators are seen in Figure 2 (Wang et al., 2019).

Figure 2. Traffic signs

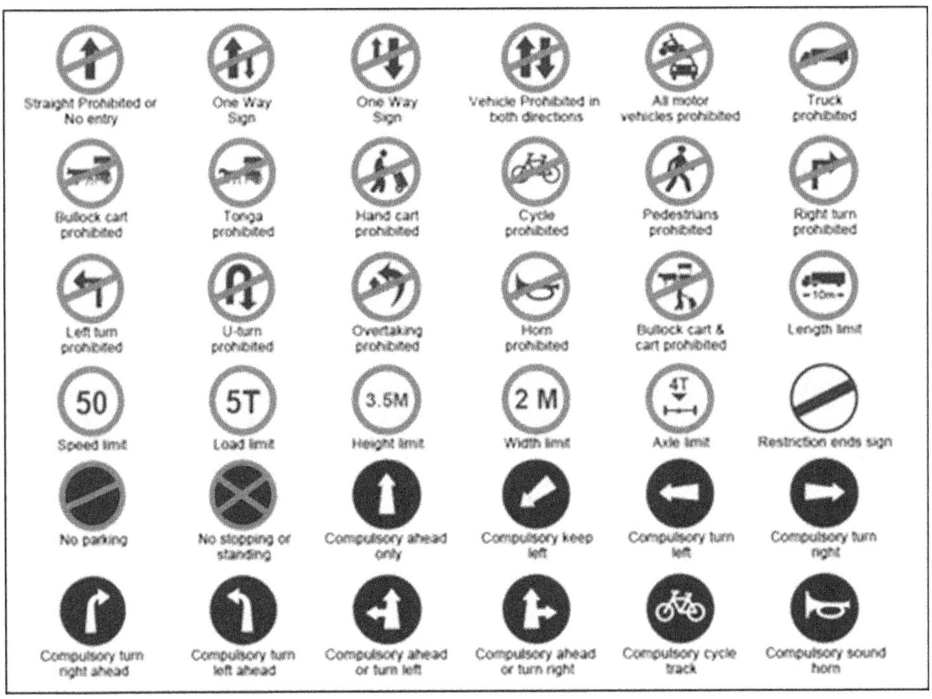

3.2. Pre-Processing

Preprocessing is an important stage in the TSD procedure because it improves the quality of the resulting image as well as the identification system's accurateness. TS images can be pre processed in a variety of ways, based on the particular properties of the visual information and the recognition system parameters. These strategies, such as reducing the noise, standardising raw data, and improving image quality relevant to categorization, are dependent entirely on the system specifications and data properties.

The following are some of the most common TSD preprocessing methods:

- During image cropping, background knowledge is removed by choosing a region of interest (ROI) that only includes the TS. This may help to reduce the amount of extraneous data that the DL system must assess, resulting in a faster and more accurate classification stage.
- Because some DL algorithms demand constant input dimensions, cropping images to a specific size can facilitate better the data. It can also minimise computational load and increase system speed.
- The pixel value in an image are scaled to a standard range, such as [0, 1] or [-1, 1], to normalise it. This improves system accuracy by decreasing the effect of lighting fluctuations and enhancing visual contrast.
- Image smoothing methodology such as Gaussian blur can be used to reduce image noise and highlight the TS margins. This has the potential to increase important classification characteristics and make edge detection systems more precise.
- To separate the colour and brightness information, use a color image converter such as HSV or YUV. This can help to improve the relevant relevant features and make color-based segmentation more accurate.

3.3. Proposed Classification

The HDBN model was developed to forecast the flow of traffic in future years. DBN is a crucial deep learning process. A stochastic ANN called a Restricted Boltzmann Machine (RBM) unit can learn a confidence interval from a set of input data. They are similar to Boltzmann machines. It directs the direction of all activities and resource use in order to produce effective results. The RBM improves accountability and transparency by allowing for concurrent interventions and reducing duplication. Each visible layer (VL) neural unit was connected to every hidden layer (HL) neural unit, although there is no link between the neural units across all tiers of the RBM technique. Additionally, the output of each RBM layer served as an input for the subsequent layer. A layered RBM infrastructure is deployed in the DBN approach's bottom layer. Layer after layer, the example data was learned using a greedy method. Figure 3 illustrates the DBN infrastructure. The DBN is made up of many different layers of latent variables, with only connections among units in different layers. The parameters of the intermediate layer RBM were derived analogously, utilizing the parameters obtained from the trained primary layer RBM as input. The training technique involves unsupervised learning. The following examples demonstrate the combined configuration energy of VL and HL in RBM:

$$E(v,h|\theta) = -\sum_{i=1}^{n} a_i v_i - \sum_{j=1}^{m} b_i h_i - \sum_{i=1}^{n}\sum_{j=1}^{m} v_i w_{ij} h_j \tag{1}$$

where θ={w_{ij}, a_i, b_j}, it might do this by connecting the visible unit's weighted value i and hidden unit *j*, a_i refers to the VL neurons' intrinsic bias and b_j can be HLs. Equation (2) yields the conditional probability of VL and HL, which is linked to Equation (3) as described in the following if the parameter θ is specified, depending on the energy function:

$$P(v,h|\theta) = \frac{e^{-E(v,h|\theta)}}{Z^{\theta}} \tag{2}$$

$$Z^{\theta} = \sum_{v,h} e^{-E(v,h|\theta)} \tag{3}$$

Figure 3. Structure of DBN

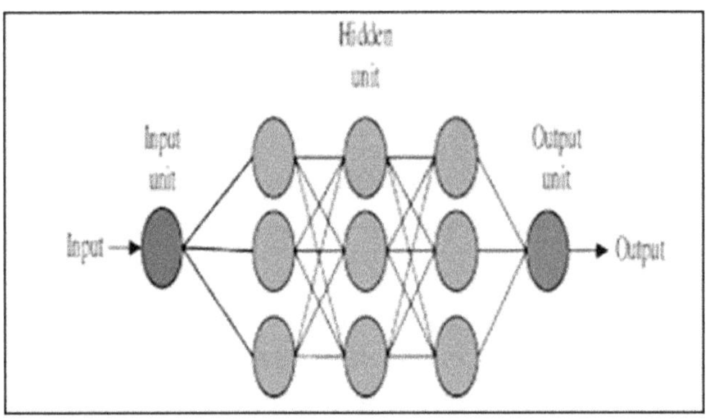

The activation probability of the jth neural unit of HL h is computed if the state of VLv is determined:

$$P(h_j = 1|v,\theta) = \sigma\left(b_j + \sum_i v_i w_{ij}\right) \tag{4}$$

The probability of the ith neuronal unit of VLv activating is attained if the HL state h can be identified:

$$P\left(v_i = 1|h,\theta\right) = \sigma\left(a_i + \sum_i h_i w_{ij}\right) \tag{5}$$

In which, $\sigma\left(x\right) = \frac{1}{1+\exp\left(-x\right)}$ is an formula for the sigmoid function., which is the activation function. Each neuron defines its state as either one or zero with probability P. The drive of trained RBM can be used to obtain the parameter method provided by log-likelihood function for the unsupervised learning technique as follows:

$$L\left(\theta\right) = \sum_{n=1}^{N} \ln\left(v^n, h\right) \tag{6}$$

$$\theta = argmaxL\left(\theta\right) = argmax\sum_{n=1}^{N} \ln\left(v^n, h\right) \tag{7}$$

The HDBN model is provided to improve the convergence rate by combining DBN with an appropriate learning process takes.

It is challenging to train the DBN model, especially when it comes to training many RBMs. Therefore, using contrastive divergence to fix the right learning data is necessary in order to train the DBN model. A training technique that is unbalanced due to a relatively high learning rate, and one that is poorly balanced due to a low learning rate. The HDBN model is developed utilising the adaptive learning step (ALS) to compute the effective learning rate in order to address this problem. Depending on how the sign changes, the step size is altered.

$$u\gamma_{ij}^{old} if\left(v_i h_{j0} - v_i h_{jk}\right)\left(v_i h_{j0}^{old} - v_i h_{j0}^{old}\right) > 0 \tag{8}$$

$$d\gamma_{ij}^{old} if\left(v_i h_{j0} - v_i h_{jk}\right)\left(v_i h_{j0}^{old} - v_i h_{jk}^{old}\right) > 0 \tag{9}$$

where γ_{ij}^{old} was the individual learning rate and u > 1 signifies the learning step is to take incremental factor, and d 1 the learning step's decremental factor.The step size would be increased and decreased based on the direction of any two upgrades that followed one another. The problems brought on by insufficient step size are avoided. Additionally, the DBN method's convergence rate improves.

3.4. Hyperparameter Tuning Using RSFO Algorithm

The final stage necessitates the use of paper in order to achieve the most optimal TSD results. To attain this, the RSFO algorithm was employed in this study, which merges the sunflower optimization (SFO) and rider optimization algorithms, to discover a superior TSD (ROA). The suggested approach combines the advantages of both ROA and SFO, resulting in enhanced TSD solutions and expedited convergence. The functionality of ROA is contingent upon the behavior of various riders at the terminus. These riders can be classified into four distinct categories: Rider, assaulter, follower, and overtaker must all be avoided. The rotation of the sun drives the SFO. SFO emulates rotational patterns that are influenced by nature. This approach allows for the identification of areas that can be optimized for improved performance. However, due to the significant number of processing steps involved, it also incurs high computational complexity. In this paper, a hybrid ROA approach combined with SFO was employed for hyperparameter tuning, enabling the production of a globally optimal solution with reduced computational steps and enhanced performance.

The following are the steps involved in hybrid K-means-based RSFO:

Step 1. Set the starting parameters.

Step 2. Utilizing the equations' provided Lagrangian optimization principle, evaluate the fitness function:

$$M=\sum_{p=1}^{k}\sum_{q=1}^{g}\left(w_{qp}d_{p}-l_{p}^{2}+\rho\ln w_{qp}+\rho\ln d_{q}-l_{p}^{2}\right)=\sum_{q=1}^{g}\eta_{q}\left(-1\sum_{p=1}^{k}w_{qp}\right) \tag{10}$$

Here, d_q refers to a data item, and the symbol for a user-defined constant is ρ, lp relates to the traffic sign, lp is utilising fuzzy membership, g is the total data, and K denotes the total signs. The aforementioned equation will therefore be derived as follows:

$$\left(\frac{\partial M}{\partial w_{qp}}\right)=d_{q}-l_{p}^{2}+\frac{\rho}{w_{qp}}-\eta_{q}=0,$$

$$w_{qp}=\frac{\rho}{\eta_{q}-d_{q}-l_{p}^{2}} \tag{11}$$

Since $\sum_{p=1}^{K}W_{qp}=1$, which is given as

$$\sum_{p=1}^{K}\left[\frac{1}{\eta_q}-d_q-l_p^{\,2}\right]=\frac{1}{\rho}, \tag{12}$$

we should ensure $W_{qp} \geq 0 \geq 0$.

Step 3. Update the riding groups' positions. We employed bypass rider for position updating to increase success rates. Riders who use a bypass always use a well-traveled path in secret from other riders. The following equation is used to update a position depending on a bypass rider:

$$B_{t+1}(r,p)=\vartheta\left[B_t(t,p)*m(p)+B_t(\mu,p)*\left[1-m(p)\right]\right] \tag{13}$$

Here, the parameters $\vartheta, t, m,$ and μ indicate the 0–1 range of random numbers. Then, k represents the user-defined number of iterations. Assuming μ=r, *T*he equation is rewritten as follows:

$$B_{t+1}(r,p)=\vartheta\left[B_t(t,p)*m(p)+B_t(r,p)*\left[1-m(p)\right]\right] \tag{14}$$

By using the sun's rotation, SFO makes changes to the destination or solution space. Sunflower rotates like the sun all the time. Consequently, the SFO position update is provided by

$$B(r,p)=B_t(r,p)+y_r\times g_r \tag{15}$$

Here, $B_t(r,p)$ shows the current location at the time t, $B_{t+1}(r,p)$ is the position as it was updated at time t+1, $B_{t+1}(r,p)$ shows the sunflower step, and g_r refers to the sunflower's orientation.

$$B_t(r,p)=\frac{B_{t+1}(r,p)}{y_r}\times g_r \tag{16}$$

End up replacing (15), which is the rider optimization position update, with (17), which is the sunflower optimization position update.

$$B_{t+1}(r,p) = \vartheta B_t(t,p) * m(p) + \left(\frac{B_{t+1}(r,p)}{y_r} \right) \times g_r * [1 - m(p)], \tag{17}$$

$$B_{t+1}(r,p) = \vartheta B_t(t,p) * m(p) + B_{t+1}(r,p)[1 - m(p)] - y_r \times g_r * [1 - m(p)] \tag{18}$$

After that, by rearranging (17) and (18), we obtain,

$$B_{t+1}(r,p) = \vartheta B_t(t,p) * m(p) + B_{t+1}(r,p) - \frac{B_{t+1}(r,p) m(p)}{y_r g_r} + y_r g_r m(p),$$

$$\frac{B_{t+1}(r,p)}{B_{t+1}(r,p)\vartheta} + \vartheta B_{r+1}(r,p) m(p) = \vartheta \left[\frac{B_t(t,p) * m(p)}{y_r t_{r+} y_r t_r m(p)} \right],$$

$$B_{t+1}(r,p)[1 - \vartheta + \vartheta m(p)] = \vartheta \left[\frac{B_t(t,p) * m(p)}{y_r t_r + y_r t_r m(p)} \right] \tag{19}$$

The complete equation will then be formatted as,

$$B_{t+1}(r,p) = \frac{1}{[1 - \vartheta[1 - \vartheta m(p)]]} \left[\frac{\vartheta B_t(t,p) * m(p)}{y_r t_r [1 - m(p)]} \right] \tag{20}$$

Step 4. Identifying the ideal response

Here, m is the maximum fitness value is taken into account as the optimum solution and updates the rider optimization parameters.

Step 5. Termination circumstance

Until the specified number of iterations has been reached, steps 2 through 4 above are repeated.

4. RESULT AND DISCUSSION

On a computer with the following specifications, we first tested our TSD system: 8 GB RAM; Intel® Core (TM) i57200 64-bit CPU. Another test is then run on a Raspberry Pi 4 board Model B with an ARM Cortex-A72 quad-core processor running at 1.5 GHz, 64 bits of CPU, and 8 GB of RAM (Li et al., 2023).

4.1. Experimentation With ARM

We used a two-minute long video sequence that described a road scenario in order to test our TSD algorithm. From the point of the sign's detection to its classification, the processing time that was obtained was calculated. The algorithm takes an average of 0.08 seconds during this brief period to correctly identify each traffic sign that is detected..

4.2. Raspberry Pi Experimentation

The Raspberry Pi (Li et al., 2023) is widely used as a cost-effective and compact platform for the deployment of TSD systems. Its versatility in communicating with various sensors and cameras makes it an ideal choice. In the context of TSD systems, the Raspberry Pi serves as the foundation for implementing computer vision, machine learning, and deep learning algorithms. It can be seamlessly installed on roadside cameras or integrated into vehicles to offer drivers and traffic control systems RT information. Nevertheless, it is worth noting that the computing capacity of a Raspberry Pi-based system might pose certain limitations in terms of performance.

4.3. Performances Metrices

Classification models employ convolutional neural networks (R-CNN), solid-state drives (SSD), and Feature Pyramid Network (FPN) models.

The following provides a definition of the potential values that can be obtained in a categorization problem:

- True positives - TP: the total number of incidents that were rated positively;
- False negatives- FN: the percentage of positive occurrences that were labelled as negative;
- True negatives - TN: the total number of instances that were labelled as negative;
- False positives - FP: the number of negative events that were labelled as positive;

- F1, accuracy, precision, recall are utilized as measurements.

The accuracy of the classification is defined as the ratio of correct predictions to all predictions.

$$Accuracy = \frac{TP+TN}{TP+FP+TN+FN}; \tag{21}$$

The proportion of correctly identified positive instances to all correct predictions is defined as precision:

$$Precision = \frac{TP}{TP+FP}; \tag{22}$$

The proportion of correctly categorised positive instances to all positive instances is known as recall, often referred to as sensitivity or true positive rate.

$$Recall = \frac{TP}{TP+FP}; \tag{23}$$

F1 is a single value that combines recall and precision:

$$F1 = \frac{Precision * Recall}{Precision + Recall} \tag{24}$$

Table 1 shows the outcomes after the relevant characteristics have been fed into the classifier. R-CNN has 88.02%, SSD has 92.54%, FPN has 95.38% and the proposed model has 99.01% of accuracy. When the models are tested with precision, R-CNN has 63.7%, SSD has 55.2%, FP has 83.2% and the proposed model has 92.7%. When comparing with all other models, the proposed model achieved better performance in accuracy analysis with 99.01%. Figure 4 offering the graphical description of proposed model.

Table 1 shows the outcomes after the relevant characteristics have been fed into the classifier. R-CNN has 70.03%, SSD has 81.6%, FPN has 88.6% and the proposed model has 90.26% of recall. When the models are tested with F1, R-CNN has 66.8%, SSD has 78.5%, FP has 85.2% and the proposed model has 89.3%. When comparing with all other models, the proposed model achieved better performance in result analysis. Figure 5 offering the graphical description of proposed model.

Table 1. Analysis of various DL classifiers

Models	Accuracy (%)	Precision (%)	Recall (%)	F1 (%)
R-CNN	88.02	63.7	70.3	66.8
SSD	92.54	55.2	81.6	78.5
FPN	95.38	83.2	88.6	85.2
Proposed model	99.10	92.7	90.26	89.3

Figure 4. Accuracy and precision comparisons

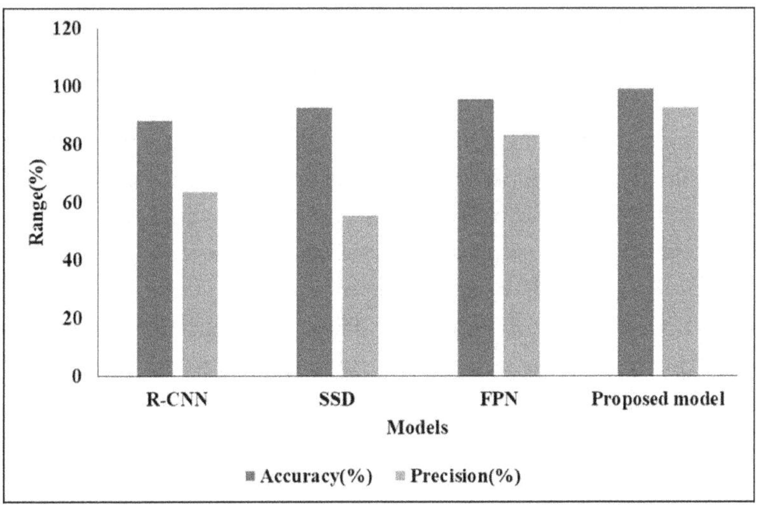

Figure 5. Recall and F1 comparisons

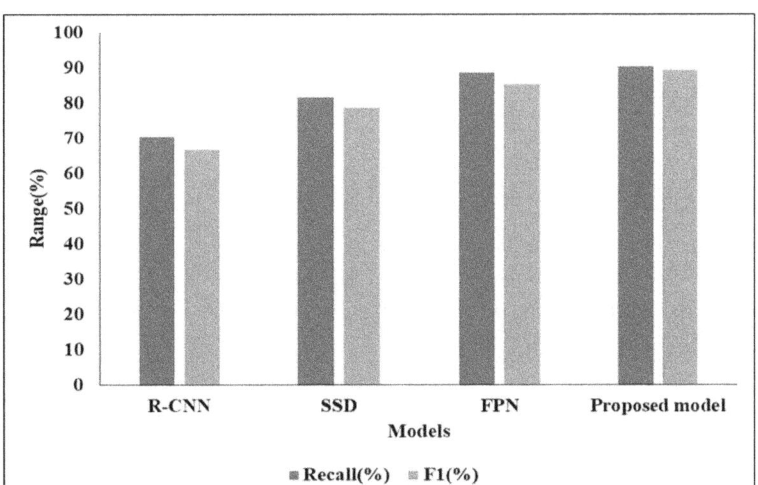

5. CONCLUSION

The products of the continual advancement of automated safety technologies are ADAS and ADS. These systems have the ability to handle all driving activities and contribute to the preservation of life and injury prevention by wanting to avoid potentially hazardous circumstances that could lead to accidents. TSD system is an important component of ADAS and ADS since it effectively and automatically recognises road signs in actual. The TSD system requires a high level of accuracy and cannot function properly if a traffic sign is ignored or misinterpreted. To ensure the accuracy of the system, this study proposes a mechanism. The recommended approach is to build a new TSD system based on the HDBN DL model classifier, with hyperparameter tuning optimized using the RSFO algorithm. The performance of the developed TSD model is evaluated on various traffic sign datasets, including the GTSRB dataset, achieving a testing accuracy rate of 99.01 percent. The system is assessed using different approaches such as video sequence analysis, RT camera monitoring, and RT embedded camera usage, all of which demonstrate strong performance. Notably, the testing step was successfully completed using our method on a CPU and a Raspberry Pi board, with processing durations of 0.06 seconds and 0.15 seconds respectively, ensuring full traffic sign recognition. In future work, we aim to expand our research by incorporating transfer learning techniques to recognize various categories of road signs used in different countries. This is necessary due to the diversity of hardware and software as well as the availability of road sign datasets. In addition, we intend to create an embedded system employing Nvidia hardware frameworks such as FPGA and Nano Jetson boards. The outcomes of these experiments will be compared with the results of actual automobile tests. Moreover, we intend to enhance the overall performance of the established TSD system by employing ensemble learning, a powerful technique in machine learning. Finally, to further validate our efforts, we are considering the possibility of collaborating with regional automakers.

REFERENCES

Babi'c, D., Babi'c, D., Fioli'c, M., & Šari'c, Ž. (2021). Analysis of Market-Ready Traffic Sign Recognition Systems in Cars: A Test Field Study. *Energies*, *14*(12), 3697. doi:10.3390/en14123697

Dalborgo, V., Murari, T. B., Madureira, V. S., Moraes, J. G. L., Bezerra, V. M. O., Santos, F. Q., Silva, A., & Monteiro, R. L. (2023). Traffic Sign Recognition with Deep Learning: Vegetation Occlusion Detection in Brazilian Environments. *Sensors (Basel)*, *23*(13), 5919. doi:10.3390/s23135919 PMID:37447772

Dang, T. P., Tran, N. T., To, V. H., & Tran Thi, M. K. (2023). Improved YOLOv5 for real-time traffic signs recognition in bad weather conditions. *The Journal of Supercomputing*, *79*(10), 1–19. doi:10.1007/s11227-023-05097-3

Fazekas, Z., Balázs, G., Gyulai, C., Potyondi, P., & Gáspár, P. (2022). Road-Type Detection Based on Traffic Sign and Lane Data. *Journal of Advanced Transportation*, *2022*, 6766455. doi:10.1155/2022/6766455

Jaya Mabel Rani, A., & Pravin, A. (2022). Clustering by Hybrid K-Means-Based Rider Sunflower Optimization Algorithm for Medical Data. *Advances in Fuzzy Systems*, *2022*, 2022. doi:10.1155/2022/7783196

Khan, M. A., Park, H., & Chae, J. (2023). A Lightweight Convolutional Neural Network (CNN) Architecture for Traffic Sign Recognition in Urban Road Networks. *Electronics (Basel)*, *12*(8), 1802. doi:10.3390/electronics12081802

Kheder, M. Q., & Mohammed, A. A. (2023). Improved traffic sign recognition system (itsrs) for autonomous vehicle based on deep convolutional neural network. *Multimedia Tools and Applications*, 1–21. doi:10.1007/s11042-023-15898-6

Kim, K., Cho, S., & Chung, W. (2021). HD map update for autonomous driving with crowdsourced data. *IEEE Robotics and Automation Letters*, *6*(2), 1895–1901. doi:10.1109/LRA.2021.3060406

Kortmann, F., Fassmeyer, P., Funk, B., & Drews, P. (2022). Watch out, pothole! featuring road damage detection in an end-to-end system for autonomous driving. *Data & Knowledge Engineering*, *142*, 102091. doi:10.1016/j.datak.2022.102091

Lambert, F. (2020). Tesla Releases New Software Update to Visually Detect Speed Limit Signs, and More. *Electrek*, *5*, 100113.

Li, Y., Li, J., & Meng, P. (2023, August 25). Attention-YOLOV4: a real-time and high-accurate traffic sign detection algorithm. Multimedia Tools and Applications. *Diagnostics (Basel)*, *13*(17), 2760. PMID:37685300

Liu, W., Ren, G., Yu, R., Guo, S., Zhu, J., & Zhang, L. (2022). Image-adaptive YOLO for object detection in adverse weather conditions. *Proceedings of the AAAI Conference on Artificial Intelligence*. 10.1609/aaai.v36i2.20072

Rajendran, S. P., Shine, L., Pradeep, R., & Vijayaraghavan, S. (2019). Real-time traffic sign recognition using YOLOv3 based detector. *Proceedings of the 2019 10th International Conference on Computing, Communication and Networking Technologies (ICCCNT)*. 10.1109/ICCCNT45670.2019.8944890

Saadna, Y., Behloul, A., & Mezzoudj, S. (2019). Speed limit sign detection and recognition system using SVM and MNIST datasets. *Neural Computing & Applications*, *31*(9), 1–11. doi:10.1007/s00521-018-03994-w

Schematic structure of DBN. (n.d.). https://www.researchgate.net/figure/Schematic-structure-of-DBN_fig2_370541543

Sharma, V., Dhiman, P., & Rout, R. K. (2023). Improved traffic sign recognition algorithm based on YOLOv4-tiny. *Journal of Visual Communication and Image Representation*, *91*, 103774. doi:10.1016/j.jvcir.2023.103774

Tabernik, D., & Skočaj, D. (2019). Deep learning for large-scale traffic-sign detection and recognition. *IEEE Transactions on Intelligent Transportation Systems*, *21*(4), 1427–1440. doi:10.1109/TITS.2019.2913588

Tao, X., Gong, Y., Shi, W., & Cheng, D. (2020). Object detection with class aware region proposal network and focused attention objective. *Pattern Recognition Letters*, *130*, 353–361. doi:10.1016/j.patrec.2018.09.025

Triki, N., Karray, M., & Ksantini, M. (2023). A real-time traffic sign recognition method using a new attention-based deep convolutional neural network for smart vehicles. *Applied Sciences (Basel, Switzerland)*, *13*(8), 4793. doi:10.3390/app13084793

Wang, G., Yuan, Y., & Guo, W. (2019). An improved rider optimization algorithm for solving engineering optimization problems. *IEEE Access : Practical Innovations, Open Solutions*, *7*, 80570–80576. doi:10.1109/ACCESS.2019.2923468

William, P., Darwante, N. K., Pawar, A. B., Jawale, M. A., & Verma, A. (2022). Framework for implementation of smart driver assistance system using augmented reality. In *Mobile Radio Communications and 5G Networks: Proceedings of Third MRCN 2022* (pp. 365-383). Singapore: Springer Nature.

Zhang, J., Xie, Z., Sun, J., Zou, X., & Wang, J. (2020). A cascaded R-CNN with multiscale attention and imbalanced samples for traffic sign detection. *IEEE Access : Practical Innovations, Open Solutions*, *8*, 29742–29754. doi:10.1109/ACCESS.2020.2972338

Zhang, P., Zhang, M., & Liu, J. (2021). Real-time HD map change detection for crowdsourcing update based on mid-to-high-end sensors. *Sensors (Basel)*, *21*(7), 2477. doi:10.3390/s21072477 PMID:33918443

Zhang, X., & Zhang, Z. (2023). Research on a traffic sign recognition method under small sample conditions. *Sensors (Basel)*, *23*(11), 5091. doi:10.3390/s23115091 PMID:37299816

Compilation of References

Abdmouleh, Z., Gastli, A., Ben-Brahim, L., Haouari, M., & Al-Emadi, N. A. (2017). Review of optimization techniques applied for the integration of distributed generation from renewable energy sources. *Renewable Energy, 113*, 266–280. doi:10.1016/j.renene.2017.05.087

Adefarati, T., & Bansal, R. C. (2019). Reliability, economic and environmental analysis of a microgrid system in the presence of renewable energy resources. *Applied Energy, 236*, 1089–1114. doi:10.1016/j.apenergy.2018.12.050

Adenle, A., Azad, H., & Joseph. (2015). *Global Assessment of Technological Innovation for Climate Change Adaptation and Mitigation in Developing World.* https://www.researchgate.net/publication/280135149_

Aggarwal, K. C., Kumar, V., & Dass, T. (1983). Snowmelt run-off for a catchment of Beas Basin. In *Proceedings of the First National Symposium on Seasonal Snowcover* (*Vol. 2*, pp. 43-63). Academic Press.

Aggarwal, A., Jain, S. K., Lohani, A. K., & Jain, N. (2013). Glacial lake outburst flood risk assessment using combined approaches of remote sensing, GIS and dam break modelling. *Geomatics, Natural Hazards & Risk.*

AGRA. (2016). *Annual Report Towards Africa Agriculture Transformation.* https://agra.org/wp-content/uploads/2017/06/AGRA-2016-AR-final.pdf

Ahamad, S., Veeraiah, V., Ramesh, J. V. N., Rajadevi, R., Reeja, S. R., Pramanik, S., & Gupta, A. (2023). *Deep Learning based Cancer Detection Technique. In Thrust Technologies' Effect on Image Processing*. IGI Global.

Ahmad, T., Madonski, R., Zhang, D., Huang, C., & Mujeeb, A. (2022). Data-driven probabilistic machine learning in sustainable smart energy/smart energy systems: Key developments, challenges, and future research opportunities in the context of smart grid paradigm. *Renewable & Sustainable Energy Reviews, 160*, 112128. doi:10.1016/j.rser.2022.112128

Ahrens, C. D., & Henson, R. (2018). *Essentials of Meteorology An Invitation to the Atmosphere* (8th ed.). Cengage Learning.

Akorede, M. F., Hizam, H., & Pouresmaeil, E. (2010). Distributed energy resources and benefits to the environment. *Renewable & Sustainable Energy Reviews*, *14*(2), 724–734. doi:10.1016/j.rser.2009.10.025

Alexander. (1977). *Introduction to Soil Microbiology*. https://scirp.org/reference/referencespapers.aspx?referenceid=1453835

Allafrica.com. (2016). *Tanzania: Irrigation Plots Increase By 233,715 Hectares in Five Years.* https://allafrica.com/stories/202008040846.html

Allen, S., Sattar, A., King, O., Zhang, G., Bhattacharya, A., Yao, T., & Bolch, T. (2021). Glacial lake outburst flood hazard under current and future conditions: First insights from a transboundary Himalayan basin. *Nat Hazards Earth Syst Sci Discuss,* 1–30.

Allen, S. K., Linsbauer, A., Randhawa, S. S., Huggel, C., Rana, P., & Kumari, A. (2016). Glacial lake outburst flood risk in Himachal Pradesh, India: An integrative and anticipatory approach considering current and future threats. *Natural Hazards*, *84*(3), 1741–1763. doi:10.1007/s11069-016-2511-x

Allen, S. K., Rastner, P., Arora, M., Huggel, C., & Stoffel, M. (2015). Lake outburst and debris flow disaster at Kedarnath, June 2013: Hydrometeorological triggering and topographic predisposition. *Landslides*, 1–13. doi:10.1007/s10346-015-0584-3

Alliance for Water Stewardship. (n.d.). https://a4ws.org/

Almeida, C. R., Teodoro, A. C., & Goncalves, A. (2021). Study of the urban heat island (UHI) using remote sensing data/techniques: A systematic review. *Environments (Basel, Switzerland)*, *8*(10), 105. doi:10.3390/environments8100105

American Water Works Association. (2016). *Water treatment plant design* (5th ed.). McGraw-Hill Education.

Anand, R., Singh, J., Pandey, D. K., Pandey, B., Nassa, V. K., & Pramanik, S. (2022). Modern Technique for Interactive Communication in LEACH-Based Ad Hoc Wireless Sensor Network. In M. M. Ghonge, S. Pramanik, & A. D. Potgantwar (Eds.), *Software Defined Networking for Ad Hoc Networks*. Springer. doi:10.1007/978-3-030-91149-2_3

Anderson, L., Ostrom, A., & Mende, G. M. (2016). *Transformative Service Research: An Agenda for The Future.* https://www.sciencedirect.com/science/article/abs/pii/S0148296312002287

Andersson, G., Donalek, P., Farmer, R., Hatziargyriou, N., Kamwa, I., Kundur, P., Martins, N., Paserba, J., Pourbeik, P., Sanchez-Gasca, J., Schulz, R., Stankovic, A., Taylor, C., & Vittal, V. (2005). Causes of the 2003 major grid blackouts in North America and Europe, and recommended means to improve system dynamic performance. *IEEE Transactions on Power Systems*, *20*(4), 1922–1928. doi:10.1109/TPWRS.2005.857942

Andrews, D. G. (2010). *An introduction to atmospheric physics*. Cambridge University Press. doi:10.1017/CBO9780511800788

Andrushia, A. D., Neebha, T. M., Patricia, A. T., Sagayam, K. M., & Pramanik, S. (2023). Capsule Network based Disease Classification for VitisVinifera Leaves. *Neural Computing & Applications*. Advance online publication. doi:10.1007/s00521-023-09058-y

Anisha, Reddy, Nguyen, Kumar, & Hanafiah. (2022). Intelligent Systems and Machine Learning for Industry: Advancements, Challenges and Practices. CRC Press, Taylor & Francis.

Apostolopoulos, I. D., & Mpesiana, T. A. (2020). Covid-19: Automatic detection from X-ray images utilizing transfer learning with convolutional neural networks. *Physical and Engineering Sciences in Medicine*, *43*(2), 635–640. doi:10.1007/s13246-020-00865-4 PMID:32524445

Arendt, A., Bliss, A., Bolch, T., Cogley, J. G., Gardner, A. S., Hagen, J.-O., Hock, R., Huss, M., Kaser, G., & Kienholz, C. (2015). *Randolph glacier inventory – a dataset of global glacier outlines. Version 5.0.* Global Land Ice Measurements from Space. Available from: https://www.glims.org/RGI/00çrgi50çTechnicalNote.pdf

AU. (2003). *Maputo Declaration Agriculture and Food Security.* https://www.nepad.org/caadp/publication/au-2003-

Aydin, H. E., & Iban, M. C. (2023). Predicting and analyzing flood susceptibility using boosting-based ensemble machine learning algorithms with SHapley Additive exPlanations. *Natural Hazards*, *116*(3), 2957–2991. doi:10.1007/s11069-022-05793-y

Babi'c, D., Babi'c, D., Fioli'c, M., & Šari'c, Ž. (2021). Analysis of Market-Ready Traffic Sign Recognition Systems in Cars: A Test Field Study. *Energies*, *14*(12), 3697. doi:10.3390/en14123697

Bagchi, A. K. (1981). *Snowmelt runoff in Beas basin using satellite images* [PhD Thesis]. Department of Civil Engineering, University of Roorkee, India.

Bakken, S. M. (2019). *The effect of dimensionality reduction on signature-based target detection for hyperspectral remote sensing. In CubeSats and SmallSats for remote sensing III.* International Society for Optics and Photonics.

Bansal, R., Jenipher, B., Nisha, V., Makhan, R., Dilip, K., Pramanik, S., Roy, S., & Gupta, A. (2022). Big Data Architecture for Network Security. In Cyber Security and Network Security. Wiley. doi:10.1002/9781119812555.ch11

Bansal, R., Obaid, A. J., Gupta, A., Singh, R., & Pramanik, S. (2021). Impact of Big Data on Digital Transformation in 5G Era. *2nd International Conference on Physics and Applied Sciences (ICPAS 2021).* 10.1088/1742-6596/1963/1/012170

Becker, F. (1987). The impact of spectral emissivity on the measurement of land surface temperature from a satellite. *International Journal of Remote Sensing*, *8*(10), 1509–1522. doi:10.1080/01431168708954793

Becker, F., & Li, Z. L. (1990). Towards a local split window method over land surfaces. *International Journal of Remote Sensing*, *11*(3), 369–393. doi:10.1080/01431169008955028

Benediktsson, J. A., Palmason, J. A., & Sveinsson, J. R. (2005). Classification of hyperspectral data from urban areas based on extended morphological profiles. *IEEE Transactions on Geoscience and Remote Sensing*, *43*(3), 480–491. doi:10.1109/TGRS.2004.842478

Bhargava, A., & Finger, S. (2017). Internet of Things: Applications and challenges in smart cities: A case study of IBM smart city projects. In *Smart Cities* (pp. 61–84). Springer.

Bhattacharya, A., Ghosal, A., Obaid, A. J., Krit, S., Shukla, V. K., Mandal, K., & Pramanik, S. (2021). Unsupervised Summarization Approach with Computational Statistics of Microblog Data. In D. Samanta, R. R. Althar, S. Pramanik, & S. Dutta (Eds.), *Methodologies and Applications of Computational Statistics for Machine Learning* (pp. 23–37). IGI Global. doi:10.4018/978-1-7998-7701-1.ch002

Big Data Analytics for Smart Water Management: A Review. (2019). *Water Resources Management*, *33*(11), 4101–4123.

Big Data Analytics for Water Demand Prediction: A Review. (2020). *Water Resources Management*, *34*(13), 4637–4659.

Big Data Analytics for Water Quality Prediction: A Review. (2021). *Water Resources Management*, *35*(15), 5591–5613.

Big Data Analytics for Water-Related Hazard Prediction: A Review. (2022). *Water Resources Management*, *36*(17), 6459–6481.

Blazquez, J., Fuentes, R., & Manzano, B. (2020). On some economic principles of the energy transition. *Energy Policy*, *147*, 111807. doi:10.1016/j.enpol.2020.111807

Bo, D. Z. (2009). A target detection algorithm for urban areas using HSRH imagery. *Joint Urban Remote Sensing Event*, 1–5.

Bodansky, D. (2016). The Paris climate change agreement: A new hope? *The American Journal of International Law*, *110*(2), 288–319. doi:10.5305/amerjintelaw.110.2.0288

Boesch, H., Liu, Y., Tamminen, J., Yang, D., Palmer, P. I., Lindqvist, H., Cai, Z., Che, K., Di Noia, A., Feng, L., Hakkarainen, J., Ialongo, I., Kalaitzi, N., Karppinen, T., Kivi, R., Kivimäki, E., Parker, R. J., Preval, S., Wang, J., ... Chen, H. (2021). Monitoring Greenhouse Gases from Space. *Remote Sensing (Basel)*, *13*(14), 2700. Advance online publication. doi:10.3390/rs13142700

Bolch, T., Buchroithner, M. F., Peters, J., Baessler, M., & Bajracharya, S. (2008). Identification of glacier motion andpotentially dangerous glacial lakes in the Mt. Everest region, Nepal using spaceborne imagery. *Natural Hazards and Earth System Sciences*, *8*(6), 1329–1340. doi:10.5194/nhess-8-1329-2008

Bolstad, P. V., Swift, L., Collins, F., & Régnière, J. (1998). Measured and predicted air temperatures at basin to regional scales in the southern Appalachian mountains. *Agricultural and Forest Meteorology*, *91*(3-4), 161–176. doi:10.1016/S0168-1923(98)00076-8

Boussouf, S., Fernández, T., & Hart, A. B. (2023). Landslide susceptibility mapping using maximum entropy (MaxEnt) and geographically weighted logistic regression (GWLR) models in the Río Aguas catchment (Almería, SE Spain). *Natural Hazards*, *117*(1), 207–235. doi:10.1007/s11069-023-05857-7

Caballero, R. (2014). *Physics of the Atmosphere*. IOP Publishing. doi:10.1088/978-0-7503-1052-9

Campbell, B. M., Vermeulen, S., Aggarwal, P., & Dolloff, C. (2016, June). *Reducing Risks to Food Security from Climate Change*. https://www.researchgate.net/publication/304619592_

Cardell, J., & Tabors, R. (1997). Operation and control in a competitive market: Distributed generation in a restructured industry. *The Energy Journal (Cambridge, Mass.)*, *18*(1_suppl), 18. doi:10.5547/ISSN0195-6574-EJ-Vol18-NoSI-6

Chandan, R. R., Soni, S., Raj, A., Veeraiah, V., Dhabliya, D., Pramanik, S., & Gupta, A. (2023). Genetic Algorithm and Machine Learning. In Advanced Bioinspiration Methods for Healthcare Standards, Policies, and Reform. IGI Global. doi:10.4018/978-1-6684-5656-9

Chisense, C. M. (2012). Classification of roof materials using hyperspectral data. *The International Archives of the Photogrammetry, Remote Sensing and Spatial Information Sciences*, *39*(B7), 103–107. doi:10.5194/isprsarchives-XXXIX-B7-103-2012

Chobanoglous, G., Burton, F. L., & Stensel, H. D. (2014). *Wastewater Treatment* (5th ed.). McGraw-Hill Education.

Choudhary, S., Narayan, V., Faiz, M., & Pramanik, S. (2022). Fuzzy Approach-Based Stable Energy-Efficient AODV Routing Protocol in Mobile Ad hoc Networks. In M. M. Ghonge, S. Pramanik, & A. D. Potgantwar (Eds.), *Software Defined Networking for Ad Hoc Networks*. Springer. doi:10.1007/978-3-030-91149-2_6

Clague, J. J., & Evans, S. G. (1992). A self-arresting moraine dam failure, St. Elias Mountains, British Columbia. Current Research, Part A. Geological Survey of Canada Paper, 92(1A), 185–188.

Clague, J. J., & Evans, S. G. (2000). A review of catastrophic drainage of moraine-dammed lakes in British Columbia. *Quaternary Science Reviews*, *19*(17-18), 1763–1783. doi:10.1016/S0277-3791(00)00090-1

Climate-ADAPT. (2022). *IPCC Sixth Assessment Report*. https://www.ipcc.ch/report/ar6/wg2/

Coll, C., & Caselles, V. (1997). A split-window algorithm for land surface temperature from advanced very high resolution radiometer data: Validation and algorithm comparison. *Journal of Geophysical Research*, *102*(D14), 16697–16713. doi:10.1029/97JD00929

Cortes, C., & Vapnik, V. (1995). Support-vector networks. *Machine Learning*, *20*(3), 273–297. doi:10.1007/BF00994018

CropMon. (2017). *Mobile Weather Services for Small Scale Farmers*. https://www.weatherimpact.com/wpcontent/uploads/2019/10

Dalal, N. a. (2005). Histograms of oriented gradients for human detection. *IEEE Computer Society Conference on Computer Vision and Pattern Recognition, 1*, 886-893.

Dalborgo, V., Murari, T. B., Madureira, V. S., Moraes, J. G. L., Bezerra, V. M. O., Santos, F. Q., Silva, A., & Monteiro, R. L. (2023). Traffic Sign Recognition with Deep Learning: Vegetation Occlusion Detection in Brazilian Environments. *Sensors (Basel), 23*(13), 5919. doi:10.3390/s23135919 PMID:37447772

Dang, T. P., Tran, N. T., To, V. H., & Tran Thi, M. K. (2023). Improved YOLOv5 for real-time traffic signs recognition in bad weather conditions. *The Journal of Supercomputing, 79*(10), 1–19. doi:10.1007/s11227-023-05097-3

De Carvalho, O. A. (2000). Spectral correlation mapper (SCM): an improvement on the spectral angle mapper (SAM). In *Summaries of the 9th JPL Airborne Earth Science Workshop*. JPL Publication.

Dekeyser. (2019). *The Food Systems Approach in Practice Our Guide for Sustainable Transformation*. https://www.researchgate.net/publication/343255882

Derdouri, A., Wang, R., Mutayama, Y., & Osaragi, T. (2021). Understanding the links between LULC changes and SUHI in cities: Insights from two-decadal studies (2001–2020). *Remote Sensing (Basel), 13*(18), 3654. doi:10.3390/rs13183654

Dhamodaran, S., Ahamad, S., Ramesh, J. V. N., Muthugurunathan, G., Manikandan, K., Pramanik, S., & Pandey, D. (2023). Food Quality Assessment using Image Processing Technique. In Thrust Technologies' Effect on Image Processing. IGI Global.

Dhamodaran, S., Ahamad, S., Ramesh, J. V. N., Sathappan, S., Namdev, A., Kanse, R. R., & Pramanik, S. (2023). *Fire Detection System Utilizing an Aggregate Technique in UAV and Cloud Computing. In Thrust Technologies' Effect on Image Processing*. IGI Global.

Dickinson, R. E. (1983). Land surface processes and climate—Surface albedos and energy balance. *Advances in Geophysics, 25*, 305–353. doi:10.1016/S0065-2687(08)60176-4

Dileep, G. J. R. E. (2020). A survey on smart grid technologies and applications. *Renewable Energy, 146*, 2589–2625. doi:10.1016/j.renene.2019.08.092

Dkhili, N., Eynard, J., Thil, S., & Grieu, S. (2020). A survey of modelling and smart management tools for power grids with prolific distributed generation. *Sustainable Energy. Grids and Networks, 21*, 100284.

Dushyant, K., Muskan, G., Gupta, A., & Pramanik, S. (2022). Utilizing Machine Learning and Deep Learning in Cyber security: An Innovative Approach. In M. M. Ghonge, S. Pramanik, R. Mangrulkar, & D. N. Le (Eds.), *Cyber security and Digital Forensics*. Wiley. doi:10.1002/9781119795667.ch12

Dutta, S., Pramanik, S., & Bandyopadhyay, S. K. (2021). Prediction of Weight Gain during COVID-19 for Avoiding Complication in Health. *International Journal of Medical Science and Current Research, 4*(3), 1042–1052.

Eckstein & Schäfer. (2021). *Global Climate Risk Index 2021*. GermanWatch. https://www.germanwatch.org/en/19777

Ehiakpor, Yussif, & Gideon. (2021). *Adoption of Interrelated Sustainable Agricultural Practices Among Smallholder Farmers in Ghana*. https://www.researchgate.net/publication/346633908

Ehsan, A., & Yang, Q. (2018). Optimal integration and planning of renewable distributed generation in the power distribution networks: A review of analytical techniques. *Applied Energy*, *210*, 44–59. doi:10.1016/j.apenergy.2017.10.106

Emerging Technologies for Wastewater Treatment and In-Plant Wet Weather Management Fact Sheet. (n.d.). Environmental Protection Agency (EPA). https://www.epa.gov/sustainable-water-infrastructure/emerging-technologies-wastewater-treatment-and-plant-wet-weather

Erenstein. (2002). *Crop Residue Mulching in Tropical and Semi-Tropical Countries: An Evaluation of Residue Availability and Other Technological Implications*. https://www.sciencedirect.com/science/article/pii/S0167198702000624

Espe, E., Potdar, V., & Chang, E. (2018). Prosumer communities and relationships in smart grids: A literature review, evolution and future directions. *Energies*, *11*(10), 2528. doi:10.3390/en11102528

Eswaran, V., & Mohan, M. (2018). Emerging technologies for wastewater treatment: A review. *Journal of Chemical Technology and Biotechnology*, *93*(10), 2747–2767.

Fadnavis, S., Mahajan, A. S., Choudhury, A. D., Roy, C., Singh, M., Biswas, M. S., & Sarin, M. M. (2020). Atmospheric Aerosols and Trace Gases. In R. Krishnan, J. Sanjay, C. Gnanaseelan, M. Mujumdar, A. Kulkarni, & S. Chakraborty (Eds.), *Assessment of Climate Change over the Indian Region: A Report of the Ministry of Earth Sciences (MoES), Government of India* (pp. 93–116). Springer Singapore. doi:10.1007/978-981-15-4327-2_5

Falkenmark, M., & Rockström, J. (2006). The new blue and green water paradigm: Breaking new ground for water resources planning and management. *Journal of Water Resources Planning and Management*, *132*(3), 129–132. doi:10.1061/(ASCE)0733-9496(2006)132:3(129)

Fang, X., Misra, S., Xue, G., & Yang, D. (2011). Smart grid—The new and improved power grid: A survey. *IEEE Communications Surveys and Tutorials*, *14*(4), 944–980. doi:10.1109/SURV.2011.101911.00087

FAO, IFAD, WHO, UNICEF, & WFP. (2019). *The State of Food Security and Nutrition in the World: Safeguarding Against Economic Slowdowns and Downturns*. https://www.fao.org/3/ca5162en/ca5162en.pdf

FAO. (2018). *Transforming Food and Agriculture to Achieve the SDGs*. https://www.fao.org/3/I9900EN/i9900en.pdf

FAO. (2020). *Climate-Smart Agriculture: capturing the synergies among mitigation, adaptation and food security*. https://www.fao.org/3/aq548e/aq548e.pdf

Farooq, A. X. (2019). Texture and Shape Features for Grass Weed Classification Using Hyperspectral Remote Sensing Images. In *IEEE International Geoscience and Remote Sensing Symposium* (pp. 7208-7211). IEEE. 10.1109/IGARSS.2019.8900132

Fathabad, A. M., Cheng, J., Pan, K., & Qiu, F. (2020). Data-driven planning for renewable distributed generation integration. *IEEE Transactions on Power Systems*, *35*(6), 4357–4368. doi:10.1109/TPWRS.2020.3001235

Fauvel, M. J., Benediktsson, J. Ó. A., Chanussot, J., & Sveinsson, J. R. (2008). Spectral and spatial classification of hyperspectral data using SVMs and morphological profiles. *IEEE Transactions on Geoscience and Remote Sensing*, *46*(11), 3804–3814. doi:10.1109/TGRS.2008.922034

Fauvel, M. J., Chanussot, J., & Benediktsson, J. A. (2009). Kernel principal component analysis for the classification of hyperspectral remote sensing data over urban areas. *EURASIP Journal on Advances in Signal Processing*, *2009*(1), 1–14. doi:10.1155/2009/783194

Fazekas, Z., Balázs, G., Gyulai, C., Potyondi, P., & Gáspár, P. (2022). Road-Type Detection Based on Traffic Sign and Lane Data. *Journal of Advanced Transportation*, *2022*, 6766455. doi:10.1155/2022/6766455

Frey, H., Haeberli, W., Linsbauer, A., Huggel, C., & Paul, F. (2010). A multi-level strategy for anticipating futureglacier lake formation and associated hazard potentials. *Natural Hazards and Earth System Sciences*, *10*(2), 339–352. doi:10.5194/nhess-10-339-2010

Friedler, E., & Wilderer, P. A. (2009). *Introduction to wastewater treatment*. John Wiley & Sons.

Gallo, K. P., Easterling, D. R., & Peterson, T. (1996). The influence of land use/land cover on climatological values of the diurnal temperature range. *Journal of Climate*, *9*(11), 2941–2944. doi:10.1175/1520-0442(1996)009<2941:TIOLUC>2.0.CO;2

Garrigues, S., Remy, S., Chimot, J., Ades, M., Inness, A., Flemming, J., Kipling, Z., Laszlo, I., Benedetti, A., Ribas, R., Jafariserajehlou, S., Fougnie, B., Kondragunta, S., Engelen, R., Peuch, V.-H., Parrington, M., Bousserez, N., Vazquez Navarro, M., & Agusti-Panareda, A. (2022). Monitoring multiple satellite aerosol optical depth (AOD) products within the Copernicus Atmosphere Monitoring Service (CAMS) data assimilation system. *Atmospheric Chemistry and Physics*, *22*(22), 14657–14692. doi:10.5194/acp-22-14657-2022

Gautier, A., Jacqmin, J., & Poudou, J. C. (2018). The prosumers and the grid. *Journal of Regulatory Economics*, *53*(1), 100–126. doi:10.1007/s11149-018-9350-5

Ghamisi, P., Bernhard, H., & Xiao, X. Z. (2016). Hyperspectral and LiDAR data fusion using extinction profiles and deep convolutional neural network. *IEEE Journal of Selected Topics in Applied Earth Observations and Remote Sensing*, *10*(6), 3011–3024. doi:10.1109/JSTARS.2016.2634863

Ghosh, R., Bhunia, R., Pramanik, S., Mohanty, S., & Patnaik, P. K. (2023). Smart City Healthcare System for Survival Forecast for Cardiac Attack Situations using Machine Learning Techniques. In *Data-Driven Mathematical Modeling in Smart Cities*. IGI Global. doi:10.4018/978-1-6684-6408-3.ch019

Gilany, N., & Iqbal, J. (2020). Geospatial analysis and simulation of glacial lake outburst flood hazard in Shyok Basin of Pakistan. *Environmental Earth Sciences*, *79*(6), 1–11. doi:10.1007/s12665-020-8867-y

Giller, Smaling, & Buresh. (1997). *The Influence of Organic and Inorganic Nutrient Inputs on Soil Organic Carbon Functional Groups Content and Maize Yields*. https://www.ncbi.nlm.nih.gov/pmc/articles/PMC8405894/

Gleick, P. H. (2003). Water use. *Annual Review of Environment and Resources*, *28*(1), 275–314. doi:10.1146/annurev.energy.28.040202.122849

Global Water Futures: Perspectives on the Future of Water Resources and Management. (2017). UNESCO.

Gowing, Mzirai, & Hatibu. (1999). *Review of Rainwater Harvesting Techniques and Evidence for their Use in Semi-Arid Tanzania*. https://www.ajol.info/index.php/tjags/article/view/115921/105468

GSMA. (2016). *Weather Forecasting and Monitoring Mobile Solutions for Climate Resilience*. https://www.gsma.com/mobilefordevelopment/wp-content/uploads/2016/02/

Guardieiro, V., Raimundo, M. M., & Poco, J. (2023). Enforcing fairness using ensemble of diverse Pareto-optimal models. *Data Mining and Knowledge Discovery*, *37*(5), 1930–1958. doi:10.1007/s10618-023-00922-y

Guo, G., Wu, Z., Cao, Z., Chen, Y., & Yang, Z. (2020). A multilevel statistical technique to identify the dominant landscape metrics of greenspace for determining land surface temperature. *Sustainable Cities and Society*, *61*, 102263. doi:10.1016/j.scs.2020.102263

Gupta, A., Verma, A., & Pramanik, S. (2022). Security Aspects in Advanced Image Processing Techniques for COVID-19. In S. Pramanik, A. Sharma, S. Bhatia, & D. N. Le (Eds.), *An Interdisciplinary Approach to Modern Network Security*. CRC Press.

Hahnel, U. J., Herberz, M., Pena-Bello, A., Parra, D., & Brosch, T. (2020). Becoming prosumer: Revealing trading preferences and decision-making strategies in peer-to-peer energy communities. *Energy Policy*, *137*, 111098. doi:10.1016/j.enpol.2019.111098

Hall, D. K., Riggs, G. A., & Salomonson, V. V. (1995). Development V.V. Salomonson, I. Appel Remote Sensing of Environment 89 (2004) 351–360 359 of methods for mapping global snow cover using moderate resolution imaging spectroradiometer data. *Remote Sensing of Environment*, *54*(2), 127–140. doi:10.1016/0034-4257(95)00137-P

Hall, D. K., Riggs, G. A., Salomonson, V. V., DeGirolamo, N. E., Bayr, K. J., & Jin, J. M. (2002). MODIS Snow-cover products. *Remote Sensing of Environment*, *83*(1-2), 181–194. doi:10.1016/S0034-4257(02)00095-0

Hall, J. W. (2014). Robust climate policies under uncertainty: A comparison of robust decision making and info-gap methods. *Risk Analysis*, *34*(4), 673–683. PMID:22519664

Han, Y. H., Kim, H., Choi, J., & Kim, Y. (2012). A shape–size index extraction for classification of high resolution multispectral satellite images. *International Journal of Remote Sensing*, *33*(6), 1682–1700. doi:10.1080/01431161.2011.599348

Haralick, R. M., Shanmugam, K., & Dinstein, I. H. (1973). Textural features for image classification. *IEEE Transactions on Systems, Man, and Cybernetics*, *6*(6), 610–621. doi:10.1109/TSMC.1973.4309314

Hardwick, S. R., Toumi, R., Pfeifer, M., Turner, E. C., Nilus, R., & Ewers, R. M. (2015). The relationship between leaf area index and microclimate in tropical forest and oil palm plantation: Forest disturbance drives changes in microclimate. *Agricultural and Forest Meteorology*, *201*, 187–195. doi:10.1016/j.agrformet.2014.11.010 PMID:28148995

Haritashya, U. K. (2005). *Glacio-Hydrological Studies of the Gangotri Glacier Basin, Himalayas* [Doctoral dissertation]. Indian Institute of Technology, Roorkee, India.

Hassani, H., Razavi-Far, R., & Saif, M. (2022). Real-time out-of-step prediction control to prevent emerging blackouts in power systems: A reinforcement learning approach. *Applied Energy*, *314*, 118861. doi:10.1016/j.apenergy.2022.118861

Heiden, U. K., Segl, K., Roessner, S., & Kaufmann, H. (2007). Determination of robust spectral features for identification of urban surface materials in hyperspectral remote sensing data. *Remote Sensing of Environment*, *111*(4), 537–552. doi:10.1016/j.rse.2007.04.008

Hemaja, Charan, & Charan. (2023). *Weather forecasting using digital image processing*. Academic Press.

Huggel, C., Kaab, A., Haeberli, W., Teysseire, P., & Paul, F. (2002). Remote sensing based assessment of hazards from glacier lake outbursts: A case study in the Swiss Alps. *Canadian Geotechnical Journal*, *39*(2), 316–330. doi:10.1139/t01-099

Huntingford, C., Jeffers, E. S., Bonsall, M. B., Christensen, H. M., Lees, T., & Yang, H. (2019). Machine learning and artificial intelligence to aid climate change research and preparedness. *Environmental Research Letters*, *14*(12), 124007. Advance online publication. doi:10.1088/1748-9326/ab4e55

Huntley, B. J. (2023). Solar Energy, Temperature and Rainfall. In B. J. Huntley (Ed.), *Ecology of Angola: Terrestrial Biomes and Ecoregions* (pp. 95–125). Springer. doi:10.1007/978-3-031-18923-4_5

IFDC. (2021). *IFDC Annual Report 2021*. https://ifdc.org/wpcontent/uploads/2022/06/IFDC-Annual-Report-2021.pdf

IFPRI. (2016). *Food Systems.* International Food Policy Research Institute. https://www.ifpri.org/topic/food-systems

Igamberdiev, R. M., Grenzdoerffer, G., Bill, R., Schubert, H., Bachmann, M., & Lennartz, B. (2011). Determination of chlorophyll content of small water bodies (kettle holes) using hyperspectral airborne data. *International Journal of Applied Earth Observation and Geoinformation, 13*(6), 912–921. doi:10.1016/j.jag.2011.04.001

IPCC. (2019). *Food Security.* https://www.ipcc.ch/site/assets/uploads/2019/08/2f.-Chapter-5_FINAL.pdf

Ito, A., & Hajima, T. (2020). Biogeophysical and biogeochemical impacts of land-use change simulated by MIROC-ES2L. *Progress in Earth and Planetary Science, 7*(45), 1–15. doi:10.1186/s40645-020-00372-w

Izuka, K., Itoh, M., Shiodera, S., Matsubara, T., Dohar, M., & Watanabe, K. (2018). Advantages of unmanned aerial vehicle (UAV) photogrammetry for landscape analysis compared with satellite data: A case study of postmining sites in Indonesia. *Cogent Geoscience, 4*(1), 1498180. doi:10.1080/23312041.2018.1498180

Jain, S. K. (2001). *Snowmelt runoff modeling and sedimentation studies in Satluj basin using remote sensing and GIS* [Unpublished PhD thesis]. University of Roorkee.

Jain, S. K., Goswami, A., & Saraf, K. (2008). Determination of Land Surface Temperature and Its Lapse Rate in the Satluj River Basin Using NOAA Data. *International Journal of Remote Sensing, 29*(11), 3091–3103. doi:10.1080/01431160701468992

Jaya Mabel Rani, A., & Pravin, A. (2022). Clustering by Hybrid K-Means-Based Rider Sunflower Optimization Algorithm for Medical Data. *Advances in Fuzzy Systems, 2022*, 2022. doi:10.1155/2022/7783196

Jayasingh, R., & Kumar, J. (2022). Speckle noise removal by SORAMA segmentation in Digital Image Processing to facilitate precise robotic surgery. *International Journal of Reliable and Quality E-Healthcare, 11*(1), 1–19. Advance online publication. doi:10.4018/IJRQEH.295083

Jeruto & Ouma. (2010). *Sustainable Horticultural Crop Production Through Intercropping: The case of fruits and vegetable crops: A review.* https://www.researchgate.net/publication/273009988_Sustainable_horticultural_crop_production_through_intercropping_The_case_of_fruits_and_vegetable_crops_A_review

Jeyram, A., Tiwari, R. S., & Sharma, K. P. (1983). Snowmelt Runoff Using Landsat Imageries. *Proceedings of the First National Symposium on Snowcover, 2*, 11–21.

Jiang, Cheng, & Gao. (2022). CSIP-Net: Convolutional Satellite Image Prediction Network for Meteorological Satellite Infrared Observation Imaging. *Atmosphere, 14*(1), 2. DOI: doi:10.3390/atmos14010025

Jiang, G. M., Zhou, W., & Liu, R. (2013). Development of split-window algorithm for land surface temperature estimation from the VIRR/FY-3A measurements. *IEEE Geoscience and Remote Sensing Letters*, *10*(4), 952–956. doi:10.1109/LGRS.2013.2255859

Jiménez-Muñoz, J. C., Sobrino, J. A., Skoković, D., Mattar, C., & Cristóbal, J. (2014). Land surface temperature retrieval methods from Landsat-8 thermal infrared sensor data. *IEEE Geoscience and Remote Sensing Letters*, *11*(10), 1840–1843. doi:10.1109/LGRS.2014.2312032

Judd, S. J. (2017). The future of desalination: Technology, economics, and the environment. *Science*, *357*(6347), aad3527. https://www.science.org/doi/10.1126/science.1200488

Kang, G. H., & Lawson, T. (1985). *Alley Cropping Sequentially Cropped Maize and Cowpea with Leucaena on A Sandy Soil in Southern Nigeria*. https://www.jstor.org/stable/42934819

Kassie, M., Jaleta, M., Shiferaw, B., & Frank, M. (2013). *Adoption of Interrelated Sustainable Agricultural Practices in Smallholder Systems Evidence from Rural Tanzania*. https://www.researchgate.net/publication/256859403_

Key, J. R., Collins, J. B., Fowler, C., & Stone, R. S. (1997). High-latitude surface temperature estimates from thermal satellite data. *Remote Sensing of Environment*, *61*(2), 302–309. doi:10.1016/S0034-4257(97)89497-7

Key, J., & Haefliger, M. (1992). Arctic ice surface temperature retrieval from AVHRR thermal channels. *Journal of Geophysical Research*, *97*(D5), 5885–5893. doi:10.1029/92JD00348

Khandelwal, S., Goyal, R., Kaul, N., & Mathew, A. (2018). Assessment of land surface temperature variation due to change in elevation of area surrounding Jaipur, India. *The Egyptian Journal of Remote Sensing and Space Sciences*, *21*(1), 87–94. doi:10.1016/j.ejrs.2017.01.005

Khanh, P. T., Ngọc, T. H., & Pramanik, S. (2023). Future of Smart Agriculture Techniques and Applications. In A. Khang & I. G. I. Global (Eds.), *Advanced Technologies and AI-Equipped IoT Applications in High Tech Agriculture*. doi:10.4018/978-1-6684-9231-4.ch021

Khan, M. A., Park, H., & Chae, J. (2023). A Lightweight Convolutional Neural Network (CNN) Architecture for Traffic Sign Recognition in Urban Road Networks. *Electronics (Basel)*, *12*(8), 1802. doi:10.3390/electronics12081802

Kheder, M. Q., & Mohammed, A. A. (2023). Improved traffic sign recognition system (itsrs) for autonomous vehicle based on deep convolutional neural network. *Multimedia Tools and Applications*, 1–21. doi:10.1007/s11042-023-15898-6

Khodadadzadeh, M. J.-D., Jun Li, Plaza, A., Ghassemian, H., Bioucas-Dias, J. M., & Xia Li. (2014). Spectral–spatial classification of hyperspectral data using local and global probabilities for mixed pixel characterization. *IEEE Transactions on Geoscience and Remote Sensing*, *52*(10), 6298–6314. doi:10.1109/TGRS.2013.2296031

Khvostikov, I. A. (1965). *The upper Layers of the Atmosphere*. National Aeronautics and Space Administration.

Kim, K., Cho, S., & Chung, W. (2021). HD map update for autonomous driving with crowdsourced data. *IEEE Robotics and Automation Letters*, *6*(2), 1895–1901. doi:10.1109/LRA.2021.3060406

Kishor Kumar Reddy, Anisha, Khan, Hanafiah, P., & R. (2024). Sustainability in Industry 5.0: Theory and Applications. CRC Press, Taylor & Francis.

Kishor Kumar Reddy, Anisha, Marlia Mohd Hanafah, Pragathi, Ramana Murthy, & Madana Mohana. (2023). *An intelligent optimized cyclone intensity prediction framework using satellite images*. Springer Earth Science Informatics. https://doi.org/10.1007/s12145-023-00983-z

Kortmann, F., Fassmeyer, P., Funk, B., & Drews, P. (2022). Watch out, pothole! featuring road damage detection in an end-to-end system for autonomous driving. *Data & Knowledge Engineering*, *142*, 102091. doi:10.1016/j.datak.2022.102091

Kowe, P., Dube, T., Mushore, T. D., Ncube, A., Nyenda, T., Mutowo, G., & Kizilirmak, G. (2022). Impacts of the spatial configuration of built-up areas and urban vegetation on land surface temperature using spectral and local spatial autocorrelation indices. *Remote Sensing Letters*, *13*(12), 1222–1235. doi:10.1080/2150704X.2022.2142073

Krishna, D. K., Watham, T., Padalia, H., Srinet, R., & Nandy, S. (2023). Improved gross primary productivity estimation using semi empirical (PRELES) model for moist Indian sal forest. *Ecological Modelling*, *475*, 110175. doi:10.1016/j.ecolmodel.2022.110175

Krishnan, R., Gnanaseelan, C., Sanjay, J., Swapna, P., Dhara, C., Sabin, T. P., & Niyogi, D. (2020). Introduction to Climate Change Over the Indian Region. In R. Krishnan, J. Sanjay, C. Gnanaseelan, M. Mujumdar, A. Kulkarni, & S. Chakraborty (Eds.), *Assessment of Climate Change over the Indian Region: A Report of the Ministry of Earth Sciences (MoES), Government of India* (pp. 1–20). Springer Singapore. doi:10.1007/978-981-15-4327-2_1

Kulkarni, A. V., Rathore, B. P., Randhawa, S. S., Sood, R. K., & Kaul, M. (2005). Glacier Atlas Chenab Basin (H. P.). RSAM/SAC/RSA/RESIPA/MWRG-GLI/ SN22.

Kulkarni, A. V., Rathore, B. P., Mahajan, S., & Mathur, P. (2005). Alarming retreat of Parbati Glacier, Beas basin, Himachal Pradesh. *Current Science*, *88*(11), 1844–1850.

Kulkarni, A. V., Singh, S. K., Mathur, P., & Mishra, V. D. (2006). Algorithm to monitor snow cover using AWiFS data of RESOURCESAT for the Himalayan region. *International Journal of Remote Sensing*, *27*(12), 2449–2457. doi:10.1080/01431160500497820

Kumar, T. S., Panigrahy, S., Kumar, P., & Parihar, J. S. (2013). Classification of floristic composition of mangrove forests using hyperspectral data: Case study of Bhitarkanika National Park. *Journal of Coastal Conservation*, *17*(1), 121–132. doi:10.1007/s11852-012-0223-2

Kuo, B.-C. H.-H.-H.-C.-S., Ho, H.-H., Li, C.-H., Hung, C.-C., & Taur, J.-S. (2013). A kernel-based feature selection method for SVM with RBF kernel for hyperspectral image classification. *IEEE Journal of Selected Topics in Applied Earth Observations and Remote Sensing*, *7*(1), 317–326. doi:10.1109/JSTARS.2013.2262926

Kuzilwa, J., & Mpeta, D. M., & Niel, F. (2017). *Strategies and Policies for Agricultural Development in Tanzania*. https://www.taylorfrancis.com/chapters/edit/10.4324/9781315651422-3/strategies-policies-agricultural-development-tanzania-joseph-andrew-kuzilwa-mpeta-daniel-marianne-nylandsted-larsen-niels-fold

Lakshmi, V., Czajkowski, K., Dubayah, R., & Susskind, J. (2001). Land surface air temperature mapping using TOVS and AVHRR. *International Journal of Remote Sensing*, *22*(4), 643–662. doi:10.1080/01431160050505900

Lambert, F. (2020). Tesla Releases New Software Update to Visually Detect Speed Limit Signs, and More. *Electrek*, *5*, 100113.

Lamnatou, C., Chemisana, D., & Cristofari, C. (2022). Smart grids and smart technologies in relation to photovoltaics, storage systems, buildings and the environment. *Renewable Energy*, *185*, 1376–1391. doi:10.1016/j.renene.2021.11.019

Lauwaet, D., De Ridder, K., Saeed, S., Brisson, E., Chatterjee, F., Van, L. N., & Hooyberghs, H. (2016). Assessing the current and future urban heat island of Brussels. *Urban Climate*, *15*, 1–15. doi:10.1016/j.uclim.2015.11.008

Lee, J., Esmaeili, A., & Showkatfar, M. (2018). Emerging technologies in wastewater treatment: A review. *Chemosphere*, *199*, 2740. https://www.sciencedirect.com/topics/earth-and-planetary-sciences/wastewater-treatment

Lee, K. H., Li, Z., Kim, Y. J., & Kokhanovsky, A. (2009). Atmospheric Aerosol Monitoring from Satellite Observations: A History of Three Decades. In Y. J. Kim, U. Platt, U. Platt, M. B. Gu, & H. Iwahashi (Eds.), *Atmospheric and Biological Environmental Monitoring* (pp. 13–38). Springer Netherlands. doi:10.1007/978-1-4020-9674-7_2

Levira. (2009). *Climate Change Impact in Agriculture Sector in Tanzania and Its Mitigation Measure*. https://iopscience.iop.org/article/10.1088/1755-1307/6/37/372049/pdf

Li, J., Carlson, B. E., Yung, Y. L., Lv, D., Hansen, J., Penner, J. E., Liao, H., Ramaswamy, V., Kahn, R. A., Zhang, P., Dubovik, O., Ding, A., Lacis, A. A., Zhang, L., & Dong, Y. (2022). Scattering and absorbing aerosols in the climate system. *Nature Reviews. Earth & Environment*, *3*(6), 363–379. doi:10.1038/s43017-022-00296-7

Lin, B., Xiaoming, T., Mai, X., Linhao, D., & Jianhua, L. (2017). Bayesian hyperspectral and multispectral image fusions via double matrix factorization. *IEEE Transactions on Geoscience and Remote Sensing*, *10*(10), 5666–5678. doi:10.1109/TGRS.2017.2711640

Liu, W., Ren, G., Yu, R., Guo, S., Zhu, J., & Zhang, L. (2022). Image-adaptive YOLO for object detection in adverse weather conditions. *Proceedings of the AAAI Conference on Artificial Intelligence*. 10.1609/aaai.v36i2.20072

Li, Y., Li, J., & Meng, P. (2023, August 25). Attention-YOLOV4: a real-time and high-accurate traffic sign detection algorithm. Multimedia Tools and Applications. *Diagnostics (Basel)*, *13*(17), 2760. PMID:37685300

Lou, Q. S. (2008). The application and prospecting of the hyperspectral remote sensing technology in marine research. *Trans Oceanol Limno*, 168-173.

Luo, G. G.-E., Chen, G., Tian, L., Qin, K., & Qian, S.-E. (2016). Minimum noise fraction versus principal component analysis as a preprocessing step for hyperspectral imagery denoising. *Canadian Journal of Remote Sensing*, *42*(2), 106–116. doi:10.1080/07038992.2016.1160772

Lu, Y., Lu, S., Horton, R., & Ren, T. (2014). An empirical model for estimating soil thermal conductivity from texture, water content, and bulk density. *Soil Science Society of America Journal*, *78*(6), 1859–1868. doi:10.2136/sssaj2014.05.0218

Lyatuu, P. M., & Urasa. (2016). *Land Access, Livelihood Strategies and Rural Household Well-being in Mvomero District, Tanzania*. https://journals.ub.uni-koeln.de/index.php/JNRD/article/view/739/762

Ma, Liu, Jiang, Jiang, Feng, & Hu. (2023). *A Novel ST-ViBe Algorithm for Satellite Fog Detection at Dawn and Dusk*. https://doi.org/10.3390/rs15092331

Mahmud, N., & Zahedi, A. (2016). Review of control strategies for voltage regulation of the smart distribution network with high penetration of renewable distributed generation. *Renewable & Sustainable Energy Reviews*, *64*, 582–595. doi:10.1016/j.rser.2016.06.030

Majeed, U., Rashid, I., Sattar, A., Allen, S., Stoffel, M., Nüsser, M., & Schmidt, S. (2021). Recession of Gya Glacier and the 2014 glacial lake outburst flood in the Trans-Himalayan region of Ladakh, India. *The Science of the Total Environment*, *756*, 144008. doi:10.1016/j.scitotenv.2020.144008 PMID:33293089

Makoi, J., Martin, V., & Wiebe, K. (2017). *Can Tanzania Feed Itself 2050 Estimating Cereal, Self Sufficiency 2050*. International Food Policy Research Institute. https://www.ifpri.org/publication/

Malhi, G. S., Kaur, M., & Kaushik, P. (2021). Impact of Climate Change on Agriculture and Its Mitigation Strategies: A Review. *Sustainability (Basel)*, *13*(3), 1318. Advance online publication. doi:10.3390/su13031318

Mall, P. K., Pramanik, S., Srivastava, S., Faiz, M., Sriramulu, S., & Kumar, M. N. (2023). FuzztNet-Based Modelling Smart Traffic System in Smart Cities Using Deep Learning Models. In *Data-Driven Mathematical Modeling in Smart Cities*. IGI Global. doi:10.4018/978-1-6684-6408-3.ch005

Mandal, A., Dutta, S., & Pramanik, S. (2021). Machine Intelligence of Pi from Geometrical Figures with Variable Parameters using SCILab. In D. Samanta, R. R. Althar, S. Pramanik, & S. Dutta (Eds.), *Methodologies and Applications of Computational Statistics for Machine Learning* (pp. 38–63). IGI Global. doi:10.4018/978-1-7998-7701-1.ch003

Manditereza, P. T., & Bansal, R. (2016). Renewable distributed generation: The hidden challenges–A review from the protection perspective. *Renewable & Sustainable Energy Reviews*, *58*, 1457–1465. doi:10.1016/j.rser.2015.12.276

Markard, J. (2018). The next phase of the energy transition and its implications for research and policy. *Nature Energy*, *3*(8), 628–633. doi:10.1038/s41560-018-0171-7

Mashingo. (2010). *Vulnerability Assessment on Pastoral Livestock and Pasture Development to the Extreme Climatic Events in Tanzania*. Dar es Salaam.

Melgani, F., & Bruzzone, L. (2004). Classification of hyperspectral remote sensing images with support vector machines. *IEEE Transactions on Geoscience and Remote Sensing*, *42*(8), 1778–1790. doi:10.1109/TGRS.2004.831865

Meslie, Y., Enbeyle, W., Pandey, B. K., Pramanik, S., Pandey, D., Dadeech, P., Belay, A., & Saini, A. (2021). Machine Intelligence-based Trend Analysis of COVID-19 for Total Daily Confirmed Cases in Asia and Africa. In D. Samanta, R. R. Althar, S. Pramanik, & S. Dutta (Eds.), *Methodologies and Applications of Computational Statistics for Machine Learning* (pp. 164–185). IGI Global. doi:10.4018/978-1-7998-7701-1.ch009

Metcalf & Eddy, Inc. (2019). *Wastewater Engineering: Treatment and Resource Recovery* (6th ed.). McGraw-Hill Education.

Metcalf & Eddy. (2013). *Wastewater engineering: Treatment and reuse* (5th ed.). McGraw-Hill Education.

Ministry of Jal Shakti, Government of India. (2021). National *Water Policy*. Author.

Ministry of Urban Development, Government of India. (2015). *Atal Mission for Rejuvenation and Urban Transformation*. AMRUT.

Ministry of Urban Development, Government of India. (2015). *Smart Cities Mission*. Author.

Mkonda, M., & He, X. (2016). Production Trends of Food Crops: Opportunities, Challenges and Prospects to Improve Tanzanian Rural Livelihoods. *Horizon Research Publishing Coperation*, *4*(4), 51–59. doi:10.13189/nrc.2016.040402

Mohammadi, M. (2012). Road classification and condition determination using hyperspectral imagery. *The International Archives of the Photogrammetry, Remote Sensing and Spatial Information Sciences*, *39*(B7). doi:10.5194/isprsarchives-XXXIX-B7-141-2012

Mohammad, P., Goswami, A., Chauhan, S., & Nayak, S. (2022). Machine learning algorithm based prediction of land use land cover and land surface temperature changes to characterize the surface urban heat island phenomena over Ahmedabad city, India. *Urban Climate*, *42*, 101–116. doi:10.1016/j.uclim.2022.101116

Molnia, B. F. (2009). Inventorying and monitoring the recent behavior of Afghanistan's glaciers, debris-covered glaciers, supraglacial lakes, and the potential for catastrophic flooding (jo ¨kulhlaups). In European Geosciences Union General Assembly 2009, Geophysical Research Abstracts (vol. 11). Academic Press.

Mondal, D., Ratnaparkhi, A., Deshpande, A., Deshpande, V., Kshirsagar, A. P., & Pramanik, S. (2023). Applications, Modern Trends and Challenges of Multiscale Modelling in Smart Cities. In *Data-Driven Mathematical Modeling in Smart Cities*. IGI Global. doi:10.4018/978-1-6684-6408-3.ch001

Monela. (2014). *Access to and Adoption of Improved Seeds by Smallholder Farmers in Tanzania: Cases of Maize and Rice Seeds in Mbeya and Morogoro Regions Sokoine University of Agriculture*. https://www.suaire.sua.ac.tz/handle/123456789/601?show=full

Montoya-Bueno, S., Muñoz-Hernández, J. I., & Contreras, J. (2016). Uncertainty management of renewable distributed generation. *Journal of Cleaner Production*, *138*, 103–118. doi:10.1016/j.jclepro.2016.02.135

Mool, P. K. (2005). *Monitoring of glaciers and glacial lakes from 1970 s to 2000 in Poiqu Basin, Tibet Autonomous Region, PR China*. Unpublished report prepared by ICIMOD, Kathmandu, Nepal, and CAREERI, Gansu, China.

Moskolaï, Abdou, & Dipanda, & Kolyang. (2021). Application of Deep Learning Architectures for Satellite Image Time Series Prediction. *Remote Sensing*, *13*(23), 4822. doi:10.3390/rs13234822

Mushtaq, Z., Sharma, M., Bangotra, P., Gautam, A. S., & Gautam, S. (2022). Atmospheric Aerosols: Some Highlights and Highlighters, Past to Recent Years. *Aerosol Science and Engineering*, *6*(2), 135–145. doi:10.1007/s41810-022-00133-w

Ndiritu, S. W., Bekele, S., & Kassie, M. (2014). *Are There Systematic Gender Differences in the Adoption of Sustainable Agricultural Intensification Practices Evidence from Kenya*. https://www.researchgate.net/publication/264242026

New Technology Trends in The Wastewater Management Industry. (2022). Forbes. https://www.forbes.com/sites/theyec/2022/10/25/new-technology-trends-in-the-wastewater-management-industry/

Ngọc, T. H., Khanh, P. T., & Pramanik, S. (2023). Smart Agriculture using a Soil Monitoring System. In A. Khang *Advanced Technologies and AI-Equipped IoT Applications in High Tech Agriculture*. doi:10.4018/978-1-6684-9231-4.ch011

NgocT. T. H.PramanikS.KhanhP. T. (2023). *The Relationship between Gender and Climate Change in Vietnam*. The Seybold Report. DOI doi:10.17605/OSF.IO/KJBPT

Nhemachena & Hassan. (2007). *Micro-level Analysis of Farmers' Adaptation to Climate Change in Southern Africa*. International Food Policy Research Institute.

Ogemah, V. (2017). *Effect Of Modified Spacing Arrangements, Fertilizer Use and Legume Intercrop on Prevalence of Cassava Brown Streak Disease in Western Kenya*. https://academicjournals.org/journal/AJAR/article-stat/6D2B91066572

Ojo, Ogundeji, & Belle. (2022, May). *Climate Change Perception and Impact of On-Farm Demonstration on Intensity of Adoption of Adaptation Strategies Among Smallholder Farmers in South Africa*. https://www.sciencedirect.com/science/article/pii/S0040162521004637

Onyeneke. (2022, November 9). *Impact of Climate Information Services on Crop Yield in Ebonyi State, Nigeria*. https://www.mdpi.com/2225-1154/11/1/7/pdf

Paavola. (2008). *Livelihood Vulnerability and Adaptation to Climate Change in Morogoro Tanzania*. https://www.researchgate.net/publication/254354050

Pandey, B. K., Pandey, D., Nassa, V. K., George, A. S., Pramanik, S., & Dadheech, P. (2023). Applications for the Text Extraction Method of Complex Degraded Images. In The Impact of Thrust Technologies on Image Processing. Nova Publishers.

Pandey, B. K., Pandey, D., Wairya, S., Agarwal, G., Dadeech, P., Dogiwal, S. R., & Pramanik, S. (2022). Application of Integrated Steganography and Image Compressing Techniques for Confidential Information Transmission. In Cyber Security and Network Security. doi:10.1002/9781119812555.ch8

Pandey, B. K., Pandey, D., Nassa, V. K., Hameed, A. S., George, A. S., Dadheech, P., & Pramanik, S. (2023). A Review of Various Text Extraction Algorithms for Images. In *The Impact of Thrust Technologies on Image Processing*. Nova Publishers. doi:10.52305/ATJL4552

Parag, Y., & Sovacool, B. K. (2016). Electricity market design for the prosumer era. *Nature Energy*, *1*(4), 1–6. doi:10.1038/nenergy.2016.32

Paul, F., Huggel, C., & Kääb, A. (2004). Combining satellite multispectral image data and a digital elevation model for mapping debris-covered glaciers. *Remote Sensing of Environment*, *89*(4), 510–518. doi:10.1016/j.rse.2003.11.007

Pei, C. a. (2007). Status and prospect of hyperspectral remote sensing technique in rock and mineral identification. *World Nuclear Geoscience*, *24*(1), 32–38.

Peng, J. H., Shen, H., He, S. W., & Wu, J. S. (2013). Soil moisture retrieving using hyperspectral data with the application of wavelet analysis. *Environmental Earth Sciences*, *69*(1), 279–288. doi:10.1007/s12665-012-1955-x

Phahlane, Setshedi, & Phuthi. (2016). *Communication of Seasonal Weather Forecast Information in two Selected Provinces of South Africa*. Agricultural Research Council and Department of Africulture, Fisheries and Forestry.

Phuangpornpitak, N., & Tia, S. (2013). Opportunities and challenges of integrating renewable energy in smart grid system. *Energy Procedia*, *34*, 282–290. doi:10.1016/j.egypro.2013.06.756

Pinstrup & Watson. (2011). *Food Policy for Developing Countries the Role of Government in Global National and Local Food Systems*. https://www.researchgate.net/publication/239522608

Policy Reforms for Sustainable Water Management. (2020). World Resources Institute.

Pramanik, S., Joardar, S., Jena, O. P., & Obaid, A. J. (2023). An Analysis of the Operations and Confrontations of Using Green IT in Sustainable Farming. *Al-Kadhum 2nd International Conference on Modern Applications of Information and Communication Technology, AIP Conference Proceedings, 2591*. 10.1063/5.0119513

Pramanik, S., Obaid, A. J., N M, & Bandyopadhyay, S.K. (2023). Applications of Big Data in Clinical Applications. *Al-Kadhum 2nd International Conference on Modern Applications of Information and Communication Technology, AIP Conference Proceedings, 2591*. 10.1063/5.0119414

Pramanik, S. (2022). An Effective Secured Privacy-Protecting Data Aggregation Method in IoT. In *Achieving Full Realization and Mitigating the Challenges of the Internet of Things*. doi:10.4018/978-1-7998-9312-7.ch008

Pramanik, S. (2022). Carpooling Solutions using Machine Learning Tools. In *Handbook of Research on Evolving Designs and Innovation in ICT and Intelligent Systems for Real-World Applications*. IGI Global. doi:10.4018/978-1-7998-9795-8.ch002

Pramanik, S. (2023). Intelligent Farming Utilizing a Soil Tracking Device. In *Convergence of Cloud Computing, AI and Agricultural Science*. doi:10.4018/979-8-3693-0200-2.ch009

Pramanik, S., & Bandyopadhyay, S. (2023). Analysis of Big Data. In *Encyclopedia of Data Science and Machine Learning*. Global. doi:10.4018/978-1-7998-9220-5.ch006

Pramanik, S., & Bandyopadhyay, S. (2023). Identifying Disease and Diagnosis in Females using Machine Learning. In I. G. I. John Wang (Ed.), *Encyclopedia of Data Science and Machine Learning*. Global. doi:10.4018/978-1-7998-9220-5.ch187

Pramanik, S., Galety, M. G., Samanta, D., & Joseph, N. P. (2022). Data Mining Approaches for Decision Support Systems. *3rd International Conference on Emerging Technologies in Data Mining and Information Security*.

Pramanik, S., Sagayam, K. M., & Jena, O. P. (2021). *Machine Learning Frameworks in Cancer Detection*. ICCSRE. doi:10.1051/e3sconf/202129701073

Pramanik, S., Singh, R. P., & Ghosh, R. (2020). Application of Bi-orthogonal Wavelet Transform and Genetic Algorithm in Image Steganography. *Multimedia Tools and Applications*, *79*(25-26), 17463–17482. Advance online publication. doi:10.1007/s11042-020-08676-1

Pramanik, S., & Suresh Raja, S. (2020). A Secured Image Steganography using Genetic Algorithm. *Advances in Mathematics: Scientific Journal*, *9*(7), 4533–4541. doi:10.37418/amsj.9.7.22

Praveenkumar, S., Veeraiah, V., Pramanik, S., Basha, S. M., Lira Neto, A. V., De Albuquerque, V. H. C., & Gupta, A. (2023). *Prediction of Patients' Incurable Diseases Utilizing Deep Learning Approaches, ICICC 2023*. Springer. doi:10.1007/978-981-99-3315-0_4

Processes. (n.d.). *Special Issue: Emerging Technologies for Water and Wastewater Treatment*. https://www.mdpi.com/journal/water/special_issues/Wastewater_Techniques

Puttamadappa, C., & Parameshachari, B. D. (2019). Demand side management of small scale loads in a smart grid using glow-worm swarm optimization technique. *Microprocessors and Microsystems*, *71*, 102886. doi:10.1016/j.micpro.2019.102886

Quincey, D. J., Richardson, S. D., Luckman, A., Lucas, R. M., Reynolds, J. M., Hambrey, M. J., & Glasser, N. F. (2007). Early recognition of glacial lake hazards in the Himalaya usingremote sensing datasets. *Global and Planetary Change*, *56*(1–2), 137–152. doi:10.1016/j.gloplacha.2006.07.013

Rajaram & Van Ginkel. (2019). *Prioritizing Achievable Goals for Food Security in the Developing World.* https://www.researchgate.net/publication/351250964

Rajendran, S. P., Shine, L., Pradeep, R., & Vijayaraghavan, S. (2019). Real-time traffic sign recognition using YOLOv3 based detector. *Proceedings of the 2019 10th International Conference on Computing, Communication and Networking Technologies (ICCCNT).* 10.1109/ICCCNT45670.2019.8944890

Rajeshwari, A., & Mani, N. D. (2014). Estimation of land surface temperature of Dindigul district using Landsat 8 data. *International Journal of Research in Engineering and Technology*, *3*(5), 122–126. doi:10.15623/ijret.2014.0305025

Ranjan, A. K., Patra, A. K., & Gorai, A. K. (2021). A Review on Estimation of Particulate Matter from Satellite-Based Aerosol Optical Depth: Data, Methods, and Challenges. *Asia-Pacific Journal of Atmospheric Sciences*, *57*(3), 679–699. doi:10.1007/s13143-020-00215-0

Reepu, S. Kumar, Chaudhary, M. G., Gupta, K. G., Pramanik, S., & Gupta, A. (2023). Information Security and Privacy in IoT. In Handbook of Research in Advancements in AI and IoT Convergence Technologies. IGI Global.

Reynolds, J.M. (2000). On the formation of supraglacial lakes on debris-covered glaciers. *IAHS Publ*, *264*, 153–161.

Ridd, M. K. (1995). Exploring a VIS (vegetation-impervious surface-soil) model for urban ecosystem analysis through remote sensing: Comparative anatomy for cities. *International Journal of Remote Sensing*, *16*(12), 2165–2185. doi:10.1080/01431169508954549

RLDC. (2009). *Farmers Adoption of Recommended Rice Varieties: A Case of Kilombero District of Morogoro Region, Tanzania.* https://www.ajol.info/index.php/sajae/article/view/121487/110948

Rogers. (2003). *Detailed Review of Rogers' Diffusion of Innovations Theory and Educational Technology-Related Studies Based on Rogers' Theory.* https://files.eric.ed.gov/fulltext/ED501453.pdf

Roggatz, C., Power, M., & Singh, N. (2020). Power system restoration: Meeting the challenge to resiliency from distributed generation. *IEEE Power & Energy Magazine*, *18*(4), 31–40. doi:10.1109/MPE.2020.2985438

Rosen, C., & Madlener, R. (2016). Regulatory options for local reserve energy markets: Implications for prosumers, utilities, and other stakeholders. *The Energy Journal (Cambridge, Mass.)*, *37*(2_suppl), 37. doi:10.5547/01956574.37.SI2.cros

Roy, A., & Pramanik, S. (2023). A Review of the Hydrogen Fuel Path to Emission Reduction in the Surface Transport Industry. *International Journal of Hydrogen Energy.*

Ruescas, A. B., Danne, O., Fomferra, N., & Brockmann, C. (2016). The Land Surface Temperature Synergistic Processor in BEAM: A Prototype towards Sentinel-3. *Data, 1*(3), 18. doi:10.3390/data1030018

Saadna, Y., Behloul, A., & Mezzoudj, S. (2019). Speed limit sign detection and recognition system using SVM and MNIST datasets. *Neural Computing & Applications, 31*(9), 1–11. doi:10.1007/s00521-018-03994-w

Sakai, A., Chikita, K., & Yamada, T. (2000). Expansion of a moraine-dammed glacial lake, Tsho Rolpa, inRolwaling Himal, Nepal Himalaya. *Limnology and Oceanography, 45*(6), 1401–1408. doi:10.4319/lo.2000.45.6.1401

Sanjay, Lohani, Singh, Chaudhary, & Thakural. (2012). Glacial lakes and glacial lake outburst flood in a Himalayan basin using remote sensing and GIS. *Natural Hazards*.

Saraf, A. K., Mishra, B. P., Choudhury, S., & Ghosh, P. (2005). Digital elevation model (DEM) generation from NOAA–AVHRR night-time data and its comparison with USGS-DEM. *International Journal of Remote Sensing, 26*(18), 3879–3887.

Sarr, M., Ayele, B., Kimani, M., & Ruhindoka, R. (2021). *Who Benefits from Climate Friendly Agriculture, The Marginal Returns to a Rainfed System of Rice Intensification in Tanzania*. https://www.researchgate.net/publication/347879128

Sattar, A., Goswami, A., & Kulkarni, A. V. (2019). Application of 1D and 2D hydrodynamic modeling to study glacial lake outburst flood (GLOF) and its impact on a hydropower station in Central Himalaya. *Natural Hazards, 97*(2), 535–553. doi:10.1007/s11069-019-03657-6

Sattari, F., & Hashim, M. (2014). A brief review of land surface temperature retrieval methods from thermal satellite sensors. *Middle East Journal of Scientific Research, 22*(5), 757–768.

Schematic structure of DBN. (n.d.). https://www.researchgate.net/figure/Schematic-structure-of-DBN_fig2_370541543

Segl, K. U. (2003). Endmember detection in urban environments using hyperspectral HyMap data. *Third EARSeL Workshop on Imaging Spectroscopy.*

Segl, K. S., Roessner, S., Heiden, U., & Kaufmann, H. (2003). Fusion of spectral and shape features for identification of urban surface cover types using reflective and thermal hyperspectral data. *ISPRS Journal of Photogrammetry and Remote Sensing, 58*(1-2), 99–112. doi:10.1016/S0924-2716(03)00020-0

Seifi Majdar, R., & Ghassemian, H. (2017). A probabilistic SVM approach for hyperspectral image classification using spectral and texture features. *International Journal of Remote Sensing, 38*(15), 4265–4284. doi:10.1080/01431161.2017.1317941

Seo, Han, Gong, Bai, Ling, Luo, Chen, Ma, Zhang, Su, Ci, Li, Yang, & Ouyang. (2023). *Pushing the skillfull global medium-range weather forecast beyond 10 days lead*. ariXiv:2304.02948 [cs.AI]

Seo, Peterson, Landsfeld, Pedreros, Verdin, Shukla, Husak, Rowland, Harrison, Hoell, & Michaelsen. (2015). *The climate hazards infrared precipitation with stations a new nvironmental record for monitoring extremes.* Academic Press.

Seo, L., Zhang, Z., & Luo, Y. (2020). Combining Optical, Fluorescence, Thermal Satellite, and Environmental Data to Predict County-Level Maize Yield in China Using Machine Learning Approaches. *Remote Sensing*, *12*(1), 21. doi:10.3390/rs12010021

Seo, M. E.-C., & Taormina, R. (1922). The Potential of Deep Learning for Satellite Rainfall Detection over. Data-Scarce Regions, the West African Savanna. *Remote Sensing*, *15*(7). https://doi.org/10.3390/rs15071922

SeoM.ChoiY.RyH.ParkH.BaeH.LeeH.SeoW. (2023). Intermediate and Future Frame Prediction of Geostationary Satellite Imagery. *arXiv:2303.04405.*

Seo, Z. Z., Tang, W., Li, M., Cao, W., & Yuan, Z. (2023). A Novel Hybrid Intelligent SOPDEL Model with Comprehensive Data Preprocessing for Long-Time-Series Climate Prediction. *Remote Sensing*, *15*(7), 1951. doi:10.3390/rs15071951

Seth, S. M. (1983). Modelling of daily snowmelt runoff during pre-monsoon month for Beas basin up to Manali. In *Proceedings of the First National Symposium on Seasonal Snowcover* (*Vol. 2*, pp. 104-115). Academic Press.

Shafri, H. Z. (2012). Hyperspectral remote sensing of urban areas: An overview of techniques and applications. *Research Journal of Applied Sciences, Engineering and Technology*, *4*(11), 1557–1565.

Shannon, M. A., Bohn, P. W., Elimelech, M., Georgiou, J. G., Gitis, V., Jones, R. R., & Xu, Y. (2008). Science and technology for water purification in the coming decades. *Nature*, *452*(7185), 301–310.

Sharma, A., & Malano, H. (2009). Evolution of urban water demand in a rapidly growing low-income semi-arid city. *Water Resources Management*, *23*(8), 1477–1494.

Sharma, V., Dhiman, P., & Rout, R. K. (2023). Improved traffic sign recognition algorithm based on YOLOv4-tiny. *Journal of Visual Communication and Image Representation*, *91*, 103774. doi:10.1016/j.jvcir.2023.103774

Shaukat, N., Ali, S. M., Mehmood, C. A., Khan, B., Jawad, M., Farid, U., Ullah, Z., Anwar, S. M., & Majid, M. (2018). A survey on consumers empowerment, communication technologies, and renewable generation penetration within Smart Grid. *Renewable & Sustainable Energy Reviews*, *81*, 1453–1475. doi:10.1016/j.rser.2017.05.208

Singh, P. (1991). A temperature lapse rate study in Western Himalayas. *Hydrology Journal Indian Association of Hydrologists*, *14*, 156–163.

Singh, R. P. (2014). Dust Storms and Their Influence on Atmospheric Parameters over the Indo-Gangetic Plains. In J. Sundaresan, K. M. Santosh, A. Déri, R. Roggema, & R. Singh (Eds.), *Geospatial Technologies and Climate Change* (pp. 21–35). Springer International Publishing. doi:10.1007/978-3-319-01689-4_2

Sinha, M., Chacko, E., Makhija, P., & Pramanik, S. (2021). Energy Efficient Smart Cities with Green IoT. In C. Chakrabarty (Ed.), *Green Technological Innovation for Sustainable Smart Societies: Post Pandemic Era*. Springer. doi:10.1007/978-3-030-73295-0_16

Skoković, D., Sobrino, J. A., Jimenez-Munoz, J. C., Soria, G., Julien, Y., Mattar, C., & Cristóbal, J. (2014). *Calibration and Validation of land surface temperature for Landsat8-TIRS sensor*. Land Product Validation and Evolution.

Smil, V. (2017). *Harvesting the Biosphere: What We Have Taken from Nature*. MIT Press.

SNV. (2019). *Horticulture Cooperatives Final Report 2019*. https://snv.org/assets/explore/download/profiling_horticulture_cooperatives_final_report.pdf

Sobrino, J. A., El Kharraz, J., & Li, Z. L. (2003). Surface temperature and water vapour retrieval from MODIS data. *International Journal of Remote Sensing*, *24*(24), 5161–5182. doi:10.1080/0143116031000102502

Sobrino, J. A., Li, Z. L., Stoll, M. P., & Becker, F. (1996). Multi-channel and multi-angle algorithms for estimating sea and land surface temperature with ATSR data. *International Journal of Remote Sensing*, *17*(11), 2089–2114. doi:10.1080/01431169608948760

Sobrino, J. A., & Raissouni, N. (2000). Toward remote sensing methods for land cover dynamic monitoring: Application to Morocco. *International Journal of Remote Sensing*, *21*(2), 353–366. doi:10.1080/014311600210876

Solomon, B. D., & Krishna, K. (2011). The coming sustainable energy transition: History, strategies, and outlook. *Energy Policy*, *39*(11), 7422–7431. doi:10.1016/j.enpol.2011.09.009

Sorooshian, Asanjan, Yang, Hsu, Lin, & Peng. (2018). *Short-Term Precipitation Forecast Based on the PERSIANN System and LSTM Recurrent Neural Networks*. AGU. https://doi.org/doi:10.1029/2018JD028375

Springmann, M., Clarck, M., Wiebe, K., & Calson, K. (2018, October 10). *Options For Keeping the Food System Within Environmental Limits*. https://www.nature.com/articles/s41586-018-0594-0

Stroeve, J., & Steffen, K. (1998). Variability of AVHRR-derived clear-sky surface temperature over the Greenland ice sheet. *Journal of Applied Meteorology*, *37*(1), 23–31. doi:10.1175/1520-0450(1998)037<0023:VOADCS>2.0.CO;2

Sun, Y., Yin, H., Wang, W., Shan, C., Notholt, J., Palm, M., Liu, K., Chen, Z., & Liu, C. (2022). Monitoring greenhouse gases (GHGs) in China: Status and perspective. *Atmospheric Measurement Techniques*, *15*(16), 4819–4834. doi:10.5194/amt-15-4819-2022

Tabernik, D., & Skočaj, D. (2019). Deep learning for large-scale traffic-sign detection and recognition. *IEEE Transactions on Intelligent Transportation Systems*, *21*(4), 1427–1440. doi:10.1109/TITS.2019.2913588

Tak, S., & Keshari, A. K. (2020). Investigating mass balance of Parvati glacier in Himalaya using satellite imagery based model. *Scientific Reports*, *10*(1), 1–16. doi:10.1038/s41598-020-69203-8 PMID:32699284

Tan, C. L., Wong, N. H., Tan, P. Y., Jusuf, S. K., & Chiam, Z. Q. (2015). Impact of plant evapotranspiration rate and shrub albedo on temperature reduction in the tropical outdoor environment. *Building and Environment*, *94*, 206–217. doi:10.1016/j.buildenv.2015.08.001

Tang, H., & Li, Z. L. (2014). Land surface temperature retrieval from thermal infrared data. In *Quantitative Remote Sensing in Thermal Infrared* (pp. 93–143). Springer. doi:10.1007/978-3-642-42027-6_5

Tao, X., Gong, Y., Shi, W., & Cheng, D. (2020). Object detection with class aware region proposal network and focused attention objective. *Pattern Recognition Letters*, *130*, 353–361. doi:10.1016/j.patrec.2018.09.025

Tao, Z., Santanello, J. A., Chin, M., Zhou, S., Tan, Q., Kemp, E. M., & Peters-Lidard, C. D. (2013). Effect of land cover on atmospheric processes and air quality over the continental United States–a NASA Unified WRF (NU-WRF) model study. *Atmospheric Chemistry and Physics*, *13*(13), 6207–6226. doi:10.5194/acp-13-6207-2013

Taylor, L., Rezai, A., & Foley, D. (2016). *An Integrated Approach to Climate Change, Income Distribution, Employment and Economic Growth*. https://www.sciencedirect.com/science/article/abs/pii/S0921800915002323

Tchobanoglous, G., Burton, F. L., & Stensel, H. D. (2014). *Wastewater treatment* (2nd ed.). McGraw-Hill Education.

Teng, A. G. (2009). Progress of hyperspectral technology application research in agriculture. *Jiangsu Nongye Kexue*, 8–11.

Thapa, K. B. (1980). *Analysis for snow-melt runoff during premonsoon months in Beas basin using satellite imageries* [ME Dissertation].

The Future of Water Management: Challenges and Opportunities. (2020). World Resources Institute.

The Nature Conservancy. (n.d.). https://www.nature.org/en-us/what-we-do/our-insights/water-security/

The United States Environmental Protection Agency (EPA). (n.d.). https://www.epa.gov/ground-water-and-drinking-water/basic-information-about-your-drinking-water

The Water Environment Federation (WEF). (n.d.). https://www.wef.org/

The World Bank. (2020, July 15). *Water and Sanitation*. https://www.worldbank.org/en/topic/water

The World Health Organization (WHO). (n.d.). https://www.who.int/teams/environment-climate-change-and-health/water-sanitation-and-health

Thinda, Ogundeji, & Belle. (2021). *Determinants of Relevant Constraints Inhibiting Farmers' Adoption of Climate Change Adaptation Strategies in South Africa*. https://www.researchgate.net/publication/342569250

Thompson, R. (1998). *Atmospheric Processes and Systems* (1st ed.). Routledge. doi:10.4324/9780203268643

Thompson, S., Benn, D., Luckman, A., & Kulessa, B. (2010). The evolution of supraglacial lakes, NgozumpaGlacier, Khumbu Himalaya, Nepal. *Geophysical Research Abstracts*, *12*, EGU2010–11396-1.

Tiwari, K. C., Arora, M. K., & Singh, D. (2011). An assessment of independent component analysis for detection of military targets from hyperspectral images. *International Journal of Applied Earth Observation and Geoinformation*, *13*(5), 730–740. doi:10.1016/j.jag.2011.03.007

Triki, N., Karray, M., & Ksantini, M. (2023). A real-time traffic sign recognition method using a new attention-based deep convolutional neural network for smart vehicles. *Applied Sciences (Basel, Switzerland)*, *13*(8), 4793. doi:10.3390/app13084793

Truslow, E. D., Manolakis, D., Pieper, M., Cooley, T., & Brueggeman, M. (2013). Performance prediction of matched filter and adaptive cosine estimator hyperspectral target detectors. *IEEE Journal of Selected Topics in Applied Earth Observations and Remote Sensing*, *7*(6), 2337–2350. doi:10.1109/JSTARS.2013.2272697

Ukita, J., Narama, C., Tadono, T., Yamanokuchi, T., Tomiyama, N., Kawamoto, S., Abe, C., Uda, T., Yabuki, H., Fujita, K., & Nishimura, K. (2011). Glacial lake inventory of Bhutan using ALOS data: Part 1, methods andpreliminary results. *Annals of Glaciology*, *52*(58), 2011. doi:10.3189/172756411797252293

Ulivieri, C. M. M. A., Castronuovo, M. M., Francioni, R., & Cardillo, A. (1994). A split window algorithm for estimating land surface temperature from satellites. *Advances in Space Research*, *14*(3), 59–65. doi:10.1016/0273-1177(94)90193-7

United Nations Water. (n.d.). https://www.unwater.org/

United Nations. (2018). *Sustainable Development Goal 6 Synthesis Report on Water and Sanitation*. Author.

United States Environmental Protection Agency (EPA). (2010). https://www.epa.gov/

V. I., Allender, S., & Aktins, V. J. (2019). *The Global Syndemic of Obesity, Undernutrition and Climate Change the Lancet Commission Report*. https://www.researchgate.net/publication/330676764_

Van Berkum, D. (2018). *Food Systems Approach Sustainable Solutions for a Sufficient*. Wageningen University and Research. https://research.wur.nl/en/publications/

Veeraiah, V., Talukdar, V., Manikandan, K., Talukdar, S. B., Solavande, V. D., Pramanik, S., & Gupta, A. (2023). Machine Learning Frameworks in Carpooling. In Handbook of Research on AI and Machine Learning Applications in Customer Support and Analytics. IGI Global. doi:10.4018/978-1-6684-7105-0.ch009

Veeraiah, V., Shiju, D. J., Ramesh, J. V. N., Ganesh, K. R., Pramanik, S., & Pandey, D. (2023). *Healthcare Cloud Services in Image Processing. In Thrust Technologies' Effect on Image Processing*. IGI Global.

Velasco, E., Roth, M., Norford, L., & Molina, L. T. (2016). Does urban vegetation enhance carbon sequestration? *Landscape and Urban Planning*, *148*, 99–107. doi:10.1016/j.landurbplan.2015.12.003

Verma, S., Prakash, D., Soni, M., & Ram, K. (2019). Atmospheric Aerosols Monitoring: Ground and Satellite-Based Instruments. In S. Sarvajayakesavalu (Ed.), *Advances in Environmental Monitoring and Assessment*. IntechOpen. doi:10.5772/intechopen.80489

Vidya Chellam, V., Veeraiah, V., Khanna, A., Sheikh, T. H., Pramanik, S., & Dhabliya, D. (2023). *A Machine Vision-based Approach for Tuberculosis Identification in Chest X-Rays Images of Patients, ICICC 2023*. Springer. doi:10.1007/978-981-99-3315-0_3

Vivone, G., Rocco, R., Giorgio, L., Mauro Dalla, M., & Jocelyn, C. (2014). Multiresolution analysis and component substitution techniques for hyperspectral pansharpening. *IEEE Geoscience and Remote Sensing Symposium*, 2649-26. 10.1109/IGARSS.2014.6947018

Wang, Zhang, Zhi, & Ji. (2023). Multi-Model Ensemble Forecasts of Surface Air Temperatures in Henan Province. *Atmosphere, 14*(3), 520. https://doi.org/. doi:10.3390/atmos14030520

Wang, G., Yuan, Y., & Guo, W. (2019). An improved rider optimization algorithm for solving engineering optimization problems. *IEEE Access : Practical Innovations, Open Solutions*, *7*, 80570–80576. doi:10.1109/ACCESS.2019.2923468

Wang, Q., & Abdelrahman, W. (2023). High-Precision AI-Enabled Flood Prediction Integrating Local Sensor Data and 3rd Party Weather Forecast. *Sensors (Basel)*, *23*(6), 3065.

Wan, Z. (2006). *MODIS land surface temperature products users' guide*. Institute for Computational Earth System Science, University of California.

Watanabe, T., Ives, J. D., & Hammond, J. E. (1994). Rapid growth of a glacial lake in Khumbu Himal, Nepal:prospects for a catastrophic flood. *Mountain Research and Development, 14*(4), 329–340. doi:10.2307/3673729

Water Environment Federation. (2020). *Operation of municipal wastewater treatment plants (Manual of Practice No. 11)*. Water Environment Federation.

Water Management in the 21st Century: Challenges and Opportunities. (2016). United Nations Development Programme.

WB. (2018). *Developing An Irrigation Fund in Tanzania*. World Bank Group. https://2030wrg.org/

WFP. (2021). *Global Hunger Crisis World Food Program.* https://www.wfp.org/global-hunger-crisis

White, M. A., Thornton, P. E., Running, S. W., & Nemani, R. R. (2000). Parameterization and sensitivity analysis of the BIOME–BGC terrestrial ecosystem model: Net primary production controls. *Earth Interactions*, *4*(3), 1–85. doi:10.1175/1087-3562(2000)004<0003:PASAOT>2.0.CO;2

Willett. (2022). *Food in the Anthropocene the EAT Lancet Commission on Healthy Diets from Sustainable Food Systems.* https://www.researchgate.net/publication/330443133_

William, P., Darwante, N. K., Pawar, A. B., Jawale, M. A., & Verma, A. (2022). Framework for implementation of smart driver assistance system using augmented reality. In *Mobile Radio Communications and 5G Networks: Proceedings of Third MRCN 2022* (pp. 365-383). Singapore: Springer Nature.

Wilson, T. A., Steven, K. R., & Lemuel, R. M. J. (1995). Perceptual-based hyperspectral image fusion using multiresolution analysis. *Optical Engineering (Redondo Beach, Calif.)*, *34*(11), 3154–3164. doi:10.1117/12.213617

World Health Organization. (2017). *Guidelines for Drinking-Water Quality.* Author.

World Resources Institute. (2021). *Aqueduct Water Risk Atlas.* Author.

World Resources Institute. (n.d.). https://www.wri.org/water

Yan, Y., Qian, Y., Sharif, H., & Tipper, D. (2012). A survey on smart grid communication infrastructures: Motivations, requirements and challenges. *IEEE Communications Surveys and Tutorials*, *15*(1), 5–20. doi:10.1109/SURV.2012.021312.00034

Yao, T., Li, Z. G., Yang, W., Guo, X. J., Zhu, L. P., Kang, S. C., Wu, Y. H., & Yu, W. S. (2010). Glacial distribution and mass balance in the Yarlung Zangbo River and its influence onlakes. *Chinese Science Bulletin*, *55*(20), 2072–2078. doi:10.1007/s11434-010-3213-5

Ying-Nong Seo, F. (2022). Rainfall Forecast Using Machine Learning with High Spatiotemporal Satellite Imagery Every 10 Minutes. *Remote Sensing*, *14*(23), 5950. doi:10.3390/rs14235950

Yirsaw, E., Wu, W., Shi, X., Temesgen, H., & Bekele, B. (2017). Land Use/Land Cover Change Modelling and the Prediction of Subsequent Changes in Ecosystem Service Values in a Coastal Area of China, the Su-Xi-Chang Region. *Sustainability (Basel)*, *9*(7), 1204. doi:10.3390/su9071204

Yokoya, N., Claas, G., & Jocelyn, C. (2017). Hyperspectral and multispectral data fusion: A comparative review of the recent literature. *IEEE Geoscience and Remote Sensing Magazine*, *5*(2), 29–56. doi:10.1109/MGRS.2016.2637824

Zafar, R., Mahmood, A., Razzaq, S., Ali, W., Naeem, U., & Shehzad, K. (2018). Prosumer based energy management and sharing in smart grid. *Renewable & Sustainable Energy Reviews*, *82*, 1675–1684. doi:10.1016/j.rser.2017.07.018

Zaidi, A. Z., Yasmeen, Z., & Siddiqui, M. D. (2013) glacial lake outburst flood (GLOF) Risk mapping in Hunza River Basin (Pakistan) using geospatial techniques. In *2013 6th International Conference on Recent Advances in Space Technologies (RAST)*. IEEE.

Zeweld, Guido, V., Girmay, & Speedman. (2017). *Smallholder Farmers' Behavioral Intentions Towards Sustainable Agricultural Practices*. Academic Press.

Zhang, Y., Odeh, I. O., & Ramadan, E. (2013). Assessment of land surface temperature in relation to landscape metrics and fractional vegetation cover in an urban/peri-urban region using Landsat data. *International Journal of Remote Sensing, 34*(1), 168-189. doi:10.1080/01431161.2012.712227

Zhang, H. S. (2011). Application of hyper spectral remote sensing for urban forestry monitoring in natural disaster zones. In *International Conference on Computer and Management (CAMAN)* (pp. 1-4). IEEE. 10.1109/CAMAN.2011.5778867

Zhang, J., Xie, Z., Sun, J., Zou, X., & Wang, J. (2020). A cascaded R-CNN with multiscale attention and imbalanced samples for traffic sign detection. *IEEE Access : Practical Innovations, Open Solutions*, *8*, 29742–29754. doi:10.1109/ACCESS.2020.2972338

Zhang, P., Zhang, M., & Liu, J. (2021). Real-time HD map change detection for crowdsourcing update based on mid-to-high-end sensors. *Sensors (Basel)*, *21*(7), 2477. doi:10.3390/s21072477 PMID:33918443

Zhang, X., & Zhang, Z. (2023). Research on a traffic sign recognition method under small sample conditions. *Sensors (Basel)*, *23*(11), 5091. doi:10.3390/s23115091 PMID:37299816

Zhang, Y., Li, Z., Bai, K., Wei, Y., Xie, Y., Zhang, Y., Ou, Y., Cohen, J., Zhang, Y., Peng, Z., Zhang, X., Chen, C., Hong, J., Xu, H., Guang, J., Lv, Y., Li, K., & Li, D. (2021). Satellite remote sensing of atmospheric particulate matter mass concentration: Advances, challenges, and perspectives. *Fundamental Research*, *1*(3), 240–258. doi:10.1016/j.fmre.2021.04.007

Zhao, D., Yu, C., Jian, Y., & Wang, Y. (2018). *An Overview of Climate Change Impacts on The Society in China*. https://www.sciencedirect.com/science/article/pii/S1674927821000411

Zhao, S., Qin, Q., Yang, Y., Xiong, Y., & Qiu, G. (2009). Comparison of two split-window methods for retrieving land surface temperature from MODIS data. *Journal of Earth System Science*, *118*(4), 345–353. doi:10.1007/s12040-009-0027-4

Zhao, W., & Du, S. (2016). Spectral–spatial feature extraction for hyperspectral image classification: A dimension reduction and deep learning approach. *IEEE Transactions on Geoscience and Remote Sensing*, *54*(8), 4544–4554. doi:10.1109/TGRS.2016.2543748

Zhu, M., Wang, J., Yang, X., Zhang, Y., Zhang, L., Ren, H., Wu, B., & Ye, L. (2022). A review of the application of machine learning in water quality evaluation. *Eco-Environment & Health*, *1*(2), 107-116. doi:10.1016/j.eehl.2022.06.001

Zwicker, L., Swinburn, B., Hawkes, C., & Brownel, K. D. (2015, June). *Patchy Progress on Obesity Prevention: Emerging Examples, Entrenched Barriers and New Thinking*. National Library of Medicine. https://pubmed.ncbi.nlm.nih.gov/25703111/

Related References

To continue our tradition of advancing academic research, we have compiled a list of recommended IGI Global readings. These references will provide additional information and guidance to further enrich your knowledge and assist you with your own research and future publications.

Ajiboye, O. E., & Yusuff, O. S. (2017). Foreign Land Acquisition: Food Security and Food Chains – The Nigerian Experience. In I. Management Association (Ed.), Natural Resources Management: Concepts, Methodologies, Tools, and Applications (pp. 1524-1545). Hershey, PA: IGI Global. https://doi.org/ doi:10.4018/978-1-5225-0803-8.ch072

Alapiki, H. E., & Amadi, L. A. (2018). Sustainable Food Consumption in the Neoliberal Order: Challenges and Policy Implications. In A. Obayelu (Ed.), *Food Systems Sustainability and Environmental Policies in Modern Economies* (pp. 90–123). Hershey, PA: IGI Global. doi:10.4018/978-1-5225-3631-4.ch005

Altaş, A. (2018). Geographical Information System Applications Utilized in Museums in Turkey Within the Scope of the Cultural Heritage Tourism: A Case Study of Mobile Application of Müze Asist. In S. Chaudhuri & N. Ray (Eds.), *GIS Applications in the Tourism and Hospitality Industry* (pp. 42–60). Hershey, PA: IGI Global. doi:10.4018/978-1-5225-5088-4.ch002

Andreea, I. R. (2018). Beyond Macroeconomics of Food and Nutrition Security. *International Journal of Sustainable Economies Management*, *7*(1), 13–22. doi:10.4018/IJSEM.2018010102

Anwar, J. (2017). Reproductive and Mental Health during Natural Disaster: Implications and Issues for Women in Developing Nations – A Case Example. In I. Management Association (Ed.), Gaming and Technology Addiction: Breakthroughs in Research and Practice (pp. 446-472). Hershey, PA: IGI Global. https://doi.org/ doi:10.4018/978-1-5225-0778-9.ch021

Awadh, H., Aksissou, M., Benhardouze, W., Darasi, F., & Snaiki, J. (2018). Socioeconomic Status of Artisanal Fishers in the West Part of Moroccan Mediterranean. *International Journal of Social Ecology and Sustainable Development*, *9*(1), 40–52. doi:10.4018/IJSESD.2018010104

Aye, G. C., & Haruna, R. F. (2018). Effect of Climate Change on Crop Productivity and Prices in Benue State, Nigeria: Implications for Food Security. In V. Erokhin (Ed.), *Establishing Food Security and Alternatives to International Trade in Emerging Economies* (pp. 244–268). Hershey, PA: IGI Global. doi:10.4018/978-1-5225-2733-6.ch012

Azizan, S. A., & Suki, N. M. (2017). Consumers' Intentions to Purchase Organic Food Products. In T. Esakki (Ed.), *Green Marketing and Environmental Responsibility in Modern Corporations* (pp. 68–82). Hershey, PA: IGI Global. doi:10.4018/978-1-5225-2331-4.ch005

Barakabitze, A. A., Fue, K. G., Kitindi, E. J., & Sanga, C. A. (2017). Developing a Framework for Next Generation Integrated Agro Food-Advisory Systems in Developing Countries. In I. Management Association (Ed.), Agri-Food Supply Chain Management: Breakthroughs in Research and Practice (pp. 47-67). Hershey, PA: IGI Global. doi:10.4018/978-1-5225-1629-3.ch004

Beachcroft-Shaw, H., & Ellis, D. (2018). Using Successful Cases to Promote Environmental Sustainability: A Social Marketing Approach. In I. Management Association (Ed.), Sustainable Development: Concepts, Methodologies, Tools, and Applications (pp. 936-953). Hershey, PA: IGI Global. doi:10.4018/978-1-5225-3817-2.ch042

Behnassi, M., Kahime, K., Boussaa, S., Boumezzough, A., & Messouli, M. (2017). Infectious Diseases and Climate Vulnerability in Morocco: Governance and Adaptation Options. In I. Management Association (Ed.), Public Health and Welfare: Concepts, Methodologies, Tools, and Applications (pp. 91-109). Hershey, PA: IGI Global. https://doi.org/ doi:10.4018/978-1-5225-1674-3.ch005

Bekele, F., & Bekele, I. (2017). Social and Environmental Impacts on Agricultural Development. In W. Ganpat, R. Dyer, & W. Isaac (Eds.), *Agricultural Development and Food Security in Developing Nations* (pp. 21–56). Hershey, PA: IGI Global. doi:10.4018/978-1-5225-0942-4.ch002

Benaouda, A., & García-Peñalvo, F. J. (2018). Towards an Intelligent System for the Territorial Planning: Agricultural Case. In F. García-Peñalvo (Ed.), *Global Implications of Emerging Technology Trends* (pp. 158–178). Hershey, PA: IGI Global. doi:10.4018/978-1-5225-4944-4.ch010

Bhaskar, A., Rao, G. B., & Vencatesan, J. (2017). Characterization and Management Concerns of Water Resources around Pallikaranai Marsh, South Chennai. In P. Rao & Y. Patil (Eds.), *Reconsidering the Impact of Climate Change on Global Water Supply, Use, and Management* (pp. 102–121). Hershey, PA: IGI Global. doi:10.4018/978-1-5225-1046-8.ch007

Bhyan, P., Shrivastava, B., & Kumar, N. (2022). Requisite Sustainable Development Contemplating Buildings: Economic and Environmental Sustainability. In A. Hussain, K. Tiwari, & A. Gupta (Eds.), *Addressing Environmental Challenges Through Spatial Planning* (pp. 269–288). IGI Global. https://doi.org/10.4018/978-1-7998-8331-9.ch014

Bogataj, D., & Drobne, D. (2017). Control of Perishable Goods in Cold Logistic Chains by Bionanosensors. In S. Joo (Ed.), *Applying Nanotechnology for Environmental Sustainability* (pp. 376–402). Hershey, PA: IGI Global. doi:10.4018/978-1-5225-0585-3.ch016

Bogataj, D., & Drobne, D. (2017). Control of Perishable Goods in Cold Logistic Chains by Bionanosensors. In I. Management Association (Ed.), Materials Science and Engineering: Concepts, Methodologies, Tools, and Applications (pp. 471-497). Hershey, PA: IGI Global. https://doi.org/ doi:10.4018/978-1-5225-1798-6.ch019

Bogueva, D., & Marinova, D. (2018). What Is More Important: Perception of Masculinity or Personal Health and the Environment? In D. Bogueva, D. Marinova, & T. Raphaely (Eds.), *Handbook of Research on Social Marketing and Its Influence on Animal Origin Food Product Consumption* (pp. 148–162). Hershey, PA: IGI Global. doi:10.4018/978-1-5225-4757-0.ch010

Bouzid, M. (2017). Waterborne Diseases and Climate Change: Impact and Implications. In M. Bouzid (Ed.), *Examining the Role of Environmental Change on Emerging Infectious Diseases and Pandemics* (pp. 89–108). Hershey, PA: IGI Global. doi:10.4018/978-1-5225-0553-2.ch004

Bowles, D. C. (2017). Climate Change-Associated Conflict and Infectious Disease. In M. Bouzid (Ed.), *Examining the Role of Environmental Change on Emerging Infectious Diseases and Pandemics* (pp. 68–88). Hershey, PA: IGI Global. doi:10.4018/978-1-5225-0553-2.ch003

Buck, J. J., & Lowry, R. K. (2017). Oceanographic Data Management: Quills and Free Text to the Digital Age and "Big Data". In P. Diviacco, A. Leadbetter, & H. Glaves (Eds.), *Oceanographic and Marine Cross-Domain Data Management for Sustainable Development* (pp. 1–22). Hershey, PA: IGI Global. doi:10.4018/978-1-5225-0700-0.ch001

Buse, C. G. (2017). Are Climate Change Adaptation Policies a Game Changer?: A Case Study of Perspectives from Public Health Officials in Ontario, Canada. In M. Bouzid (Ed.), *Examining the Role of Environmental Change on Emerging Infectious Diseases and Pandemics* (pp. 230–257). Hershey, PA: IGI Global. doi:10.4018/978-1-5225-0553-2.ch010

Calderon, F. A., Giolo, E. G., Frau, C. D., Rengel, M. G., Rodriguez, H., Tornello, M., ... Gallucci, R. (2018). Seismic Microzonation and Site Effects Detection Through Microtremors Measures: A Review. In N. Ceryan (Ed.), *Handbook of Research on Trends and Digital Advances in Engineering Geology* (pp. 326–349). Hershey, PA: IGI Global. doi:10.4018/978-1-5225-2709-1.ch009

Carfì, D., Donato, A., & Panuccio, D. (2018). A Game Theory Coopetitive Perspective for Sustainability of Global Feeding: Agreements Among Vegan and Non-Vegan Food Firms. In I. Management Association (Ed.), Game Theory: Breakthroughs in Research and Practice (pp. 71-104). Hershey, PA: IGI Global. https://doi.org/doi:10.4018/978-1-5225-2594-3.ch004

Castagnolo, V. (2018). Analyzing, Classifying, Safeguarding: Drawing for the Borgo Murattiano Neighbourhood of Bari. In G. Carlone, N. Martinelli, & F. Rotondo (Eds.), *Designing Grid Cities for Optimized Urban Development and Planning* (pp. 93–108). Hershey, PA: IGI Global. doi:10.4018/978-1-5225-3613-0.ch006

Chekima, B. (2018). The Dilemma of Purchase Intention: A Conceptual Framework for Understanding Actual Consumption of Organic Food. *International Journal of Sustainable Economies Management*, *7*(2), 1–13. doi:10.4018/IJSEM.2018040101

Chen, Y. (2017). Sustainable Supply Chains and International Soft Landings: A Case of Wetland Entrepreneurship. In B. Christiansen & F. Kasarcı (Eds.), *Corporate Espionage, Geopolitics, and Diplomacy Issues in International Business* (pp. 232–247). Hershey, PA: IGI Global. doi:10.4018/978-1-5225-1031-4.ch013

Çıtak, L., Akel, V., & Ersoy, E. (2018). Investors' Reactions to the Announcement of New Constituents of BIST Sustainability Index: An Analysis by Event Study and Mean-Median Tests. In M. Risso & S. Testarmata (Eds.), *Value Sharing for Sustainable and Inclusive Development* (pp. 270–289). Hershey, PA: IGI Global. doi:10.4018/978-1-5225-3147-0.ch012

D'Aleo, V., D'Aleo, F., & Bonanno, R. (2018). New Food Industries Toward a New Level of Sustainable Supply: Success Stories, Business Models, and Strategies. In V. Erokhin (Ed.), *Establishing Food Security and Alternatives to International Trade in Emerging Economies* (pp. 74–97). Hershey, PA: IGI Global. doi:10.4018/978-1-5225-2733-6.ch004

Dagevos, H., & Reinders, M. J. (2018). Flexitarianism and Social Marketing: Reflections on Eating Meat in Moderation. In D. Bogueva, D. Marinova, & T. Raphaely (Eds.), *Handbook of Research on Social Marketing and Its Influence on Animal Origin Food Product Consumption* (pp. 105–120). Hershey, PA: IGI Global. doi:10.4018/978-1-5225-4757-0.ch007

Danisman, G. O. (2022). What Drives Eco-Design Innovations in European SMEs? In U. Akkucuk (Ed.), *Disruptive Technologies and Eco-Innovation for Sustainable Development* (pp. 191–206). IGI Global. https://doi.org/10.4018/978-1-7998-8900-7.ch011

Deenapanray, P. N., & Ramma, I. (2017). Adaptations to Climate Change and Climate Variability in the Agriculture Sector in Mauritius: Lessons from a Technical Needs Assessment. In I. Management Association (Ed.), Natural Resources Management: Concepts, Methodologies, Tools, and Applications (pp. 655-680). Hershey, PA: IGI Global. https://doi.org/ doi:10.4018/978-1-5225-0803-8.ch030

Deenapanray, P. N., & Ramma, I. (2017). Adaptations to Climate Change and Climate Variability in the Agriculture Sector in Mauritius: Lessons from a Technical Needs Assessment. In I. Management Association (Ed.), Natural Resources Management: Concepts, Methodologies, Tools, and Applications (pp. 655-680). Hershey, PA: IGI Global. https://doi.org/ doi:10.4018/978-1-5225-0803-8.ch030

Deshpande, S., Basu, S. K., Li, X., & Chen, X. (2017). Smart, Innovative and Intelligent Technologies Used in Drug Designing. In I. Management Association (Ed.), Pharmaceutical Sciences: Breakthroughs in Research and Practice (pp. 1175-1191). Hershey, PA: IGI Global. https://doi.org/ doi:10.4018/978-1-5225-1762-7.ch045

Dlamini, P. N. (2017). Use of Information and Communication Technologies Tools to Capture, Store, and Disseminate Indigenous Knowledge: A Literature Review. In P. Ngulube (Ed.), *Handbook of Research on Theoretical Perspectives on Indigenous Knowledge Systems in Developing Countries* (pp. 225–247). Hershey, PA: IGI Global. doi:10.4018/978-1-5225-0833-5.ch010

Dolejsova, M., & Kera, D. (2017). The Fermentation GutHub Project and the Internet of Microbes. In S. Konomi & G. Roussos (Eds.), *Enriching Urban Spaces with Ambient Computing, the Internet of Things, and Smart City Design* (pp. 25–46). Hershey, PA: IGI Global. doi:10.4018/978-1-5225-0827-4.ch002

Dolunay, O. (2018). A Paradigm Shift: Empowering Farmers to Eliminate the Waste in the Form of Fresh Water and Energy Through the Implementation of 4R+T. In I. Management Association (Ed.), Sustainable Development: Concepts, Methodologies, Tools, and Applications (pp. 882-892). Hershey, PA: IGI Global. https://doi.org/doi:10.4018/978-1-5225-3817-2.ch039

Dube, P., Heijman, W. J., Ihle, R., & Ochieng, J. (2018). The Potential of Traditional Leafy Vegetables for Improving Food Security in Africa. In V. Erokhin (Ed.), *Establishing Food Security and Alternatives to International Trade in Emerging Economies* (pp. 220–243). Hershey, PA: IGI Global. doi:10.4018/978-1-5225-2733-6.ch011

Duruji, M. M., & Urenma, D. F. (2017). The Environmentalism and Politics of Climate Change: A Study of the Process of Global Convergence through UNFCCC Conferences. In I. Management Association (Ed.), *Natural Resources Management: Concepts, Methodologies, Tools, and Applications* (pp. 77-108). Hershey, PA: IGI Global. 10.4018/978-1-5225-0803-8.ch004

Dutta, U. (2017). Agro-Geoinformatics, Potato Cultivation, and Climate Change. In S. Londhe (Ed.), *Sustainable Potato Production and the Impact of Climate Change* (pp. 247–271). Hershey, PA: IGI Global. doi:10.4018/978-1-5225-1715-3.ch012

Edirisinghe, R., Stranieri, A., & Wickramasinghe, N. (2017). A Taxonomy for mHealth. In N. Wickramasinghe (Ed.), *Handbook of Research on Healthcare Administration and Management* (pp. 596–615). Hershey, PA: IGI Global. doi:10.4018/978-1-5225-0920-2.ch036

Ekpeni, N. M., & Ayeni, A. O. (2018). Global Natural Hazard and Disaster Vulnerability Management. In A. Eneanya (Ed.), *Handbook of Research on Environmental Policies for Emergency Management and Public Safety* (pp. 83–104). Hershey, PA: IGI Global. doi:10.4018/978-1-5225-3194-4.ch005

Ene, C., Voica, M. C., & Panait, M. (2017). Green Investments and Food Security: Opportunities and Future Directions in the Context of Sustainable Development. In M. Mieila (Ed.), *Measuring Sustainable Development and Green Investments in Contemporary Economies* (pp. 163–200). Hershey, PA: IGI Global. doi:10.4018/978-1-5225-2081-8.ch007

Escamilla, I., Ruíz, M. T., Ibarra, M. M., Soto, V. L., Quintero, R., & Guzmán, G. (2018). Geocoding Tweets Based on Semantic Web and Ontologies. In M. Lytras, N. Aljohani, E. Damiani, & K. Chui (Eds.), *Innovations, Developments, and Applications of Semantic Web and Information Systems* (pp. 372–392). Hershey, PA: IGI Global. doi:10.4018/978-1-5225-5042-6.ch014

Escribano, A. J. (2018). Marketing Strategies for Trendy Animal Products: Sustainability as a Core. In F. Quoquab, R. Thurasamy, & J. Mohammad (Eds.), *Driving Green Consumerism Through Strategic Sustainability Marketing* (pp. 169–203). Hershey, PA: IGI Global. doi:10.4018/978-1-5225-2912-5.ch010

Eudoxie, G., & Roopnarine, R. (2017). Climate Change Adaptation and Disaster Risk Management in the Caribbean. In W. Ganpat & W. Isaac (Eds.), *Environmental Sustainability and Climate Change Adaptation Strategies* (pp. 97–125). Hershey, PA: IGI Global. doi:10.4018/978-1-5225-1607-1.ch004

Farmer, L. S. (2017). Data Analytics for Strategic Management: Getting the Right Data. In V. Wang (Ed.), *Encyclopedia of Strategic Leadership and Management* (pp. 810–822). Hershey, PA: IGI Global. doi:10.4018/978-1-5225-1049-9.ch056

Fattal, L. R. (2017). Catastrophe: An Uncanny Catalyst for Creativity. In R. Shin (Ed.), *Convergence of Contemporary Art, Visual Culture, and Global Civic Engagement* (pp. 244–262). Hershey, PA: IGI Global. doi:10.4018/978-1-5225-1665-1.ch014

Forti, I. (2017). A Cross Reading of Landscape through Digital Landscape Models: The Case of Southern Garda. In A. Ippolito (Ed.), *Handbook of Research on Emerging Technologies for Architectural and Archaeological Heritage* (pp. 532–561). Hershey, PA: IGI Global. doi:10.4018/978-1-5225-0675-1.ch018

Gharbi, A., De Runz, C., & Akdag, H. (2017). Urban Development Modelling: A Survey. In S. Faiz & K. Mahmoudi (Eds.), *Handbook of Research on Geographic Information Systems Applications and Advancements* (pp. 96–124). Hershey, PA: IGI Global. doi:10.4018/978-1-5225-0937-0.ch004

Ghosh, I., & Ghoshal, I. (2018). Implications of Trade Liberalization for Food Security Under the ASEAN-India Strategic Partnership: A Gravity Model Approach. In V. Erokhin (Ed.), *Establishing Food Security and Alternatives to International Trade in Emerging Economies* (pp. 98–118). Hershey, PA: IGI Global. doi:10.4018/978-1-5225-2733-6.ch005

Glaves, H. M. (2017). Developing a Common Global Framework for Marine Data Management. In P. Diviacco, A. Leadbetter, & H. Glaves (Eds.), *Oceanographic and Marine Cross-Domain Data Management for Sustainable Development* (pp. 47–68). Hershey, PA: IGI Global. doi:10.4018/978-1-5225-0700-0.ch003

Godulla, A., & Wolf, C. (2018). Future of Food: Transmedia Strategies of National Geographic. In R. Gambarato & G. Alzamora (Eds.), *Exploring Transmedia Journalism in the Digital Age* (pp. 162–182). Hershey, PA: IGI Global. doi:10.4018/978-1-5225-3781-6.ch010

Gomes, P. P. (2018). Food and Environment: A Review on the Sustainability of Six Different Dietary Patterns. In A. Obayelu (Ed.), *Food Systems Sustainability and Environmental Policies in Modern Economies* (pp. 15–31). Hershey, PA: IGI Global. doi:10.4018/978-1-5225-3631-4.ch002

Gonzalez-Feliu, J. (2018). Sustainability Evaluation of Green Urban Logistics Systems: Literature Overview and Proposed Framework. In A. Paul, D. Bhattacharyya, & S. Anand (Eds.), *Green Initiatives for Business Sustainability and Value Creation* (pp. 103–134). Hershey, PA: IGI Global. doi:10.4018/978-1-5225-2662-9.ch005

Goodland, R. (2017). A Fresh Look at Livestock Greenhouse Gas Emissions and Mitigation. In I. Management Association (Ed.), Natural Resources Management: Concepts, Methodologies, Tools, and Applications (pp. 124-139). Hershey, PA: IGI Global. doi:10.4018/978-1-5225-0803-8.ch006

Goundar, S., & Appana, S. (2018). Mainstreaming Development Policies for Climate Change in Fiji: A Policy Gap Analysis and the Role of ICTs. In I. Management Association (Ed.), Sustainable Development: Concepts, Methodologies, Tools, and Applications (pp. 402-432). Hershey, PA: IGI Global. doi:10.4018/978-1-5225-3817-2.ch020

Granell-Canut, C., & Aguilar-Moreno, E. (2018). Geospatial Influence in Science Mapping. In M. Khosrow-Pour, D.B.A. (Ed.), Encyclopedia of Information Science and Technology, Fourth Edition (pp. 3473-3483). Hershey, PA: IGI Global. doi:10.4018/978-1-5225-2255-3.ch302

Grigelis, A., Blažauskas, N., Gelumbauskaitė, L. Ž., Gulbinskas, S., Suzdalev, S., & Ferrarin, C. (2017). Marine Environment Data Management Related to the Human Activity in the South-Eastern Baltic Sea (The Lithuanian Sector). In P. Diviacco, A. Leadbetter, & H. Glaves (Eds.), *Oceanographic and Marine Cross-Domain Data Management for Sustainable Development* (pp. 282–302). Hershey, PA: IGI Global. doi:10.4018/978-1-5225-0700-0.ch012

Guma, I. P., Rwashana, A. S., & Oyo, B. (2018). Food Security Policy Analysis Using System Dynamics: The Case of Uganda. *International Journal of Information Technologies and Systems Approach, 11*(1), 72–90. doi:10.4018/IJITSA.2018010104

Guma, I. P., Rwashana, A. S., & Oyo, B. (2018). Food Security Indicators for Subsistence Farmers Sustainability: A System Dynamics Approach. *International Journal of System Dynamics Applications, 7*(1), 45–64. doi:10.4018/IJSDA.2018010103

Gupta, P., & Goyal, S. (2017). Wildlife Habitat Evaluation. In A. Santra & S. Mitra (Eds.), *Remote Sensing Techniques and GIS Applications in Earth and Environmental Studies* (pp. 258–264). Hershey, PA: IGI Global. doi:10.4018/978-1-5225-1814-3.ch013

Hanson, T., & Hildebrand, E. (2018). GPS Travel Diaries in Rural Transportation Research: A Focus on Older Drivers. In I. Management Association (Ed.), Intelligent Transportation and Planning: Breakthroughs in Research and Practice (pp. 609-625). Hershey, PA: IGI Global. doi:10.4018/978-1-5225-5210-9.ch027

Hartman, M. B. (2017). Research-Based Climate Change Public Education Programs. In I. Management Association (Ed.), Natural Resources Management: Concepts, Methodologies, Tools, and Applications (pp. 992-1003). Hershey, PA: IGI Global. doi:10.4018/978-1-5225-0803-8.ch046

Hashim, N. (2017). Zanzibari Seaweed: Global Climate Change and the Promise of Adaptation. In I. Management Association (Ed.), Natural Resources Management: Concepts, Methodologies, Tools, and Applications (pp. 365-391). Hershey, PA: IGI Global. https://doi.org/ doi:10.4018/978-1-5225-0803-8.ch019

Herrera, J. E., Argüello, L. V., Gonzalez-Feliu, J., & Jaimes, W. A. (2017). Decision Support System Design Requirements, Information Management, and Urban Logistics Efficiency: Case Study of Bogotá, Colombia. In G. Jamil, A. Soares, & C. Pessoa (Eds.), *Handbook of Research on Information Management for Effective Logistics and Supply Chains* (pp. 223–238). Hershey, PA: IGI Global. doi:10.4018/978-1-5225-0973-8.ch012

Huizinga, T., Ayanso, A., Smoor, M., & Wronski, T. (2017). Exploring Insurance and Natural Disaster Tweets Using Text Analytics. *International Journal of Business Analytics*, *4*(1), 1–17. doi:10.4018/IJBAN.2017010101

Hung, K., Kalantari, M., & Rajabifard, A. (2017). An Integrated Method for Assessing the Text Content Quality of Volunteered Geographic Information in Disaster Management. *International Journal of Information Systems for Crisis Response and Management*, *9*(2), 1–17. doi:10.4018/IJISCRAM.2017040101

Husnain, A., & Avdic, A. (2018). Identifying the Contemporary Status of E-Service Sustainability Research. In I. Management Association (Ed.), Sustainable Development: Concepts, Methodologies, Tools, and Applications (pp. 467-485). Hershey, PA: IGI Global. https://doi.org/ doi:10.4018/978-1-5225-3817-2.ch022

Iarossi, M. P., & Ferro, L. (2017). "The Past is Never Dead. It's Not Even Past": Virtual Archaeological Promenade. In A. Ippolito & M. Cigola (Eds.), *Handbook of Research on Emerging Technologies for Digital Preservation and Information Modeling* (pp. 228–255). Hershey, PA: IGI Global. doi:10.4018/978-1-5225-0680-5.ch010

Ignjatijević, S., & Cvijanović, D. (2018). Analysis of Serbian Honey Production and Exports. In *Exploring the Global Competitiveness of Agri-Food Sectors and Serbia's Dominant Presence: Emerging Research and Opportunities* (pp. 109–139). Hershey, PA: IGI Global. doi:10.4018/978-1-5225-2762-6.ch005

Jana, S. K., & Karmakar, A. K. (2017). Globalization, Governance, and Food Security: The Case of BRICS. In I. Management Association (Ed.), Natural Resources Management: Concepts, Methodologies, Tools, and Applications (pp. 692-712). Hershey, PA: IGI Global. https://doi.org/ doi:10.4018/978-1-5225-0803-8.ch032

Jana, S. K., & Karmakar, A. K. (2017). Food Security in Asia: Is There Convergence? In I. Management Association (Ed.), Natural Resources Management: Concepts, Methodologies, Tools, and Applications (pp. 109-123). Hershey, PA: IGI Global. https://doi.org/ doi:10.4018/978-1-5225-0803-8.ch005

John, J., & Kumar, S. (2018). A Locational Decision Making Framework for Shipbreaking Under Multiple Criteria. In I. Management Association (Ed.), Operations and Service Management: Concepts, Methodologies, Tools, and Applications (pp. 504-527). Hershey, PA: IGI Global. doi:10.4018/978-1-5225-3909-4.ch024

John, J., & Srivastava, R. K. (2018). Decision Insights for Shipbreaking using Environmental Impact Assessment: Review and Perspectives. *International Journal of Strategic Decision Sciences*, *9*(1), 45–62. doi:10.4018/IJSDS.2018010104

Joshi, Y., & Rahman, Z. (2018). Determinants of Sustainable Consumption Behaviour: Review and Conceptual Framework. In A. Paul, D. Bhattacharyya, & S. Anand (Eds.), *Green Initiatives for Business Sustainability and Value Creation* (pp. 239–262). Hershey, PA: IGI Global. doi:10.4018/978-1-5225-2662-9.ch011

Juma, D. W., Reuben, M., Wang, H., & Li, F. (2018). Adaptive Coevolution: Realigning the Water Governance Regime to the Changing Climate. In I. Management Association (Ed.), Hydrology and Water Resource Management: Breakthroughs in Research and Practice (pp. 346-357). Hershey, PA: IGI Global. https://doi.org/ doi:10.4018/978-1-5225-3427-3.ch014

K., S., & Tripathy, B. K. (2018). Neighborhood Rough-Sets-Based Spatial Data Analytics. In M. Khosrow-Pour, D.B.A. (Ed.), *Encyclopedia of Information Science and Technology, Fourth Edition* (pp. 1835-1844). Hershey, PA: IGI Global. https:// doi.org/ doi:10.4018/978-1-5225-2255-3.ch160

Kabir, F. (2018). Towards a More Gender-Inclusive Climate Change Policy. In N. Mahtab, T. Haque, I. Khan, M. Islam, & I. Wahid (Eds.), *Handbook of Research on Women's Issues and Rights in the Developing World* (pp. 354–369). Hershey, PA: IGI Global. doi:10.4018/978-1-5225-3018-3.ch022

Kabir, F. (2018). Towards a More Gender-Inclusive Climate Change Policy. In I. Management Association (Ed.), Climate Change and Environmental Concerns: Breakthroughs in Research and Practice (pp. 525-540). Hershey, PA: IGI Global. https://doi.org/ doi:10.4018/978-1-5225-5487-5.ch027

Kanyamuka, J. S., Jumbe, C. B., & Ricker-Gilbert, J. (2018). Making Agricultural Input Subsidies More Effective and Profitable in Africa: The Role of Complementary Interventions. In A. Obayelu (Ed.), *Food Systems Sustainability and Environmental Policies in Modern Economies* (pp. 172–187). Hershey, PA: IGI Global. doi:10.4018/978-1-5225-3631-4.ch008

Karimi, H., & Gholamrezafahimi, F. (2017). Study of Integrated Coastal Zone Management and Its Environmental Effects: A Case of Iran. In R. Singh, A. Singh, & V. Srivastava (Eds.), *Environmental Issues Surrounding Human Overpopulation* (pp. 64–88). Hershey, PA: IGI Global. doi:10.4018/978-1-5225-1683-5.ch004

Kaya, I. R., Hutabarat, J., & Bambang, A. N. (2018). "Sasi": A New Path to Sustain Seaweed Farming From Up-Stream to Down-Stream in Kotania Bay, Molucass. *International Journal of Social Ecology and Sustainable Development, 9*(2), 28–36. doi:10.4018/IJSESD.2018040103

Khader, V. (2018). Technologies for Food, Health, Livelihood, and Nutrition Security. In I. Management Association (Ed.), Food Science and Nutrition: Breakthroughs in Research and Practice (pp. 94-112). Hershey, PA: IGI Global. https://doi.org/doi:10.4018/978-1-5225-5207-9.ch005

Kocadağlı, A. Y. (2017). The Temporal and Spatial Development of Organic Agriculture in Turkey. In W. Ganpat, R. Dyer, & W. Isaac (Eds.), *Agricultural Development and Food Security in Developing Nations* (pp. 130–156). Hershey, PA: IGI Global. doi:10.4018/978-1-5225-0942-4.ch006

Koundouri, P., Giannouli, A., & Souliotis, I. (2017). An Integrated Approach for Sustainable Environmental and Socio-Economic Development Using Offshore Infrastructure. In I. Management Association (Ed.), Renewable and Alternative Energy: Concepts, Methodologies, Tools, and Applications (pp. 1581-1601). Hershey, PA: IGI Global. doi:10.4018/978-1-5225-1671-2.ch056

Kumar, A., & Dash, M. K. (2017). Sustainability and Future Generation Infrastructure on Digital Platform: A Study of Generation Y. In N. Ray (Ed.), *Business Infrastructure for Sustainability in Developing Economies* (pp. 124–142). Hershey, PA: IGI Global. doi:10.4018/978-1-5225-2041-2.ch007

Kumar, A., Mukherjee, A. B., & Krishna, A. P. (2017). Application of Conventional Data Mining Techniques and Web Mining to Aid Disaster Management. In A. Kumar (Ed.), *Web Usage Mining Techniques and Applications Across Industries* (pp. 138–167). Hershey, PA: IGI Global. doi:10.4018/978-1-5225-0613-3.ch006

Kumar, C. P. (2017). Impact of Climate Change on Groundwater Resources. In I. Management Association (Ed.), Natural Resources Management: Concepts, Methodologies, Tools, and Applications (pp. 1094-1120). Hershey, PA: IGI Global. doi:10.4018/978-1-5225-0803-8.ch052

Kumari, S., & Patil, Y. (2017). Achieving Climate Smart Agriculture with a Sustainable Use of Water: A Conceptual Framework for Sustaining the Use of Water for Agriculture in the Era of Climate Change. In P. Rao & Y. Patil (Eds.), *Reconsidering the Impact of Climate Change on Global Water Supply, Use, and Management* (pp. 122–143). Hershey, PA: IGI Global. doi:10.4018/978-1-5225-1046-8.ch008

Kumari, S., & Patil, Y. (2018). Achieving Climate Smart Agriculture With a Sustainable Use of Water: A Conceptual Framework for Sustaining the Use of Water for Agriculture in the Era of Climate Change. In I. Management Association (Ed.), Climate Change and Environmental Concerns: Breakthroughs in Research and Practice (pp. 111-133). Hershey, PA: IGI Global. doi:10.4018/978-1-5225-5487-5.ch006

Kursah, M. B. (2017). Least-Cost Pipeline using Geographic Information System: The Limit to Technicalities. *International Journal of Applied Geospatial Research, 8*(3), 1–15. doi:10.4018/ijagr.2017070101

Lahiri, S., Ghosh, D., & Bhakta, J. N. (2017). Role of Microbes in Eco-Remediation of Perturbed Aquatic Ecosystem. In J. Bhakta (Ed.), *Handbook of Research on Inventive Bioremediation Techniques* (pp. 70–107). Hershey, PA: IGI Global. doi:10.4018/978-1-5225-2325-3.ch004

Lallo, C. H., Smalling, S., Facey, A., & Hughes, M. (2018). The Impact of Climate Change on Small Ruminant Performance in Caribbean Communities. In I. Management Association (Ed.), Climate Change and Environmental Concerns: Breakthroughs in Research and Practice (pp. 193-218). Hershey, PA: IGI Global. doi:10.4018/978-1-5225-5487-5.ch010

Laurini, R. (2017). Nature of Geographic Knowledge Bases. In S. Faiz & K. Mahmoudi (Eds.), *Handbook of Research on Geographic Information Systems Applications and Advancements* (pp. 29–60). Hershey, PA: IGI Global. doi:10.4018/978-1-5225-0937-0.ch002

Lawrence, J., Simpson, L., & Piggott, A. (2017). Protected Agriculture: A Climate Change Adaptation for Food and Nutrition Security. In I. Management Association (Ed.), Natural Resources Management: Concepts, Methodologies, Tools, and Applications (pp. 140-158). Hershey, PA: IGI Global. doi:10.4018/978-1-5225-0803-8.ch007

Leadbetter, A., Cheatham, M., Shepherd, A., & Thomas, R. (2017). Linked Ocean Data 2.0. In P. Diviacco, A. Leadbetter, & H. Glaves (Eds.), *Oceanographic and Marine Cross-Domain Data Management for Sustainable Development* (pp. 69–99). Hershey, PA: IGI Global. doi:10.4018/978-1-5225-0700-0.ch004

Lucas, M. R., Rego, C., Vieira, C., & Vieira, I. (2017). Proximity and Cooperation for Innovative Regional Development: The Case of the Science and Technology Park of Alentejo. In L. Carvalho (Ed.), *Handbook of Research on Entrepreneurial Development and Innovation Within Smart Cities* (pp. 199–228). Hershey, PA: IGI Global. doi:10.4018/978-1-5225-1978-2.ch010

Ma, X., Beaulieu, S. E., Fu, L., Fox, P., Di Stefano, M., & West, P. (2017). Documenting Provenance for Reproducible Marine Ecosystem Assessment in Open Science. In P. Diviacco, A. Leadbetter, & H. Glaves (Eds.), *Oceanographic and Marine Cross-Domain Data Management for Sustainable Development* (pp. 100–126). Hershey, PA: IGI Global. doi:10.4018/978-1-5225-0700-0.ch005

Mabe, L. K., & Oladele, O. I. (2017). Application of Information Communication Technologies for Agricultural Development through Extension Services: A Review. In T. Tossy (Ed.), *Information Technology Integration for Socio-Economic Development* (pp. 52–101). Hershey, PA: IGI Global. doi:10.4018/978-1-5225-0539-6.ch003

Malomo, B. I. (2018). A Review of Psychological Resilience as a Response to Natural Hazards in Nigeria. In A. Eneanya (Ed.), *Handbook of Research on Environmental Policies for Emergency Management and Public Safety* (pp. 147–165). Hershey, PA: IGI Global. doi:10.4018/978-1-5225-3194-4.ch008

Manchiraju, S. (2018). Predicting Behavioral Intentions Toward Sustainable Fashion Consumption: A Comparison of Attitude-Behavior and Value-Behavior Consistency Models. In I. Management Association (Ed.), Fashion and Textiles: Breakthroughs in Research and Practice (pp. 1-21). Hershey, PA: IGI Global. doi:10.4018/978-1-5225-3432-7.ch001

Manzella, G. M., Bartolini, R., Bustaffa, F., D'Angelo, P., De Mattei, M., Frontini, F., ... Spada, A. (2017). Semantic Search Engine for Data Management and Sustainable Development: Marine Planning Service Platform. In P. Diviacco, A. Leadbetter, & H. Glaves (Eds.), *Oceanographic and Marine Cross-Domain Data Management for Sustainable Development* (pp. 127–154). Hershey, PA: IGI Global. doi:10.4018/978-1-5225-0700-0.ch006

Mbonigaba, J. (2018). Comparing the Effects of Unsustainable Production and Consumption of Food on Health and Policy Across Developed and Less Developed Countries. In A. Obayelu (Ed.), *Food Systems Sustainability and Environmental Policies in Modern Economies* (pp. 124–158). Hershey, PA: IGI Global. doi:10.4018/978-1-5225-3631-4.ch006

McKeown, A. E. (2017). Nurses, Healthcare, and Environmental Pollution and Solutions: Breaking the Cycle of Harm. In I. Management Association (Ed.), Natural Resources Management: Concepts, Methodologies, Tools, and Applications (pp. 392-415). Hershey, PA: IGI Global. https://doi.org/ doi:10.4018/978-1-5225-0803-8.ch020

Mili, B., Barua, A., & Katyaini, S. (2017). Climate Change and Adaptation through the Lens of Capability Approach: A Case Study from Darjeeling, Eastern Himalaya. In I. Management Association (Ed.), Natural Resources Management: Concepts, Methodologies, Tools, and Applications (pp. 1351-1365). Hershey, PA: IGI Global. https://doi.org/ doi:10.4018/978-1-5225-0803-8.ch064

Mir, S. A., Shah, M. A., Mir, M. M., & Iqbal, U. (2017). New Horizons of Nanotechnology in Agriculture and Food Processing Industry. In B. Nayak, A. Nanda, & M. Bhat (Eds.), *Integrating Biologically-Inspired Nanotechnology into Medical Practice* (pp. 230–258). Hershey, PA: IGI Global. doi:10.4018/978-1-5225-0610-2.ch009

Moallem, M., Sterrett, W. L., Gordon, C. R., Sukhera, S. M., Mahmood, A., & Bashir, A. (2021). An Investigation of the Effects of Integrating Computing and Project- or Problem-Based Learning in the Context of Environmental Sciences: A Case of Pakistani STEM Teachers. In S. Schroth & J. Daniels (Eds.), *Building STEM Skills Through Environmental Education* (pp. 49–89). IGI Global. https://doi.org/10.4018/978-1-7998-2711-5.ch003

Mujere, N., & Moyce, W. (2017). Climate Change Impacts on Surface Water Quality. In W. Ganpat & W. Isaac (Eds.), *Environmental Sustainability and Climate Change Adaptation Strategies* (pp. 322–340). Hershey, PA: IGI Global. doi:10.4018/978-1-5225-1607-1.ch012

Mukherjee, A. B., Krishna, A. P., & Patel, N. (2018). Geospatial Technology for Urban Sciences. In *Geospatial Technologies in Urban System Development: Emerging Research and Opportunities* (pp. 99–120). Hershey, PA: IGI Global. doi:10.4018/978-1-5225-3683-3.ch005

Nagarajan, S. K., & Sangaiah, A. K. (2017). Vegetation Index: Ideas, Methods, Influences, and Trends. In N. Kumar, A. Sangaiah, M. Arun, & S. Anand (Eds.), *Advanced Image Processing Techniques and Applications* (pp. 347–386). Hershey, PA: IGI Global. doi:10.4018/978-1-5225-2053-5.ch016

Naraine, L., & Meehan, K. (2017). Strengthening Food Security with Sustainable Practices by Smallholder Farmers in Lesser Developed Economies. In W. Ganpat, R. Dyer, & W. Isaac (Eds.), *Agricultural Development and Food Security in Developing Nations* (pp. 57–81). Hershey, PA: IGI Global. doi:10.4018/978-1-5225-0942-4.ch003

Naraine, L., & Meehan, K. (2017). Strengthening Food Security with Sustainable Practices by Smallholder Farmers in Lesser Developed Economies. In W. Ganpat, R. Dyer, & W. Isaac (Eds.), *Agricultural Development and Food Security in Developing Nations* (pp. 57–81). Hershey, PA: IGI Global. doi:10.4018/978-1-5225-0942-4.ch003

Naraine, L., & Meehan, K. (2017). Strengthening Food Security with Sustainable Practices by Smallholder Farmers in Lesser Developed Economies. In W. Ganpat, R. Dyer, & W. Isaac (Eds.), *Agricultural Development and Food Security in Developing Nations* (pp. 57–81). Hershey, PA: IGI Global. doi:10.4018/978-1-5225-0942-4.ch003

Nikolaou, K., Tsakiridou, E., Anastasiadis, F., & Mattas, K. (2017). Exploring Alternative Distribution Channels of Agricultural Products. *International Journal of Food and Beverage Manufacturing and Business Models*, *2*(2), 36–66. doi:10.4018/IJFBMBM.2017070103

Nishat, K. J., & Rahman, M. S. (2018). Disaster, Vulnerability, and Violence Against Women: Global Findings and a Research Agenda for Bangladesh. In N. Mahtab, T. Haque, I. Khan, M. Islam, & I. Wahid (Eds.), *Handbook of Research on Women's Issues and Rights in the Developing World* (pp. 235–250). Hershey, PA: IGI Global. doi:10.4018/978-1-5225-3018-3.ch014

O'Hara, S., Jones, D., & Trobman, H. B. (2018). Building an Urban Food System Through UDC Food Hubs. In A. Burtin, J. Fleming, & P. Hampton-Garland (Eds.), *Changing Urban Landscapes Through Public Higher Education* (pp. 116–143). Hershey, PA: IGI Global. doi:10.4018/978-1-5225-3454-9.ch006

Obayelu, A. E. (2018). Integrating Environment, Food Systems, and Sustainability in Feeding the Growing Population in Developing Countries. In A. Obayelu (Ed.), *Food Systems Sustainability and Environmental Policies in Modern Economies* (pp. 1–14). Hershey, PA: IGI Global. doi:10.4018/978-1-5225-3631-4.ch001

Othman, R., Nath, N., & Laswad, F. (2018). Environmental Reporting and Accounting: Sustainability Hybridisation. In G. Azevedo, J. da Silva Oliveira, R. Marques, & A. Ferreira (Eds.), *Handbook of Research on Modernization and Accountability in Public Sector Management* (pp. 130–158). Hershey, PA: IGI Global. doi:10.4018/978-1-5225-3731-1.ch007

Ouadi, A., & Zitouni, A. (2021). Phasor Measurement Improvement Using Digital Filter in a Smart Grid. In A. Recioui & H. Bentarzi (Eds.), *Optimizing and Measuring Smart Grid Operation and Control* (pp. 100–117). IGI Global. https://doi.org/10.4018/978-1-7998-4027-5.ch005

Padigala, B. S. (2018). Traditional Water Management System for Climate Change Adaptation in Mountain Ecosystems. In I. Management Association (Ed.), Climate Change and Environmental Concerns: Breakthroughs in Research and Practice (pp. 630-655). Hershey, PA: IGI Global. doi:10.4018/978-1-5225-5487-5.ch033

Panda, C. K. (2018). Mobile Phone Usage in Agricultural Extension in India: The Current and Future Perspective. In F. Mtenzi, G. Oreku, D. Lupiana, & J. Yonazi (Eds.), *Mobile Technologies and Socio-Economic Development in Emerging Nations* (pp. 1–21). Hershey, PA: IGI Global. doi:10.4018/978-1-5225-4029-8.ch001

Pandian, S. L., Yarrakula, K., & Chaudhury, P. (2018). GIS-Based Decision Support System for Village Level: A Case Study in Andhra Pradesh. In S. Chaudhuri & N. Ray (Eds.), *GIS Applications in the Tourism and Hospitality Industry* (pp. 275–295). Hershey, PA: IGI Global. doi:10.4018/978-1-5225-5088-4.ch012

Quaranta, G., & Salvia, R. (2017). Social-Based Product Innovation and Governance in The Milk Sector: The Case of Carciocacio and Innonatura. In T. Tarnanidis, M. Vlachopoulou, & J. Papathanasiou (Eds.), *Driving Agribusiness With Technology Innovations* (pp. 293–310). Hershey, PA: IGI Global. doi:10.4018/978-1-5225-2107-5.ch015

Rahman, M. K., Schmidlin, T. W., Munro-Stasiuk, M. J., & Curtis, A. (2017). Geospatial Analysis of Land Loss, Land Cover Change, and Landuse Patterns of Kutubdia Island, Bangladesh. *International Journal of Applied Geospatial Research*, *8*(2), 45–60. doi:10.4018/IJAGR.2017040104

Rajack-Talley, T. A. (2017). Agriculture, Trade Liberalization and Poverty in the ACP Countries. In W. Ganpat, R. Dyer, & W. Isaac (Eds.), *Agricultural Development and Food Security in Developing Nations* (pp. 1–20). Hershey, PA: IGI Global. doi:10.4018/978-1-5225-0942-4.ch001

Rajamanickam, S. (2018). Exploring Landscapes in Regional Convergence: Environment and Sustainable Development in South Asia. In I. Management Association (Ed.), Sustainable Development: Concepts, Methodologies, Tools, and Applications (pp. 1051-1087). Hershey, PA: IGI Global. https://doi.org/ doi:10.4018/978-1-5225-3817-2.ch047

Rizvi, S. M., & Dearden, A. (2018). KHETI: ICT Solution for Agriculture Extension and Its Replication in Open and Distance Learning. In U. Pandey & V. Indrakanti (Eds.), *Open and Distance Learning Initiatives for Sustainable Development* (pp. 163–174). Hershey, PA: IGI Global. doi:10.4018/978-1-5225-2621-6.ch008

Roşu, L., & Macarov, L. I. (2017). Management of Drought and Floods in the Dobrogea Region. In I. Management Association (Ed.), Agri-Food Supply Chain Management: Breakthroughs in Research and Practice (pp. 372-403). Hershey, PA: IGI Global. https://doi.org/ doi:10.4018/978-1-5225-1629-3.ch016

Rouzbehani, K., & Rouzbehani, S. (2018). Mapping Women's World: GIS and the Case of Breast Cancer in the US. *International Journal of Public Health Management and Ethics*, *3*(1), 14–25. doi:10.4018/IJPHME.2018010102

Roy, D. (2018). Success Factors of Adoption of Mobile Applications in Rural India: Effect of Service Characteristics on Conceptual Model. In M. Khosrow-Pour, D.B.A. (Ed.), Green Computing Strategies for Competitive Advantage and Business Sustainability (pp. 211-238). Hershey, PA: IGI Global. https://doi.org/ doi:10.4018/978-1-5225-5017-4.ch010

Sajeva, M., Lemon, M., & Sahota, P. S. (2017). Governance for Food Security: A Framework for Social Learning and Scenario Building. *International Journal of Food and Beverage Manufacturing and Business Models*, *2*(2), 67–84. doi:10.4018/ IJFBMBM.2017070104

Sambhanthan, A., & Potdar, V. (2017). A Study of the Parameters Impacting Sustainability in Information Technology Organizations. *International Journal of Knowledge-Based Organizations*, *7*(3), 27–39. doi:10.4018/IJKBO.2017070103

Sanga, C., Kalungwizi, V. J., & Msuya, C. P. (2017). Bridging Gender Gaps in Provision of Agricultural Extension Service Using ICT: Experiences from Sokoine University of Agriculture (SUA) Farmer Voice Radio (FVR) Project in Tanzania. In I. Management Association (Ed.), Discrimination and Diversity: Concepts, Methodologies, Tools, and Applications (pp. 682-697). Hershey, PA: IGI Global. https://doi.org/ doi:10.4018/978-1-5225-1933-1.ch031

Santra, A., & Mitra, D. (2017). Role of Remote Sensing in Potential Fishing Zone Forecast. In A. Santra & S. Mitra (Eds.), *Remote Sensing Techniques and GIS Applications in Earth and Environmental Studies* (pp. 243–257). Hershey, PA: IGI Global. doi:10.4018/978-1-5225-1814-3.ch012

Schaap, D. (2017). SeaDataNet: Towards a Pan-European Infrastructure for Marine and Ocean Data Management. In P. Diviacco, A. Leadbetter, & H. Glaves (Eds.), *Oceanographic and Marine Cross-Domain Data Management for Sustainable Development* (pp. 155–177). Hershey, PA: IGI Global. doi:10.4018/978-1-5225-0700-0.ch007

Seckin-Celik, T. (2017). Sustainability Reporting and Sustainability in the Turkish Business Context. In U. Akkucuk (Ed.), *Ethics and Sustainability in Global Supply Chain Management* (pp. 115–132). Hershey, PA: IGI Global. doi:10.4018/978-1-5225-2036-8.ch006

Segbefia, A. Y., Barnes, V. R., Akpalu, L. A., & Mensah, M. (2018). Environmental Location Assessment for Seaweed Cultivation in Ghana: A Spatial Multi-Criteria Approach. *International Journal of Applied Geospatial Research*, *9*(1), 51–64. doi:10.4018/IJAGR.2018010104

Sen, Y. (2018). How to Manage Sustainability: A Framework for Corporate Sustainability Tools. In I. Management Association (Ed.), Sustainable Development: Concepts, Methodologies, Tools, and Applications (pp. 568-589). Hershey, PA: IGI Global. doi:10.4018/978-1-5225-3817-2.ch026

Shamshiry, E., Abdulai, A. M., Mokhtar, M. B., & Komoo, I. (2015). Regional Landfill Site Selection with GIS and Analytical Hierarchy Process Techniques: A Case Study of Langkawi Island, Malaysia. In P. Thomas, M. Srihari, & S. Kaur (Eds.), *Handbook of Research on Cultural and Economic Impacts of the Information Society* (pp. 248–282). Hershey, PA: IGI Global. doi:10.4018/978-1-4666-8598-7.ch011

Sharma, Y. K., Mangla, S. K., Patil, P. P., & Uniyal, S. (2018). Analyzing Sustainable Food Supply Chain Management Challenges in India. In M. Ram & J. Davim (Eds.), *Soft Computing Techniques and Applications in Mechanical Engineering* (pp. 162–180). Hershey, PA: IGI Global. doi:10.4018/978-1-5225-3035-0.ch008

Silvestrelli, P. (2018). The Impact of Events: To Which Extent Are They Sustainable for Tourist Destinations? Some Evidences From Expo Milano 2015. In M. Risso & S. Testarmata (Eds.), *Value Sharing for Sustainable and Inclusive Development* (pp. 185–204). Hershey, PA: IGI Global. doi:10.4018/978-1-5225-3147-0.ch008

Silvius, G. (2018). Sustainability Evaluation of IT/IS Projects. In I. Management Association (Ed.), Sustainable Development: Concepts, Methodologies, Tools, and Applications (pp. 26-40). Hershey, PA: IGI Global. https://doi.org/ doi:10.4018/978-1-5225-3817-2.ch002

Singh, R., Srivastava, P., Singh, P., Upadhyay, S., & Raghubanshi, A. S. (2017). Human Overpopulation and Food Security: Challenges for the Agriculture Sustainability. In R. Singh, A. Singh, & V. Srivastava (Eds.), *Environmental Issues Surrounding Human Overpopulation* (pp. 12–39). Hershey, PA: IGI Global. doi:10.4018/978-1-5225-1683-5.ch002

Singh, R., Srivastava, P., Singh, P., Upadhyay, S., & Raghubanshi, A. S. (2017). Human Overpopulation and Food Security: Challenges for the Agriculture Sustainability. In R. Singh, A. Singh, & V. Srivastava (Eds.), *Environmental Issues Surrounding Human Overpopulation* (pp. 12–39). Hershey, PA: IGI Global. doi:10.4018/978-1-5225-1683-5.ch002

Srivastava, N. (2017). Climate Change Mitigation: Collective Efforts and Responsibly. In I. Management Association (Ed.), Natural Resources Management: Concepts, Methodologies, Tools, and Applications (pp. 64-76). Hershey, PA: IGI Global. https://doi.org/ doi:10.4018/978-1-5225-0803-8.ch003

Stanganelli, M., & Gerundo, C. (2017). Understanding the Role of Urban Morphology and Green Areas Configuration During Heat Waves. *International Journal of Agricultural and Environmental Information Systems*, *8*(2), 50–64. doi:10.4018/IJAEIS.2017040104

Stewart, M. K., Hagood, D., & Ching, C. C. (2017). Virtual Games and Real-World Communities: Environments that Constrain and Enable Physical Activity in Games for Health. *International Journal of Game-Based Learning*, *7*(1), 1–19. doi:10.4018/IJGBL.2017010101

Stone, R. J. (2017). Modelling the Frequency of Tropical Cyclones in the Lower Caribbean Region. In W. Ganpat & W. Isaac (Eds.), *Environmental Sustainability and Climate Change Adaptation Strategies* (pp. 341–349). Hershey, PA: IGI Global. doi:10.4018/978-1-5225-1607-1.ch013

Syed, A., & Jabeen, U. A. (2018). Climate Change Impact on Agriculture and Food Security. In A. Eneanya (Ed.), *Handbook of Research on Environmental Policies for Emergency Management and Public Safety* (pp. 223–237). Hershey, PA: IGI Global. doi:10.4018/978-1-5225-3194-4.ch012

Tam, G. C. (2017). The Global View of Sustainability. In *Managerial Strategies and Green Solutions for Project Sustainability* (pp. 1–24). Hershey, PA: IGI Global. doi:10.4018/978-1-5225-2371-0.ch001

Tam, G. C. (2017). Understanding Project Sustainability. In *Managerial Strategies and Green Solutions for Project Sustainability* (pp. 110–139). Hershey, PA: IGI Global. doi:10.4018/978-1-5225-2371-0.ch005

Tam, G. C. (2017). Perspectives on Sustainability. In *Managerial Strategies and Green Solutions for Project Sustainability* (pp. 53–76). Hershey, PA: IGI Global. doi:10.4018/978-1-5225-2371-0.ch003

Tang, M., & Karunanithi, A. T. (2018). Visual Logic Maps (vLms). In *Advanced Concept Maps in STEM Education: Emerging Research and Opportunities* (pp. 108–149). Hershey, PA: IGI Global. doi:10.4018/978-1-5225-2184-6.ch005

Taşçıoğlu, M., & Yener, D. (2018). The Value and Scope of GIS in Marketing and Tourism Management. In S. Chaudhuri & N. Ray (Eds.), *GIS Applications in the Tourism and Hospitality Industry* (pp. 189–211). Hershey, PA: IGI Global. doi:10.4018/978-1-5225-5088-4.ch009

ThanhTung, B. (2021). Pharmacology and Therapeutic Applications of Resveratrol. In A. Hussain & S. Behl (Eds.), *Treating Endocrine and Metabolic Disorders With Herbal Medicines* (pp. 321–333). IGI Global. https://doi.org/10.4018/978-1-7998-4808-0.ch014

Tianming, G. (2018). Food Security and Rural Development on Emerging Markets of Northeast Asia: Cases of Chinese North and Russian Far East. In V. Erokhin (Ed.), *Establishing Food Security and Alternatives to International Trade in Emerging Economies* (pp. 155–176). Hershey, PA: IGI Global. doi:10.4018/978-1-5225-2733-6.ch008

Tiftikçigil, B. Y., Yaşgül, Y. S., & Güriş, B. (2017). Sustainability of Foreign Trade Deficit in Energy: The Case of Turkey. In N. Ray (Ed.), *Business Infrastructure for Sustainability in Developing Economies* (pp. 94–109). Hershey, PA: IGI Global. doi:10.4018/978-1-5225-2041-2.ch005

Tiwari, S., Vaish, B., & Singh, P. (2018). Population and Global Food Security: Issues Related to Climate Change. In I. Management Association (Ed.), Climate Change and Environmental Concerns: Breakthroughs in Research and Practice (pp. 41-64). Hershey, PA: IGI Global. https://doi.org/ doi:10.4018/978-1-5225-5487-5.ch003

Toujani, A., & Achour, H. (2018). A Data Mining Framework for Forest Fire Mapping. In I. Management Association (Ed.), Information Retrieval and Management: Concepts, Methodologies, Tools, and Applications (pp. 771-794). Hershey, PA: IGI Global. doi:10.4018/978-1-5225-5191-1.ch033

Tripathy, B., & K., S. B. (2017). Rough Fuzzy Set Theory and Neighbourhood Approximation Based Modelling for Spatial Epidemiology. In I. Management Association (Ed.), *Public Health and Welfare: Concepts, Methodologies, Tools, and Applications* (pp. 1257-1268). Hershey, PA: IGI Global. https://doi.org/ doi:10.4018/978-1-5225-1674-3.ch058

Trukhachev, A. (2017). New Approaches to Regional Branding through Green Production and Utilization of Existing Natural Advantages. In I. Management Association (Ed.), Advertising and Branding: Concepts, Methodologies, Tools, and Applications (pp. 1758-1778). Hershey, PA: IGI Global. doi:10.4018/978-1-5225-1793-1.ch081

Tsobanoglou, G. O., & Vlachopoulou, E. I. (2017). Social-Ecological Systems in Local Fisheries Communities. In G. Korres, E. Kourliouros, & M. Michailidis (Eds.), *Handbook of Research on Policies and Practices for Sustainable Economic Growth and Regional Development* (pp. 306–316). Hershey, PA: IGI Global. doi:10.4018/978-1-5225-2458-8.ch026

Tuydes-Yaman, H., & Karatas, P. (2018). Evaluation of Walkability and Pedestrian Level of Service. In I. Management Association (Ed.), Intelligent Transportation and Planning: Breakthroughs in Research and Practice (pp. 264-291). Hershey, PA: IGI Global. doi:10.4018/978-1-5225-5210-9.ch012

Uddin, S., Chakravorty, S., Ray, A., & Sherpa, K. S. (2018). Optimal Location of Sub-Station Using Q-GIS and Multi-Criteria Decision Making Approach. *International Journal of Decision Support System Technology, 10*(2), 65–79. doi:10.4018/IJDSST.2018040104

Uddin, S., Chakravorty, S., Sherpa, K. S., & Ray, A. (2018). Power Distribution System Planning Using Q-GIS. *International Journal of Energy Optimization and Engineering, 7*(2), 61–75. doi:10.4018/IJEOE.2018040103

Uzun, F. V. (2018). Natural Resources Management. In A. Eneanya (Ed.), *Handbook of Research on Environmental Policies for Emergency Management and Public Safety* (pp. 1–21). Hershey, PA: IGI Global. doi:10.4018/978-1-5225-3194-4.ch001

V, M., Agrawal, R., Sharma, V., & T.N., K. (2018). Supply Chain Social Sustainability and Manufacturing. In I. Management Association (Ed.), *Technology Adoption and Social Issues: Concepts, Methodologies, Tools, and Applications* (pp. 226-252). Hershey, PA: IGI Global. https://doi.org/ doi:10.4018/978-1-5225-5201-7.ch011

van der Vliet-Bakker, J. M. (2017). Environmentally Forced Migration and Human Rights. In C. Akrivopoulou (Ed.), *Defending Human Rights and Democracy in the Era of Globalization* (pp. 146–180). Hershey, PA: IGI Global. doi:10.4018/978-1-5225-0723-9.ch007

Vargas-Hernández, J. G. (2022). A Comprehensive Entrepreneurship Model for the Internationalization of Green Innovation Businesses. In U. Akkucuk (Ed.), *Disruptive Technologies and Eco-Innovation for Sustainable Development* (pp. 131–149). IGI Global. https://doi.org/10.4018/978-1-7998-8900-7.ch008

Vaskov, A. G., Lin, Z. Y., Tyagunov, M. G., Shestopalova, T. A., & Deryugina, G. V. (2018). Design of Renewable Sources GIS for ASEAN Countries. In V. Kharchenko & P. Vasant (Eds.), *Handbook of Research on Renewable Energy and Electric Resources for Sustainable Rural Development* (pp. 1–25). Hershey, PA: IGI Global. doi:10.4018/978-1-5225-3867-7.ch001

Vázquez, D. G., & Gil, M. T. (2017). Sustainability in Smart Cities: The Case of Vitoria-Gasteiz (Spain) – A Commitment to a New Urban Paradigm. In L. Carvalho (Ed.), *Handbook of Research on Entrepreneurial Development and Innovation Within Smart Cities* (pp. 248–268). Hershey, PA: IGI Global. doi:10.4018/978-1-5225-1978-2.ch012

Wahab, I. N., & Soonthodu, S. (2018). Geographical Information System in Eco-Tourism. In S. Chaudhuri & N. Ray (Eds.), *GIS Applications in the Tourism and Hospitality Industry* (pp. 61–75). Hershey, PA: IGI Global. doi:10.4018/978-1-5225-5088-4.ch003

Weiss-Randall, D. (2018). Cultivating Environmental Justice. In *Utilizing Innovative Technologies to Address the Public Health Impact of Climate Change: Emerging Research and Opportunities* (pp. 110–143). Hershey, PA: IGI Global. doi:10.4018/978-1-5225-3414-3.ch004

Weiss-Randall, D. (2018). Cultivating Resilience. In *Utilizing Innovative Technologies to Address the Public Health Impact of Climate Change: Emerging Research and Opportunities* (pp. 204–235). Hershey, PA: IGI Global. doi:10.4018/978-1-5225-3414-3.ch007

Weiss-Randall, D. (2018). Climate Change Solutions: Where Do We Go From Here? In *Utilizing Innovative Technologies to Address the Public Health Impact of Climate Change: Emerging Research and Opportunities* (pp. 236–268). Hershey, PA: IGI Global. doi:10.4018/978-1-5225-3414-3.ch008

Whyte, K. P., List, M., Stone, J. V., Grooms, D., Gasteyer, S., Thompson, P. B., . . . Bouri, H. (2018). Uberveillance, Standards, and Anticipation: A Case Study on Nanobiosensors in U.S. Cattle. In I. Management Association (Ed.), Biomedical Engineering: Concepts, Methodologies, Tools, and Applications (pp. 577-596). Hershey, PA: IGI Global. https://doi.org/ doi:10.4018/978-1-5225-3158-6.ch025

Wulff, E. (2017). Data and Operational Oceanography: A Review in Support of Responsible Fisheries and Aquaculture. In P. Diviacco, A. Leadbetter, & H. Glaves (Eds.), *Oceanographic and Marine Cross-Domain Data Management for Sustainable Development* (pp. 303–324). Hershey, PA: IGI Global. doi:10.4018/978-1-5225-0700-0.ch013

Yener, D. (2017). Geographic Information Systems and Its Applications in Marketing Literature. In S. Faiz & K. Mahmoudi (Eds.), *Handbook of Research on Geographic Information Systems Applications and Advancements* (pp. 158–172). Hershey, PA: IGI Global. doi:10.4018/978-1-5225-0937-0.ch006

Yu, K., Liu, Y., & Sharma, A. (2021). Analyze the Effectiveness of the Algorithm for Agricultural Product Delivery Vehicle Routing Problem Based on Mathematical Model. *International Journal of Agricultural and Environmental Information Systems*, *12*(3), 26–38. https://doi.org/10.4018/IJAEIS.2021070103

Related References

Zhou, X., Sharma, A., & Mohindru, V. (2021). Research on Linear Programming Algorithm for Mathematical Model of Agricultural Machinery Allocation. *International Journal of Agricultural and Environmental Information Systems, 12*(3), 1–12. https://doi.org/10.4018/IJAEIS.2021070101

About the Contributors

Rubeena Vohra, Ph.D, is working as an Associate Professor in Bharati Vidyapeeth's College of Engineering, New Delhi since 2009. She has completed her Ph.D in Remote Sensing and GIS from Delhi Technological University (formerly DCE), New Delhi, India in year 2020. She has received Best Researcher Award for her contribution in publications from Bharati Vidyapeeth's College of Engineering. She has received Gold Medal in year 2011 from GGSIPU, New Delhi for being M.tech (ECE) topper of her batch. She was awarded by topper certificate in B.Tech (ECE) from LLRIET, Punjab Technical University, in year 2003. She has published 15+research papers in referred SCIE/SCI/ESCI/SCOPUS international and national journals/conferences. She has published book chapters in various Scopus indexed books. She has authored one book in Remote Sensing domain. She has edited books in AI and Remote Sensing & GIS domain. She is an active member of various technical societies and clubs. She is reviewer with many reputed international journals and conferences.

Ashish Kumar, Ph.D., is working as an associate professor with Bennett University, Greater Noida, U.P., India till date. He has worked with Bharati Vidyapeeth's College of Engineering (Affiliated to GGS Indraprastha University) from Aug 2009 to Jul 2022. He has completed his Ph.D. in Computer Science and Engineering from Delhi Technological University (formerly DCE), New Delhi, India in 2020. He received the Best Researcher award from the Delhi Technological University for his contribution to the computer vision domain. He has completed M. Tech with distinction in Computer Science and Engineering from GGS Indraprastha University, New Delhi. He has published more than 25 research papers in various reputed national and international journals and conferences. He has published 15+ book chapters in various Scopus-indexed books. He has authored/edited several books in AI, computer vision, and healthcare domains with reputed publishers. He is an active member of various international societies and clubs. He is a reviewer with many reputed journals and in the technical program committee of various national/ international conferences. Dr. Kumar also served as a session chair at many international and national conferences. His current research interests include object tracking, image processing, artificial intelligence, and medical imaging analysis.

* * *

Aravinda K. has 35 years of professional experience, with 18 years of industrial experience and 17 years of academic teaching. Served as R&D Engineer at Swede (India) Teltronics Ltd, Bangalore, as Senior Engineer at Pulsetone Industries, Bangalore, and as Lecturer at Sapthagiri College of Engineering, Bengaluru. Currently working as Professor & HoD - ECE at New Horizon College of Engineering, Bengaluru. Publications: One book, One Souvenir article, 6 Journal articles, and 16 Conference articles.

Dyvavani Krishna is a PhD scholar pursuing higher research in urban climate studies and urban biodiversity. My major research focuses are urban ecology, forest ecology and their interrelationship with climate change. I have a MTech Degree in Remote Sensing and GIS with specialization in Forest Resources and Ecosystem Analysis, a MSc Degree in Biodiversity studies and Management, and a BSc Degree in Life Sciences.

Avinash Kumar is a Ph.D. scholar pursuing higher research in assessing surface water quality parameters using Remote sensing and GIS. My major research focuses are river water quality and rainwater harvesting. I have an MTech Degree in civil engineering with a specialization in Hydraulics and water resources engineering and a BTech Degree in Civil engineering.

Manish Kumar is currently working as an Assistant Professor in Operations & IT Department, ICFAI Business School, Hyderabad, India. Before that he worked under the central government (MHRD-NPIU) project as an Assistant Professor in Bhagalpur College of Engineering, Bhagalpur, India. He obtained his PhD in Industrial Management from Indian Institute of Technology (Banaras Hindu University), Varanasi, India. He completed his MTech in Industrial Engineering and Management from Indian Institute of Technology (Indian School of Mines), Dhanbad, India and B.Tech in Mechanical Engineering from Biju Patnaik University of Technology, Orissa, India. He had successfully completed two government funded projects. His areas of interests include energy policy, sustainability, operations management and decision science.

Bharat Singh Patel is working as an Assistant Professor in Prestige University, Indore, India. He holds PhD and M. tech in Industrial Management from Indian Institute of Technology (Banaras Hindu University), Varanasi, India. His research papers have been published in ABDC listed journals and Scopus indexed journals. His research interests include supply chain management, supply chain agility, lean manufacturing and agile manufacturing.

Sabyasachi Pramanik is a professional IEEE member. He obtained a PhD in Computer Science and Engineering from Sri Satya Sai University of Technology and Medical Sciences, Bhopal, India. Presently, he is an Associate Professor, Department of Computer Science and Engineering, Haldia Institute of Technology, India. He has many publications in various reputed international conferences, journals, and book chapters (Indexed by SCIE, Scopus, ESCI, etc). He is doing research in the fields of Artificial Intelligence, Data Privacy, Cybersecurity, Network Security, and Machine Learning. He also serves on the editorial boards of several international journals. He is a reviewer of journal articles from IEEE, Springer, Elsevier, Inderscience, IET and IGI Global. He has reviewed many conference papers, has been a keynote speaker, session chair, and technical program committee member at many international conferences. He has authored a book on Wireless Sensor Network. He has edited 8 books from IGI Global, CRC Press, Springer and Wiley Publications.

Gopinadh Rongali is currently working as Project Scientist-II at NCMRWF, Ministry of Earth Sciences, Noida, U.P. Previously he worked as a Research Associate-III/Post Doctoral Fellow at the Department of Water Resources Development & Management (WRD&M), Indian Institute of Technology Roorkee (IITR). Gopinadh does research in Geoinformatics (GIS), Hydrology, Cryosphere and Remote Sensing.

Cherian Samuel obtained his degree of Bachelor of Technology in Mechanical Engineering and Master of Technology degree in Production Engineering both from Calicut University. He obtained his PhD degree from Industrial Engineering & Management department, IIT Kharagpur. He is currently working as Associate Professor in the Department of Mechanical Engineering at the Indian Institute of Technology, Banaras Hindu University, India. He has publications to his credit in national and International journals. His areas of interest include Supply Chain Management, System Dynamics, Production and Operations Management, etc.

Arunadevi Thirumalraj did her BE in Computer Science Engineering in Government college of Engineering, Bodinayakanur and she has worked as a Research associate in various industrial and academic research institutes for past 6 years. Now she is pursing ME Computer Science Engineering in K. Ramakrishnan College of Technology, (KRCT), Tiruchirappalli. She is actively involved in the field of innovative teaching and Learning. She has publishing a good number of papers in indexed journals.

Jaya Yadav is a PhD scholar pursuing higher research in climate studies. My major research focuses are rainfall prediction, machine learning and their utilization for climate studies. I have a MTech Degree in Geomatics Engineering and a BTech Degree in Electronica and Communication Engineering.

Index

I

L

M

N

O

P

Q

R

S

T

Recommended Reference Books

IGI Global's reference books are available in three unique pricing formats:
Print Only, E-Book Only, or Print + E-Book.

Order direct through IGI Global's Online Bookstore at **www.igi-global.com** or through your preferred provider.

ISBN: 9781668445488
EISBN: 9781668445501
© 2022; 334 pp.
List Price: US$ **240**

ISBN: 9781668467503
EISBN: 9781668467527
© 2022; 296 pp.
List Price: US$ **240**

ISBN: 9781799889007
EISBN: 9781799889021
© 2022; 340 pp.
List Price: US$ **215**

ISBN: 9781799898511
EISBN: 9781799898535
© 2022; 276 pp.
List Price: US$ **250**

ISBN: 9781799896913
EISBN: 9781799896937
© 2022; 219 pp.
List Price: US$ **215**

ISBN: 9781668436868
EISBN: 9781668436875
© 2022; 2,064 pp.
List Price: US$ **2,085**

Do you want to stay current on the latest research trends, product announcements, news, and special offers?
Join IGI Global's mailing list to receive customized recommendations, exclusive discounts, and more.
Sign up at: **www.igi-global.com/newsletters.**

Publisher of Timely, Peer-Reviewed Inclusive Research Since 1988

IGI Global
PUBLISHER of TIMELY KNOWLEDGE

www.igi-global.com Sign up at www.igi-global.com/newsletters facebook.com/igiglobal twitter.com/igiglobal

www.ingramcontent.com/pod-product-compliance
Lightning Source LLC
LaVergne TN
LVHW080643011026
849160LV00014B/1311
* 9 7 9 8 3 6 9 3 4 4 5 2 1 *